Statistics and Econometrics for Finance

More information about this series at http://www.springer.com/series/10377

Abdulkader Aljandali • Motasam Tatahi

Economic and Financial Modelling with EViews

A Guide for Students and Professionals

Springer

Abdulkader Aljandali
Department of Accounting, Finance
and Economics
Coventry University London
London, UK

Motasam Tatahi
Department of Economics, Finance
and Accounting
Regent's University London-European
Business School London
London, UK

ISSN 2199-093X ISSN 2199-0948 (electronic)
Statistics and Econometrics for Finance
ISBN 978-3-319-92984-2 ISBN 978-3-319-92985-9 (eBook)
https://doi.org/10.1007/978-3-319-92985-9

Library of Congress Control Number: 2018944692

This Springer imprint is published by the registered company Springer International Publishing AG part of Springer Nature.
The registered company address is: Gewerbestrasse 11, 6330 Cham, Switzerland

Contents

List of Figures

List of Tables

Chapter 1
Introduction to EViews

1.1 Introduction

EViews is an interactive computer programme for statistical and econometric analysis. With EViews you can quickly develop a statistical relation from your data and then use the relation to forecast future values of the data. Note that EViews cannot be used for financial analysis (e.g. annuities, NPV, IRR etc.), simulation and cost analysis.

EViews was developed by economists and most of its uses are in economics or financial economics. The package provides convenient ways to enter data series from the keyboard, to create new series from existing ones, to import series from Microsoft Excel, IBM SPSS Statistics and SAS to display and print series and to carry out statistical analysis of the relationships among series. Results appear in windows and can be manipulated with standard windows techniques.

EViews commands are accessed via ***menus***. Most applications contain their own set of menus, which are located on the menu bar along the top of the application window. There are generally drop-down menus associated with the items in the main menu bar. For example, the main menu bar contains the following tabs to choose from:

Selecting the **File** tab will open a drop-down menu containing additional commands. EViews uses ***dialogue boxes*** for the entry of extra information. For example, if you select the menu item to run regression, EViews opens a dialogue box prompting the user for additional information about this specification, while providing default suggestions for various options. EViews uses the "Esc" key as the break key. If you wish to cancel the current task or ongoing operation, simply press "Esc" on your keyboard.

A. Aljandali, M. Tatahi, *Economic and Financial Modelling with EViews*, Statistics and Econometrics for Finance, https://doi.org/10.1007/978-3-319-92985-9_1

Finally, when you launch the programme, you will see the EViews window of Fig. 1.1. At its very top is the ***title bar*** and when EViews is the active programme, the title has a colour and intensity that differs from other windows—it is darker. Just below the title bar is the ***main menu***. Drop-down menus appear as each item is chosen. Below the main menu is a white area called the ***command*** window.

EViews commands may be typed in here and commands are executed when the user hits the "Enter" key. This area acts very much like a word processor. Knowledge of EViews syntax is needed to use the command window. At the very bottom of the window is a ***status line*** which is divided into subsections. The left section will sometimes contain status messages sent by EViews. These can be cleared by clicking the box at the far left of the status line. The next subsection shows the default directory that EViews will use to look for data and programmes.

The dark area in the middle of the window of Fig. 1.1 is the ***work area***. Here, EViews displays the various object windows that it creates (think of these windows as similar to sheets of paper that you might place on your desk as you work).

EViews is built around the concept of **objects**. Data series, equations and systems are just a few examples of objects. Each object has its own window, its own menus, its own procedures and its own views of the data. Most statistical procedures are simply alternative views of the object. For example, a simple menu choice from a series window changes the view between a spreadsheet, line and bar graphs, a histogram-and-statistics view, a correlogram etc.

Fig. 1.1 The EViews opening window

Similarly, an equation window allows the user to switch between the display of the equation specification, basic estimation results, graphics depicting the actual, fitted and residual values for the dependent variable, tables, forecast graphs and evaluations and more than a dozen diagnostic and hypothesis tests. You can cut-and-paste any of these views into your favourite word processor. Various object windows in the **work area** will overlap each other with the foremost window being in focus or active. Only the active window has a darkened title bar. When a window is partly covered, you can bring it to the top by clicking on its title bar or on a visible portion of that window. You can also cycle through the displayed windows by pressing the F6 key.

1.2 Importing Data into EViews

The first step in any project is to get the data into the package. EViews will read data from a wide variety of common data formats, including database (*.dbf), Microsoft Excel (*.xls), HTML (*.htm; *.html), IBM SPSS Statistics (*.sav) and text files (*.csv; *.txt; *.dat) though this list is not exhaustive. Any file that is not an EViews workfile is called a **foreign file**. To load a foreign file, click:

```
File...
  Open...
    Foreign Data as Workfile
```

If you have previously saved a data file in EViews, it has the EViews default extension of (*.wf1) and may be opened via:

```
File...
  Open...
   EViews workfile
```

For example, the file used on the page overleaf contains data of the Harmon Company's monthly sales (MONSALES) in relation to the financial values of six discount offers (DISC1, DISC2, . . ., DISC6) that the firm trialled for ten consecutive months in 2015. Open this file and you will be presented with the Table read specification dialogue box of Fig. 1.2.

The **Select variables** tab permits the user to deselect any of the data series in this file, by simply clicking the ticks in Fig. 1.2. The variables are initially listed in the order in which they appear in the file. You can sort the data by clicking on the header for the column; the display will be toggled between three states—the original order, sorted ascending and sorted descending. Click the OK button and the data will be imported into EViews as shown in the ***work area*** of Fig. 1.3.

In the title bar of the workfile window (the rear of the two windows—currently inactive), you will see the **Workfile** designation followed by the workfile name. Below the toolbar and in the main portion of the window, EViews displays the

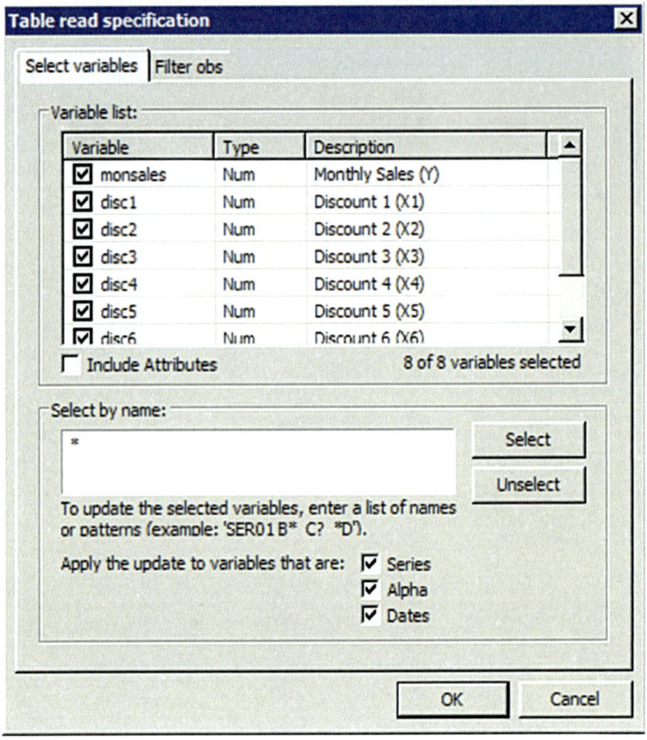

Fig. 1.2 Table read specification dialogue box

contents of the workfile. Note that EViews always adds two extra default variables. **C** stands for "constant" and permits the researcher to use this, for example, as the intercept term in a regression equation. The second default variable is named **resid** and is used to store the residuals from any working model. By making the workfile active (click on the title bar), the user may select:

```
View
  Details +/-
```

(or simply click on the Details +/– button on the toolbar) to toggle between the standard workfile display and a display that provides additional information about the date that the object was created or updated, as well as the label information that you may have attached to the object in the foreign file.

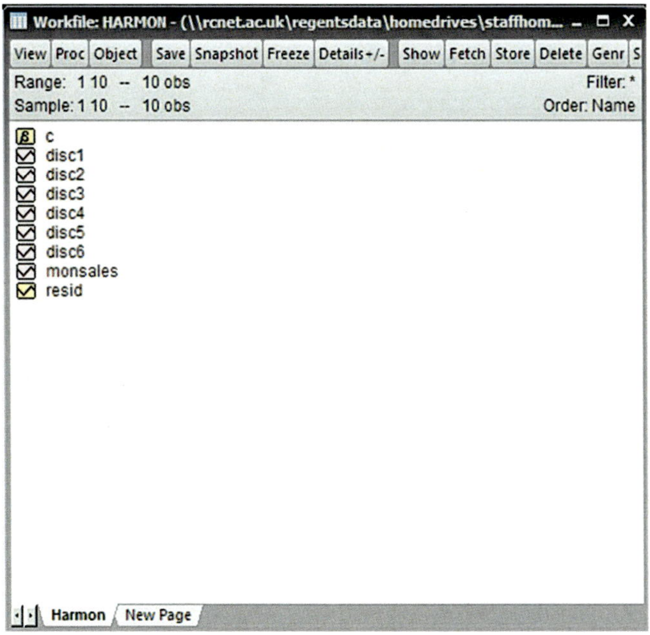

Fig. 1.3 HARMON data imported into EViews

1.2.1 Reading Excel/IBM SPSS Data Variables

If the Microsoft Excel or the IBM SPSS Statistics data file which is to be read into EViews contains a date variable that relates to annual data **only**, then that date variable will translate across with no problem. However, Microsoft Excel and IBM SPSS Statistics date variables such as years/quarters, years/months, years/months/quarters etc. do not directly translate into EViews and action needs to be taken within EViews to assign such date variables to the cases in the data file. For example, you will note in the left hand (dark grey) margin of Fig. 1.3, that the variables are not dated. The readings are simply listed as observations (Obs) from 1 to 10.

We need to replace the observation numbers (Obs) with the appropriate dates, for example, to label graphs. The first step in achieving this is to make the workfile HARMON window in Fig. 1.3 active (by clicking the title bar).

Now that it is active, click:

```
Proc
 Structure/Resize Current Page
```

Which generates the Workfile Structure dialogue box of Fig. 1.4. Under the heading ‘Workfile structure type’, click the downward pointing, black arrow and

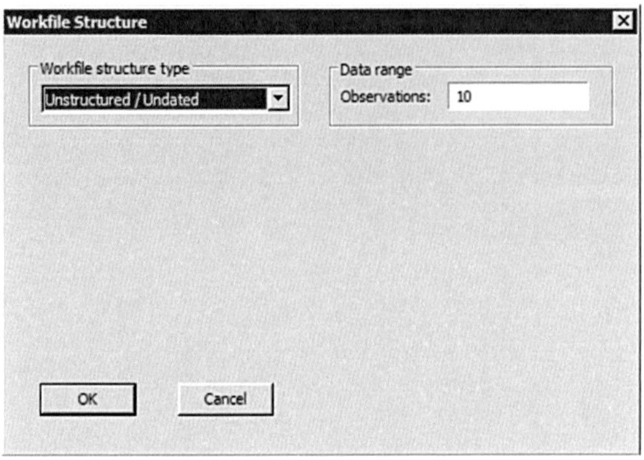

Fig. 1.4 The Workfile structure dialogue box

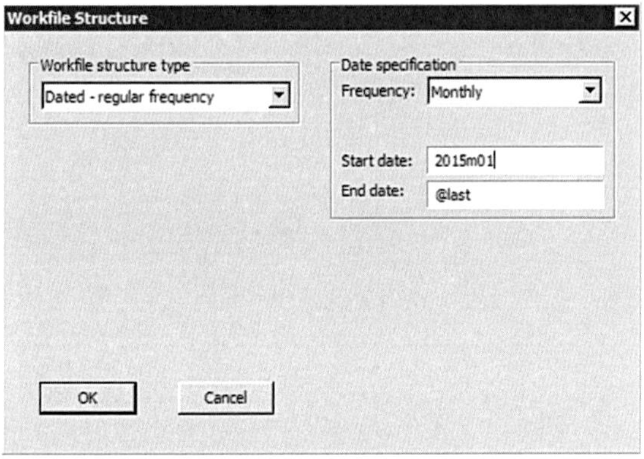

Fig. 1.5 Appropriate set up for the Workfile Structure dialogue box

select the option "Dated—regular frequency" as shown in Fig. 1.5. Under the heading 'Date specification', select:

```
Frequency: Monthly
Start date: 2015M01
End date: @last
```

Similarly, if the data were quarterly, a start date would be set as 2015Q01, for example. There is no need to type in the final date, since EViews computes it (however, you may type it in if you wish). In the above example, you could type in the End date: 2015M10 and the Start date: @first—EViews would compute the first date. Click the OK button and in the EViews ***work area***, the data file should appear as per Fig. 1.6, now appropriately dated from 2015M01 to 2015M10.

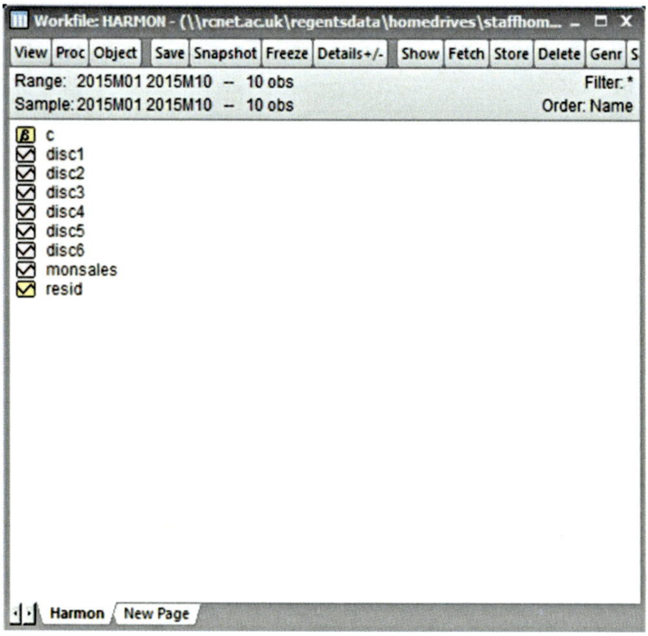

Fig. 1.6 The correctly dated data file

1.2.2 Saving and Opening an EViews Data File

You may save modified or updated versions of your named workfile by clicking:

```
File
 Save As ...
```

The EViews extension .wf1 is used for data files. If the file already exists, EViews will ask you if you want to update the version on disk. Whenever you overwrite a workfile on disk, EViews will usually keep a backup copy of the overwritten file with the first character in the extension changed to, for example MYDATA.wf1.

When you save a data file, EViews will present the default dialogue box showing the current default options for storing the data in your workfile. The first choice is whether to save the data as **single precision** or **double precision**. The former creates smaller files on disk, but saves the data with fewer digits of accuracy—7 as opposed to 16 with double precision. You may also choose to save the data in **compressed** or **non-compressed** form. Save the data file as HARMON.wf1. To load an EViews workfile, click:

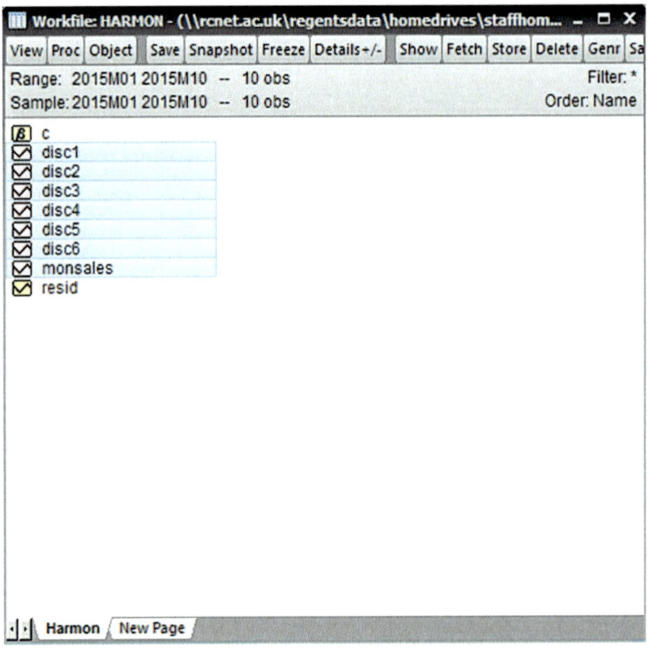

Fig. 1.7 Selecting variables

```
File
 Open
  EViews Workfile...
```

and navigate to the appropriate directory to open HARMON.wf1.The six discount variables, monthly sales and C and RESID are listed as the variables. In order to see the data values, click DISC1 and it will go dark to indicate that it is selected. Press the 'Shift' key and use the down arrow key to select the other variables up to and including MONSALES (Use the 'Shift' and 'Ctrl' keys to select multiple items that are not contiguous). All selected variables will be highlighted as per Fig. 1.7.

Next click:

```
View
 Show
```

and the user is provided with a list of all selected variables in the Show dialogue box of Fig. 1.8. Click the OK button to reveal the data which displays the values of various discounts in addition to the main variable, monthly sales as shown in Fig. 1.9.

Having previously highlighted the variables of interest, it is possible to display summary statistics for each variable. Click:

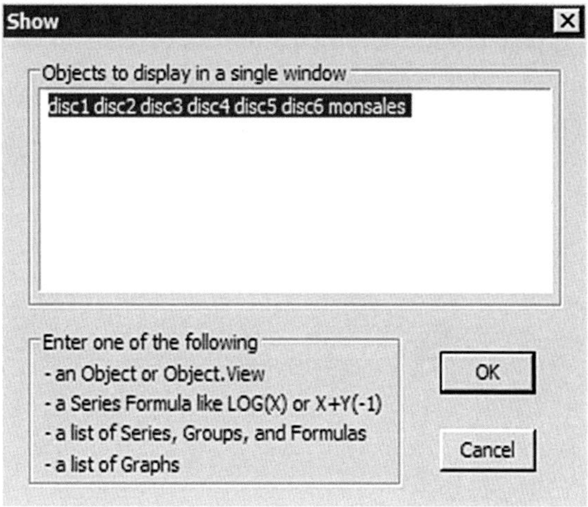

Fig. 1.8 The Show dialogue box

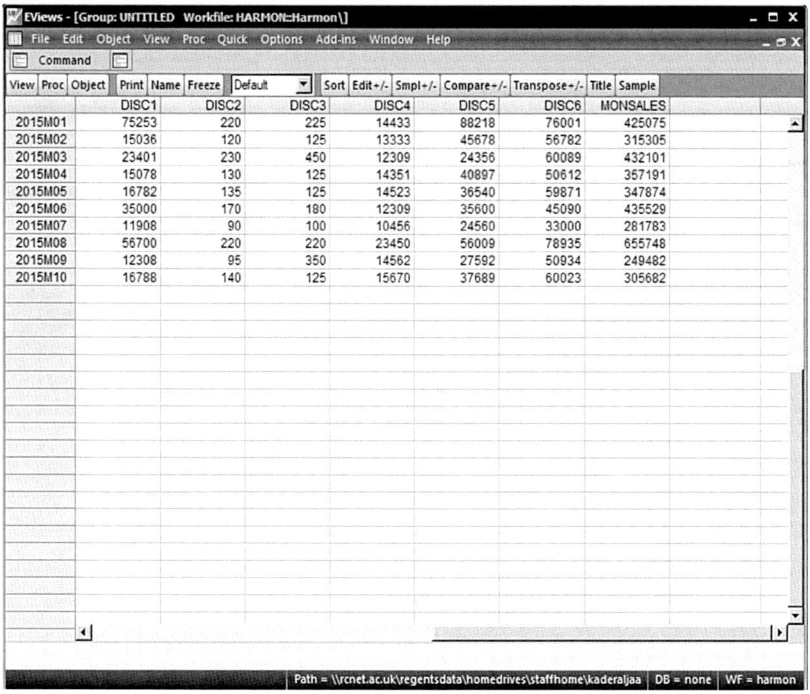

	DISC1	DISC2	DISC3	DISC4	DISC5	DISC6	MONSALES
2015M01	75253	220	225	14433	88218	76001	425075
2015M02	15036	120	125	13333	45678	56782	315305
2015M03	23401	230	450	12309	24356	60089	432101
2015M04	15078	130	125	14351	40897	50612	357191
2015M05	16782	135	125	14523	36540	59871	347874
2015M06	35000	170	180	12309	35600	45090	435529
2015M07	11908	90	100	10456	24560	33000	281783
2015M08	56700	220	220	23450	56009	78935	655748
2015M09	12308	95	350	14562	27592	50934	249482
2015M10	16788	140	125	15670	37689	60023	305682

Fig. 1.9 EViews output showing models variables

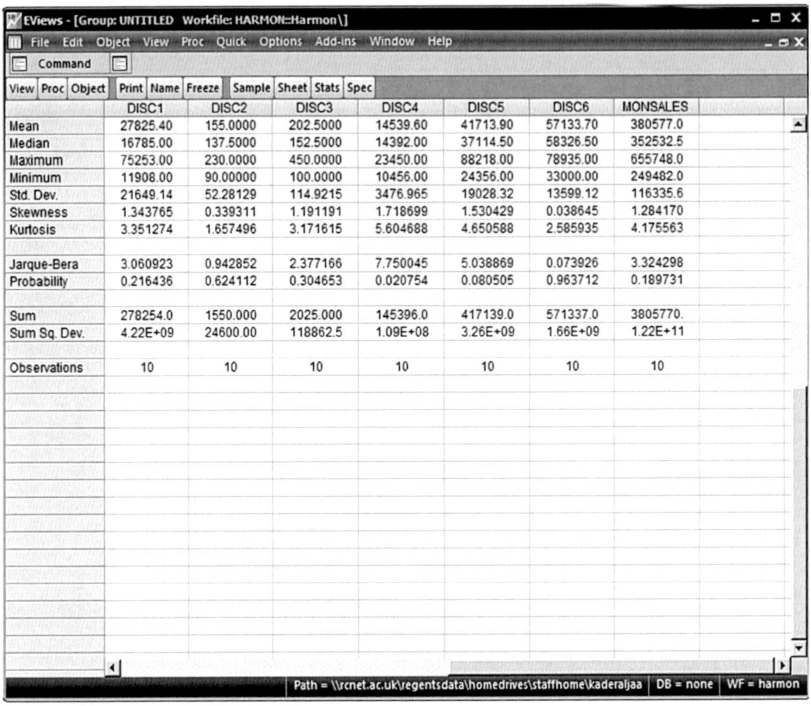

	DISC1	DISC2	DISC3	DISC4	DISC5	DISC6	MONSALES
Mean	27825.40	155.0000	202.5000	14539.60	41713.90	57133.70	380577.0
Median	16785.00	137.5000	152.5000	14392.00	37114.50	58326.50	352532.5
Maximum	75253.00	230.0000	450.0000	23450.00	88218.00	78935.00	655748.0
Minimum	11908.00	90.00000	100.0000	10456.00	24356.00	33000.00	249482.0
Std. Dev.	21649.14	52.28129	114.9215	3476.965	19028.32	13599.12	116335.6
Skewness	1.343765	0.339311	1.191191	1.718699	1.530429	0.038645	1.284170
Kurtosis	3.351274	1.657496	3.171615	5.604688	4.650588	2.585935	4.175563
Jarque-Bera	3.060923	0.942852	2.377166	7.750045	5.038869	0.073926	3.324298
Probability	0.216436	0.624112	0.304653	0.020754	0.080505	0.963712	0.189731
Sum	278254.0	1550.000	2025.000	145396.0	417139.0	571337.0	3805770.
Sum Sq. Dev.	4.22E+09	24600.00	118862.5	1.09E+08	3.26E+09	1.66E+09	1.22E+11
Observations	10	10	10	10	10	10	10

Fig. 1.10 Descriptive Statistics in EViews

```
Quick
  Group statistics
    Descriptive statistics
      Individual samples
```

Type the variables as per Fig. 1.8 which will then produce spreadsheet-type output in which summary statistics including the mean, median, standard deviation, skewness, kurtosis are displayed Fig. 1.10 also presents result of running the Jarque-Bera test which examines the null hypothesis that a particular variable is normally distributed.

Under the numerical value of the Jarque-Bera test statistic is the probability or significance associated with the test. If the significance is less than 0.05, we reject the null hypothesis that the variable in question is normally distributed. The Jarque-Bera test and other tests will be examined in the next Chap. 2.

Chapter 2
A Guideline for Running Regression

Unlike IBM SPSS Statistics, EViews does not offer the options of stepwise entry or backward removal of variables when running regression. However, you could include all the variables in EViews and eliminate the non-significant variables one by one until only the significant ones remain. The main advantage of running regression in EViews is that formal testing procedures exist for testing hypotheses concerning the residuals.

2.1 EViews Regression

Open the HARMON data file on EViews which was generated in the previous chapter. From the main EViews menu at the top, click:

```
Quick
 Estimate equation
```

to produce the Estimate Equation dialogue box of Fig. 2.1. Under the heading 'Equation specification', type in the variables that you wish to include, starting off with the dependent variable, here MONSALES. Follow this with the independent variables. Note that should we have wanted to include an intercept term, then you would add C to this list. In fact, we have established that the intercept is not significant; neither are DISC1, DISC3 and DISC5. Under the heading 'Method', choose LS for least squares.

The 'Sample' period is set to 2015M01–2015M10, but you can change this if the regression is to be run on a subset of the cases. Click the OK button to generate the dialogue box of Fig. 2.2 in which the regression results are displayed.

All the regression coefficients are statistically significant ($p < 0.025$) and the coefficient of determination has value 92.35%. Now proceed to run formal statistical tests on the residuals. Click the view button in the dialogue box of Fig. 2.2 and click:

A. Aljandali, M. Tatahi, *Economic and Financial Modelling with EViews*, Statistics and Econometrics for Finance, https://doi.org/10.1007/978-3-319-92985-9_2

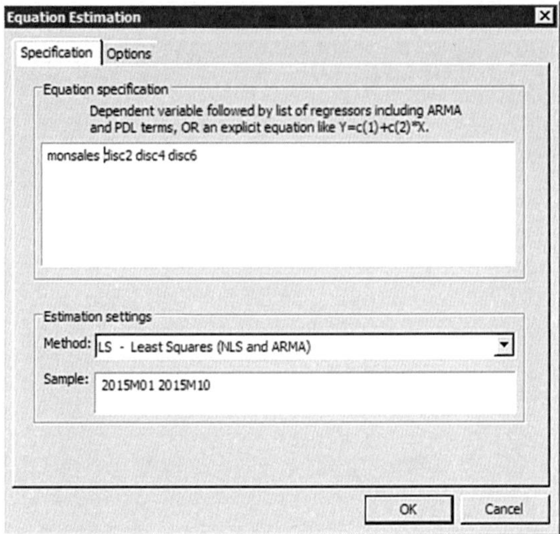

Fig. 2.1 The Equation Estimation dialogue box

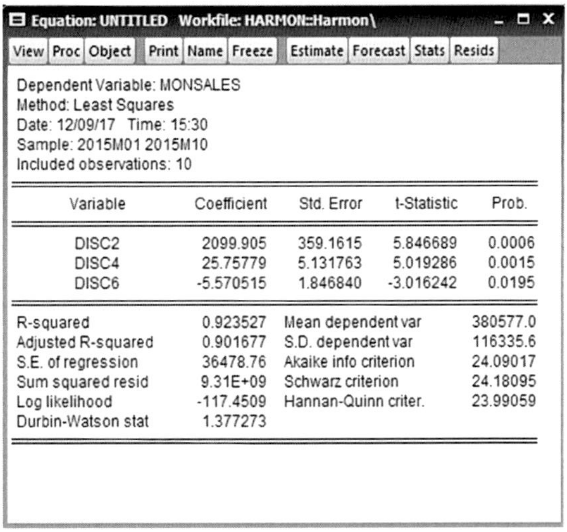

Dependent Variable: MONSALES
Method: Least Squares
Date: 12/09/17 Time: 15:30
Sample: 2015M01 2015M10
Included observations: 10

Variable	Coefficient	Std. Error	t-Statistic	Prob.
DISC2	2099.905	359.1615	5.846689	0.0006
DISC4	25.75779	5.131763	5.019286	0.0015
DISC6	-5.570515	1.846840	-3.016242	0.0195

R-squared	0.923527	Mean dependent var	380577.0
Adjusted R-squared	0.901677	S.D. dependent var	116335.6
S.E. of regression	36478.76	Akaike info criterion	24.09017
Sum squared resid	9.31E+09	Schwarz criterion	24.18095
Log likelihood	-117.4509	Hannan-Quinn criter.	23.99059
Durbin-Watson stat	1.377273		

Fig. 2.2 The EViews regression equation

```
Residual tests
 Histogram - Normality Test
```

to derive the output of Fig. 2.3. This diagram presents the results of applying the Jarque-Bera test of normality to the regression residuals discussed in Chap. 1.

The null hypothesis of the Jarque-Bera test is that the residuals are normally distributed and thus the significance level of 0.588 attached to the Jarque-Bera test statistic leads us to fail rejecting this null (**Note:** This is a one-tailed test). The graph may be edited and saved or sent to the clipboard (see the next section).

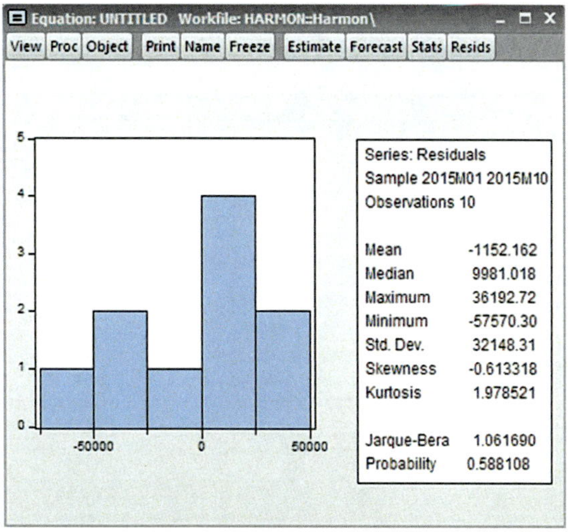

Fig. 2.3 The Jarque-Bera test of normality

We now turn to a statistical test for serial (or temporal) autocorrelation in the residuals. Click the View button on the results box of Fig. 2.3 and click:

```
Residual tests
  Serial correlation LM test
```

LM stands for Lagrange Multiplier. You will be prompted to specify how many lags you want tested for temporal autocorrelation, the default being 2. For quarterly data, it would be wise to change this to 4 or 8 (autocorrelation over 1 or 2 years' worth of data). Click the OK button to generate the output of Fig. 2.4. In essence you need only the results at the top of this output. The null is that the data (or here the residuals) exhibit no serial (temporal) autocorrelation. There are two statistics available for this test and it is most rare for them to contradict each other in terms of acceptance or reject of the null.

Firstly, the F statistic has a significance of 0.415; secondly the chi-square statistic has a significance of 0.228. Both lead us to fail to reject the null (since $p < 0.05$ for this test) and conclude that the regression residuals do not exhibit temporal autocorrelation. Lastly, we turn to a test for the homoscedasticity of the regression residuals. Click the View button in Fig. 2.4 and then click:

```
Residual tests
  White heteroscedasticity (no cross terms)
```

to derive the output of Fig. 2.5. The null for White's test is that the residuals are homoscedastic. Again two test statistics are available. The F test has a significance of 0.896 and the chi-square statistic a significance of 0.699. Both suggest that we fail to reject the null (since $p < 0.05$ for this test); the residuals are homoscedastic. At any

Equation: UNTITLED Workfile: HARMON::Harmon\

View | Proc | Object | Print | Name | Freeze | Estimate | Forecast | Stats | Resids

Breusch-Godfrey Serial Correlation LM Test:

F-statistic	1.054742	Prob. F(2,5)	0.4148
Obs*R-squared	2.967140	Prob. Chi-Square(2)	0.2268

Test Equation:
Dependent Variable: RESID
Method: Least Squares
Date: 12/09/17 Time: 15:35
Sample: 2015M01 2015M10
Included observations: 10
Presample missing value lagged residuals set to zero.

Variable	Coefficient	Std. Error	t-Statistic	Prob.
DISC2	147.7390	412.2252	0.358394	0.7347
DISC4	3.363929	5.646927	0.595710	0.5773
DISC6	-1.109658	2.079848	-0.533528	0.6165
RESID(-1)	-0.109542	0.566387	-0.193404	0.8543
RESID(-2)	-0.995518	0.719756	-1.383134	0.2252

R-squared	0.295710	Mean dependent var	-1152.162
Adjusted R-squared	-0.267721	S.D. dependent var	32148.31
S.E. of regression	36196.79	Akaike info criterion	24.13818
Sum squared resid	6.55E+09	Schwarz criterion	24.28947
Log likelihood	-115.6909	Hannan-Quinn criter.	23.97221
Durbin-Watson stat	1.621994		

Fig. 2.4 The Breusch-Godfrey test of serial autocorrelation

point, you can return to the original regression output of Fig. 2.2, by clicking the view button and selecting:

```
Estimation Output
```

from the drop-down menu. It is possible to generate a graph of the observed (or actual) data, the forecasted (or fitted) values from the regression analysis, plus the regression residuals. Click the View button and select:

```
Actual, fitted, residual
  Actual, fitted, residual graph
```

Which generates the output of Fig. 2.6. Similarly, there is the option of generating a table of these three variables.

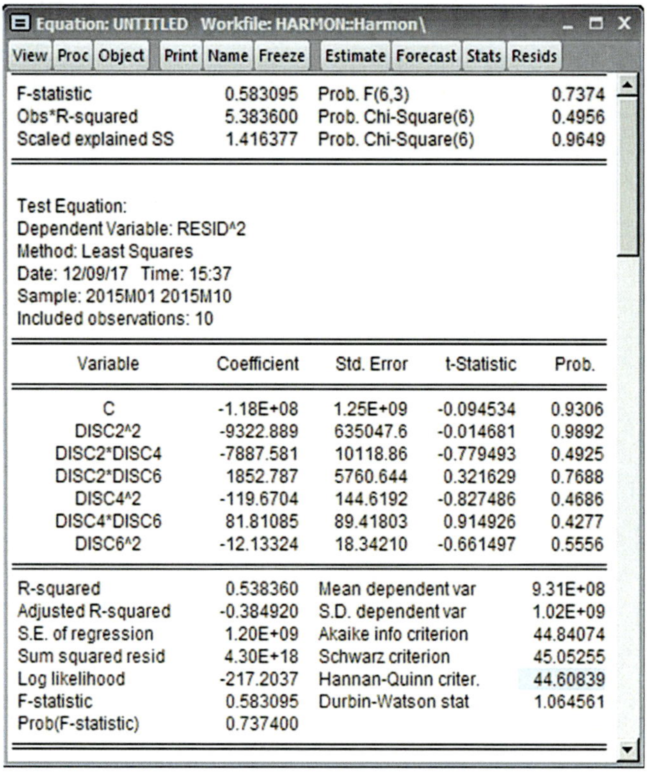
Equation: UNTITLED Workfile: HARMON::Harmon\

View | Proc | Object | Print | Name | Freeze | Estimate | Forecast | Stats | Resids

F-statistic	0.583095	Prob. F(6,3)	0.7374
Obs*R-squared	5.383600	Prob. Chi-Square(6)	0.4956
Scaled explained SS	1.416377	Prob. Chi-Square(6)	0.9649

Test Equation:
Dependent Variable: RESID^2
Method: Least Squares
Date: 12/09/17 Time: 15:37
Sample: 2015M01 2015M10
Included observations: 10

Variable	Coefficient	Std. Error	t-Statistic	Prob.
C	-1.18E+08	1.25E+09	-0.094534	0.9306
DISC2^2	-9322.889	635047.6	-0.014681	0.9892
DISC2*DISC4	-7887.581	10118.86	-0.779493	0.4925
DISC2*DISC6	1852.787	5760.644	0.321629	0.7688
DISC4^2	-119.6704	144.6192	-0.827486	0.4686
DISC4*DISC6	81.81085	89.41803	0.914926	0.4277
DISC6^2	-12.13324	18.34210	-0.661497	0.5556

R-squared	0.538360	Mean dependent var	9.31E+08
Adjusted R-squared	-0.384920	S.D. dependent var	1.02E+09
S.E. of regression	1.20E+09	Akaike info criterion	44.84074
Sum squared resid	4.30E+18	Schwarz criterion	45.05255
Log likelihood	-217.2037	Hannan-Quinn criter.	44.60839
F-statistic	0.583095	Durbin-Watson stat	1.064561
Prob(F-statistic)	0.737400		

Fig. 2.5 White's Heteroscedasticity test

2.1.1 Saving the Regression Equation

It is possible to save the regression equation in Fig. 2.2. Click the Object button and you will be asked to provide a name for this equation via 'Name to identify object'—we called it "Eq. (2.1)". The name used should be 24 characters long maximum, with a recommended length of 16 or less characters. If you then save the file (with extension .wf1), this equation is saved along with the other study variables. Upon reopening the file (remembering that it is now in EViews format) you will see:

equation1

along with the list of saved variables. Click this to restore the equation and you may proceed to examine residuals etc.

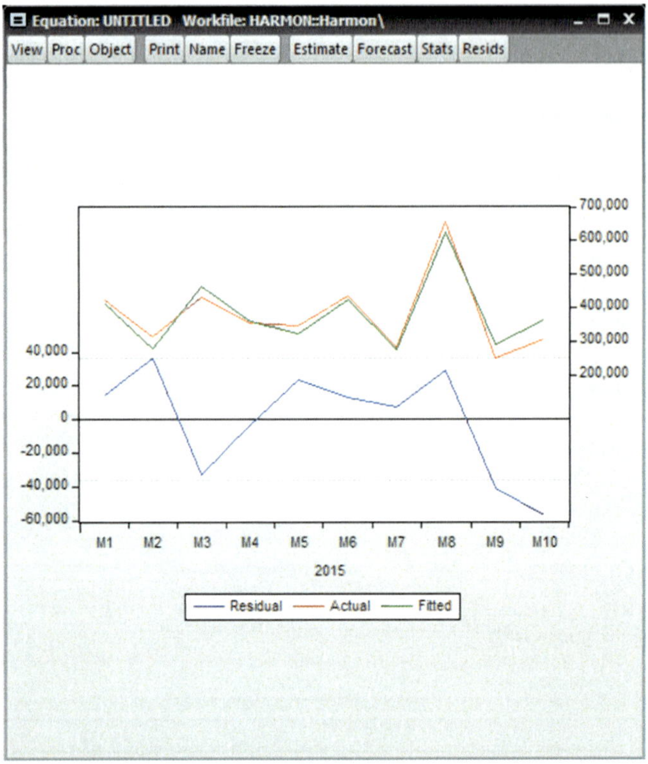

Fig. 2.6 The observed, fitted values and the residuals

2.1.2 Editing and Saving Regression Graphics

Figure 2.6 is produced using the default EViews graphical settings. However, you may wish to change the colours used (especially for black and white printing), edit the line styles (e.g. use broken lines) change the symbols used or add your own text. Graph objects are created by **freezing** a view, which is achieved by clicking the Freeze button in Fig. 2.6. Click the options tab on the top bar menu to reveal the Graph Options dialogue box of Fig. 2.7 much of which is self-explanatory.

In Fig. 2.6, the residuals were plotted in blue, the actual values in red and the fitted values in green. In Fig. 2.7, you scroll down these as required. You can change the 'Colour' and the 'Line Pattern' as required. You can change the thickness of the line via the 'Line/Symbol width' heading from its default of ¾ point. You can add a 'Symbol' to the line as per Fig. 2.8 by clicking:

```
Edit
 Copy
```

from the EViews main menu to send the edited graph to the clipboard.

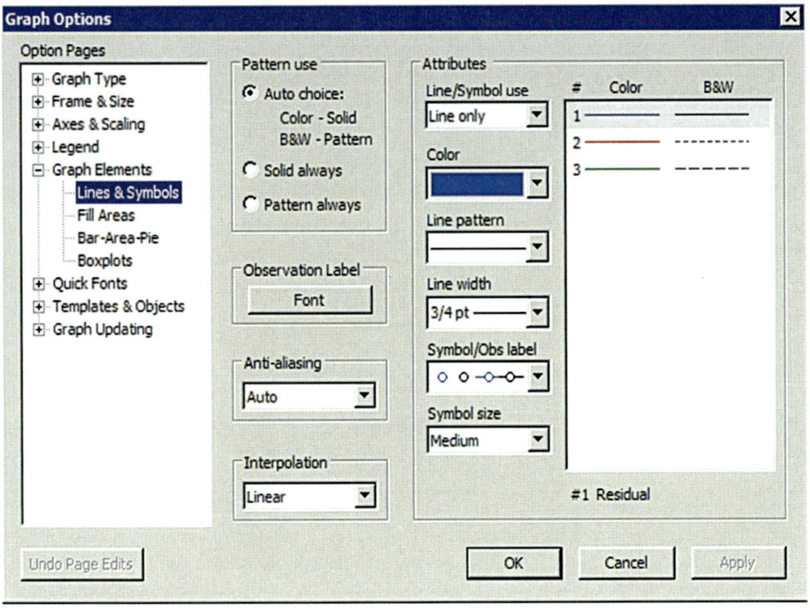

Fig. 2.7 The Graph Options dialogue box

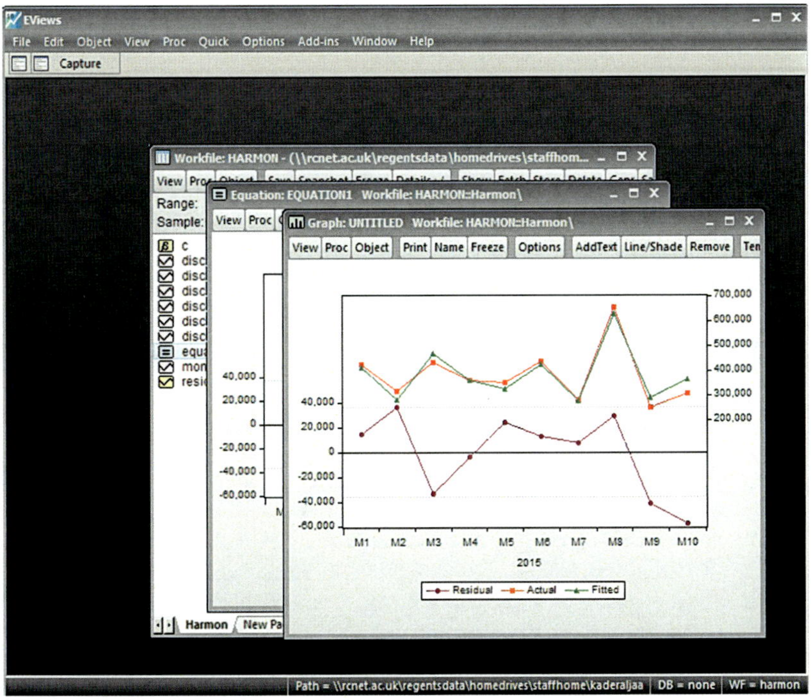

Fig. 2.8 An edited plot

A Graph Metafile dialogue box will appear. Once the graph is on the clipboard, you can open your word processor and paste it in, where it can be sized and positioned as required. To save the edited graph, either right click inside the active graph or click the Proc button. Select ***Save graph to disc*** and change the name and path as required. Next choose the file type. The default type is called Enhanced Metafile which has the extension *.emf. The regression residuals are saved by default with the name **resid**. You may remember that C and resid are set up automatically by EViews. If you wanted a plot of just the residuals, from the main EViews menu click:

```
Quick
 Graph
   Line graph
```

and enter the variable name **resid** in the resultant dialogue box. You may want a plot of the standardised residuals (i.e. mean of residuals = 0, standard deviation of residuals = 1), which will assist in the determination of **outliers**. Recall that residuals are standardised via:

$$sresid = \frac{resid - mean}{SD}$$

where **sresid** stands for the standardised residual and SD stands for the standard deviation of the original residuals. To establish the numerical values for the mean and standard deviation of the residuals, from the main EViews menu click:

```
Quick
 Series statistics
   Histogram and stats
```

You will be prompted to enter a series name, type resid and then click OK. You will see that the mean value of the residuals is -1152.162 with standard deviation of 32,148.31. To generate a data series containing the standardised residuals (variable name here is **sresid**), click:

```
Quick
 Generate Series
```

to access the ***Generate Series by Equation*** dialogue box of Fig. 2.9. Under the heading ‘Enter Equation’, type in the requirements. Leave the ‘Sample’ heading untouched, since we need the standardised residuals for all readings. Click the OK button and you will see that the variable **sresid** is now added to your list of variables. Double click **sresid** in this list of variables and you will be presented with the output of Fig. 2.10. Note that none of the 10 months is an outlier since no standardised residual lies beyond ± 2. To generate a plot of **sresid** over time, from the main EViews menu, click:

```
Quick
  Graph
```

Fig. 2.9 The Generate Series by Equation dialogue box

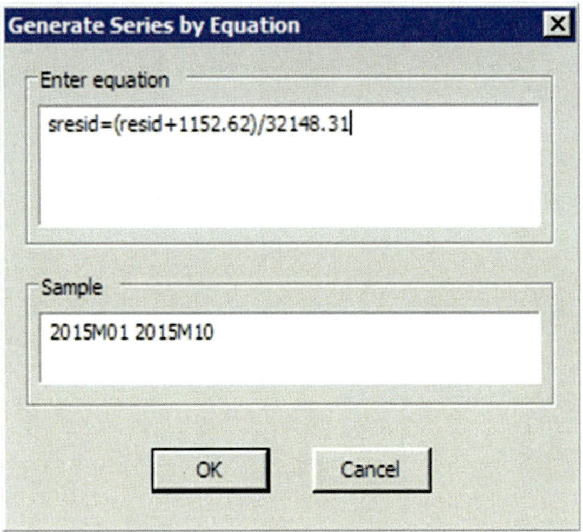

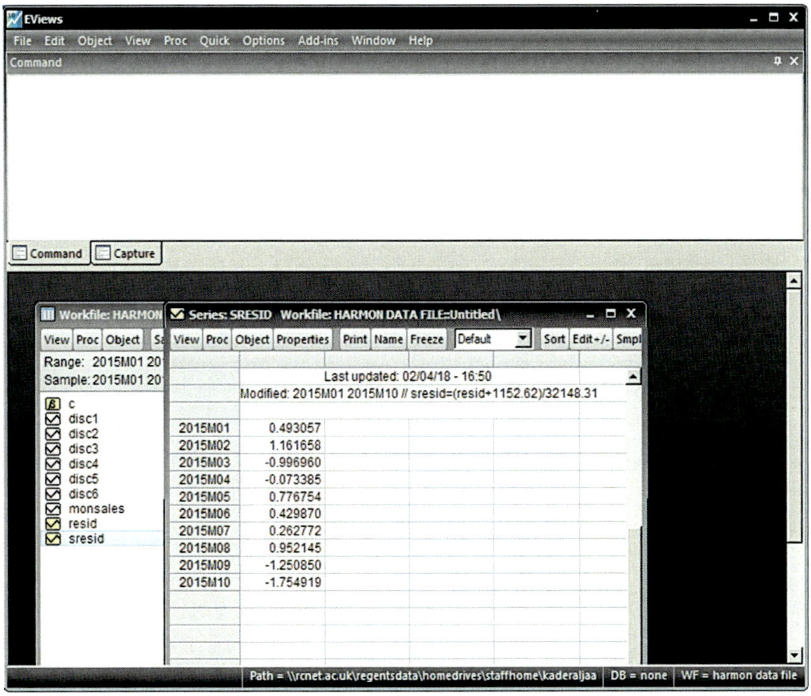

Fig. 2.10 A monthly listing of the standardised residuals

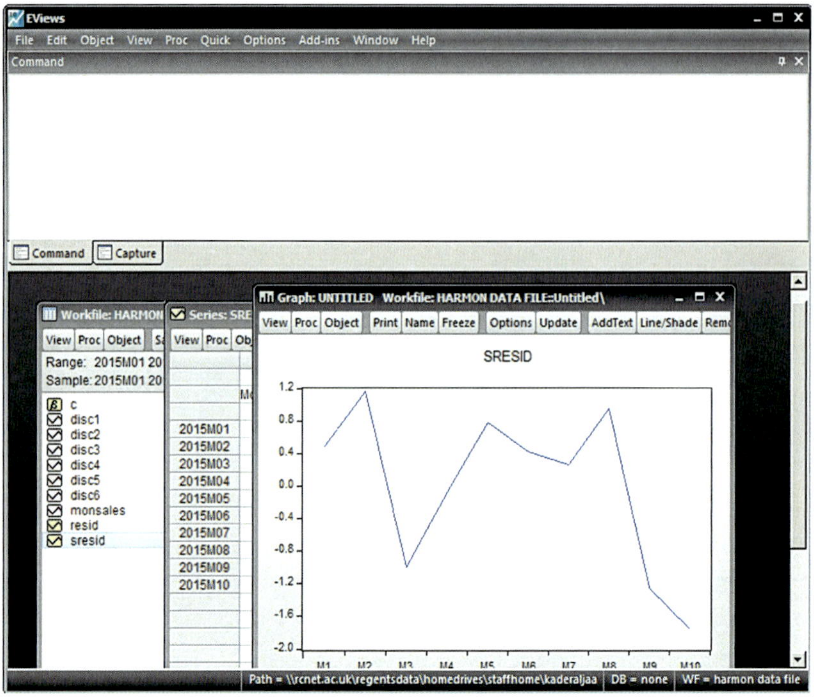

Fig. 2.11 A plot of the standardised residuals

and in the resulting Series list dialogue box, enter the name **sresid** and then choose Line & Sympbol. This will produce the results in Fig. 2.11, which may be edited by freezing the plot.

2.2 The Cobb-Douglas Function

The **Cobb–Douglas** production function is widely used to represent the relationship of an output to input. It was proposed by Knut Wicksell (1851–1926) and tested against statistical evidence by Charles Cobb and Paul Douglas in 1900–1928. For production, the function can be represented as follows:

$$Y = AL^{\alpha}K^{\beta} \tag{2.1}$$

where:

- Y = total production (the monetary value of all goods produced in a year)
- L = labor input
- K = capital input
- A = total factor productivity

- α and β are the output elastic ties of labor and capital, respectively. These values are constants determined by available technology.

Output elasticity measures the responsiveness of output to a change in levels of either labor or capital used in production. For example if $\alpha = 0.15$, a 1% increase in labor would lead to approximately a 0.15% increase in output. Further, if $\alpha + \beta = 1$, the production function has constant returns to scale. That is, if L and K are each increased by 20%, Y increases by 20%. If $\alpha + \beta < 1$, returns to scale are decreasing and if $\alpha + \beta > 1$, returns to scale are increasing. Assuming perfect competition and $\alpha + \beta = 1$, α and β can be shown to be labor and capital's share of output. Based on the above theory the model we use can be presented as follows:

$$LY = \beta_0 + \beta_1 LK + \beta_2 LL \tag{2.2}$$

To estimate the above model we collect output (Y), capital (K) and labor (L) data which we convert to logs in EViews by clicking:

```
Quick
   Generate Series
```

And then by typing LY = LOG(Y) under the heading 'Enter Equation'. Table 2.1 shows the converted log data of production, capital and labor.

2.2.1 *Estimation of the Cobb-Douglas Model*

Under ***Quick*** from the top-bar menu choose ***Graph***, write the name of one of the independent variables first and then the name of the dependent variable in the List of Series Window. The first name will be in the horizontal axis. Then from the list choose ***Scatter*** and then click OK which generates the graph in Fig. 2.12. Do the same for the second explanatory variable (LL) as shown in Fig. 2.13. Both figures indicate an increase in output (Y) when capital (K) and labor (L) increase. The next step is to run a regression with LY as the dependent variable, LK and LL the independent variables respectively. The regression output is presented in Fig. 2.14.

How to Paste EViews output on a Word Document File

1. After generating the graph click the button Print Scrn/SysRq on your keyboard.
2. Go to Start, All Programs and choose Accessories and then Paint.
3. Under Paint, choose the Edit available on the bar menu. Click Paste from Edit to have the output from EViews in the paint window.

(continued)

Table 2.1 Logs of Production, capital and labor data sets

LY	LK	LL
4.61	4.61	4.61
4.62	4.67	4.65
4.72	4.74	4.70
4.80	4.80	4.77
4.82	4.88	4.81
4.80	4.93	4.75
4.96	5.00	4.83
5.02	5.09	4.89
5.02	5.17	4.93
4.84	5.22	4.80
5.04	5.29	4.94
5.07	5.34	4.97
5.03	5.38	4.98
5.18	5.42	5.02
5.21	5.46	5.04
5.13	5.50	5.00
5.24	5.58	5.04
5.42	5.70	5.20
5.42	5.81	5.28
5.41	5.90	5.30
5.38	5.96	5.26
5.44	6.01	5.26
5.19	6.03	4.99
5.48	6.07	5.08

4. Use the Select button from the top bar menu to select the EViews output you want to cut.
5. Go to Edit and select Cut or alternatively press Ctrl + X.
6. Open a word document file and select paste which will then paste the output from EViews on the word file.

Based on Fig. 2.14, the regression line can be written as follows:

$$LY = -0.177 + 0.233LK + 0.807LL + e$$

2.2.2 *Interpret the Regression Equation*

1. As the sign of LK and LL indicate, there is a positive relationship between output, capital and labor respectively.

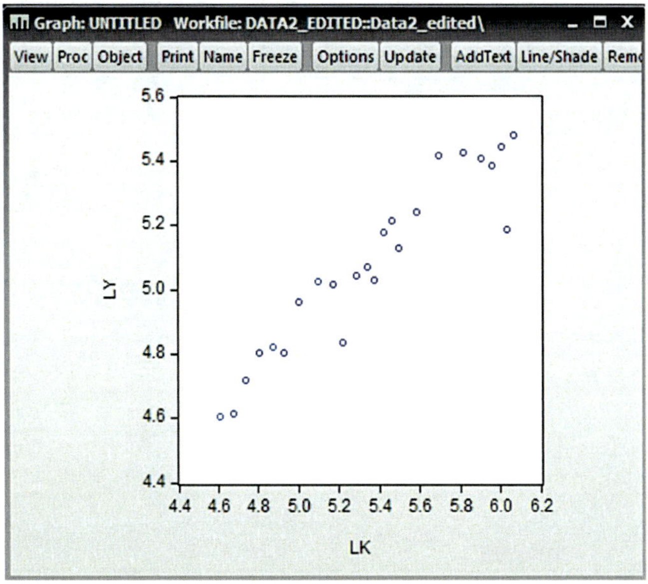

Fig. 2.12 Scatter plot of Log (Y) Log (K) data

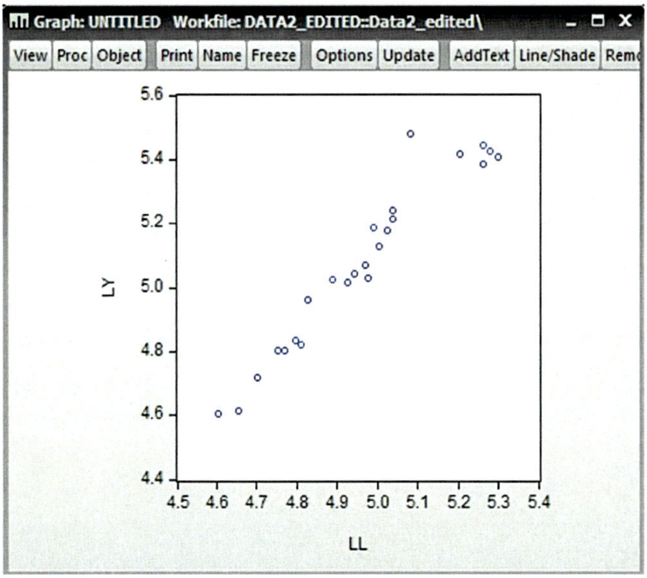

Fig. 2.13 Scatter plot of Log (Y) and Log (L) data

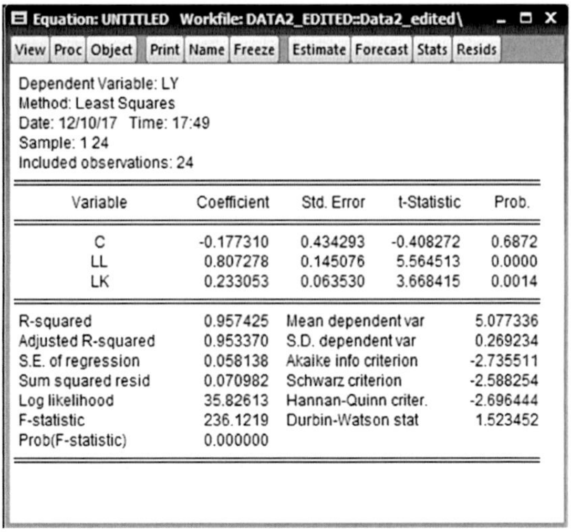

Equation: UNTITLED Workfile: DATA2_EDITED::Data2_edited\

View | Proc | Object | Print | Name | Freeze | Estimate | Forecast | Stats | Resids

Dependent Variable: LY
Method: Least Squares
Date: 12/10/17 Time: 17:49
Sample: 1 24
Included observations: 24

Variable	Coefficient	Std. Error	t-Statistic	Prob.
C	-0.177310	0.434293	-0.408272	0.6872
LL	0.807278	0.145076	5.564513	0.0000
LK	0.233053	0.063530	3.668415	0.0014

R-squared	0.957425	Mean dependent var	5.077336
Adjusted R-squared	0.953370	S.D. dependent var	0.269234
S.E. of regression	0.058138	Akaike info criterion	-2.735511
Sum squared resid	0.070982	Schwarz criterion	-2.588254
Log likelihood	35.82613	Hannan-Quinn criter.	-2.696444
F-statistic	236.1219	Durbin-Watson stat	1.523452
Prob(F-statistic)	0.000000		

Fig. 2.14 Estimate a regression equation dialogue box

2. One percent increase in capital (K) used in the production process will result in an expected increase of 0.233 of the total output (Y), leaving the labor input (L) constant.
3. Leaving capital input (K) constant, 1% increase in labor use (L) in the production process will result in an expected increase of 0.807 of the total output (Y).
4. If both capital (K) and labor (L) have not changed (meaning the value of both are zero), the output still has a value of −0.177. That said, it should be noted that the constant is not significant (Prob. 0.6872 > 0.05)

Remember, we use percentage when the data is converted to logs; otherwise we use unit change to interpret the regression equation.

2.2.3 *Testing the Coefficients*

Tests of the gradients (β) are run on the regression coefficients in order to find out whether they are reliable (i.e. statistically significant). For each regression coefficient (β), we run a hypothesis test as follows:

1. H_0: $\beta_1 = 0$
2. H_1: $\beta_1 \neq 0$
3. $\alpha = 5\%$ or 0.05
4. Calculate Test statistic = Coefficient β_1 – (β_1 * (β_1 in the H_0))/S.E. (β_1)

For LK, we reject the nil hypothesis (H_0) because the p-value of the gradient's test is 0.0014 which is smaller than 0.05. Therefore we conclude that evidence shows that the LK gradient is statistically significant. We run a similar hypothesis test for the LL gradient as well as for the intercept.

2.2.4 Comment on the Value of the R^2 and Testing the R^2

The R^2 value (coefficient of determination) suggests that approximately 95% of the total variation in output is explained by changes to capital (K) and labor (L). The F-statistic is usually used to test the significance of the R^2:

1. $H_0 : R^2 = 0.$
2. $H_1 : R^2 \neq 0.$
3. $\alpha = 5\%$ or 0.05.
4. Use calculated R^2 and find F-value through this formula.

$$F(k-1, n-k) = \frac{ESS/(k-1)}{RSS/(n-k)} = \frac{\frac{ESS}{TSS}/(k-1)}{\frac{ESS}{TSS}/(n-k)} = \frac{R^2/(k-1)}{(1-R^2)/(n-k)}$$

Test statistics derived in this way can be shown to follow a F-distribution with $v_1 = k-1$ and $v_2 = n-k$ degrees of freedom.

Testing at the 5% level of significance, one-tailed test, the critical F-value based on $3 - 1 = 2$ degree of freedom in the numerator and $24 - 3 = 21$ degrees of freedom in the denominator which shows a critical value of 3.44 based on the readings of the F-table. Therefore we reject the null hypothesis in favour of the alternative given that the F-statistic = 236.1219 which is greater than 3.44. We therefore conclude that the value of coefficient of determination is significant. The other way to make a decision is looking at the probability of F-statistic, which equals 0.0000 as reported in Fig. 2.14. As a result of that we reject the H_0 given that the F-statistic (0.0000) is smaller than 0.05 and we conclude the $\mathbf{R^2}$ is statistically significant.

2.2.5 Multicollinearity and Residual Analysis

A scatter plot of the residuals over time gives always an idea about their behaviour. Note that to obtain residual graphs and residual tests from EViews, the regression equation should be generated first. Follow steps 1–3 for a graphical examination of the residuals.

Step 1: Run the regression and generate the residuals

Under ***Quick*** choose ***Generate Series*** and in the command widow write: ***E*** = resid—then click OK to save the residuals. Double click on "**E**" from the workfile window to display the data shown in Fig. 2.15.

Step 2: Plot the residuals

Under ***Quick*** choose ***Graph*** and write **E** in the command window and then under specific select ***Dot plot***. Under ***Options***, select ***Axes & Scalling*** then ***data axis labels***. Under ***Axes ticks & lines***, choose ***zero line background*** as shown in Fig. 2.16. Click OK to generate the output of Fig. 2.17. It can be said that residual observations are serially correlated with one outlier.

Step 3: Generate a table of Actual, Fitted and Residuals

After running the regression, from the output window select: ***View—Actual, Fitted, Residual—Actual Fitted Residual Table*** then click OK to generate the output shown in Fig. 2.18.

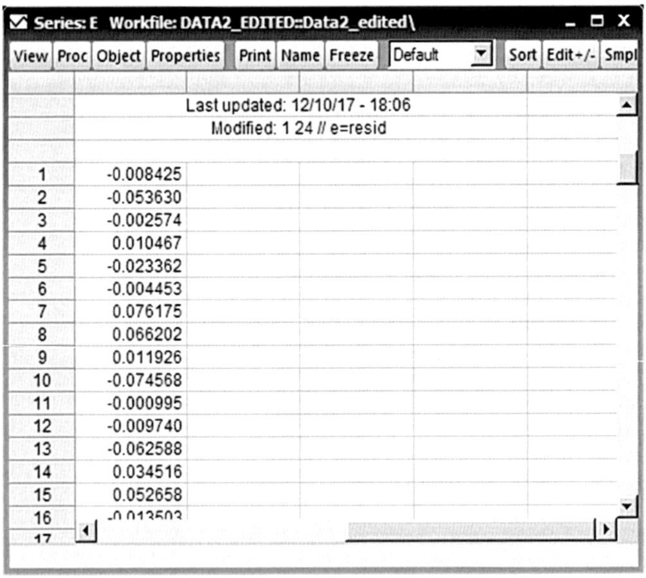

Fig. 2.15 Regression equation residuals

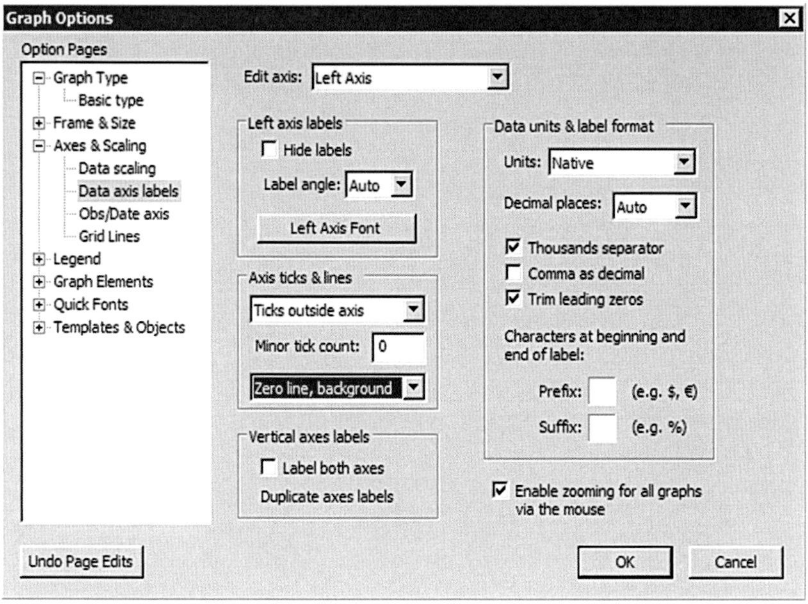

Fig. 2.16 Graph Options dialogue box

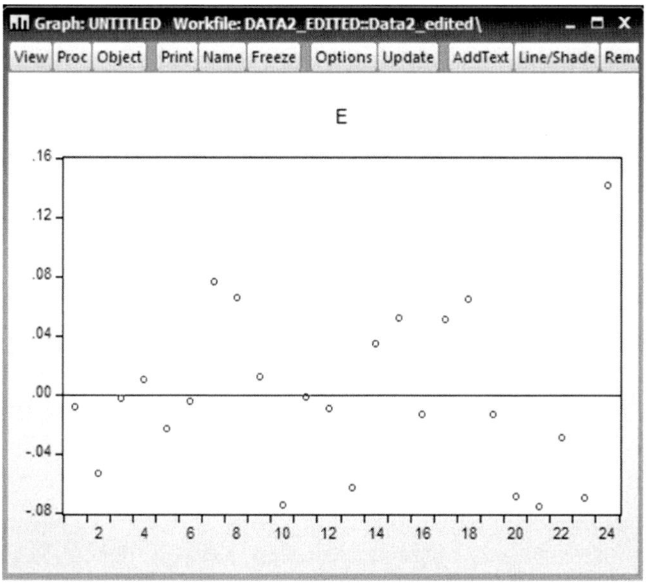

Fig. 2.17 Plot of the E residuals

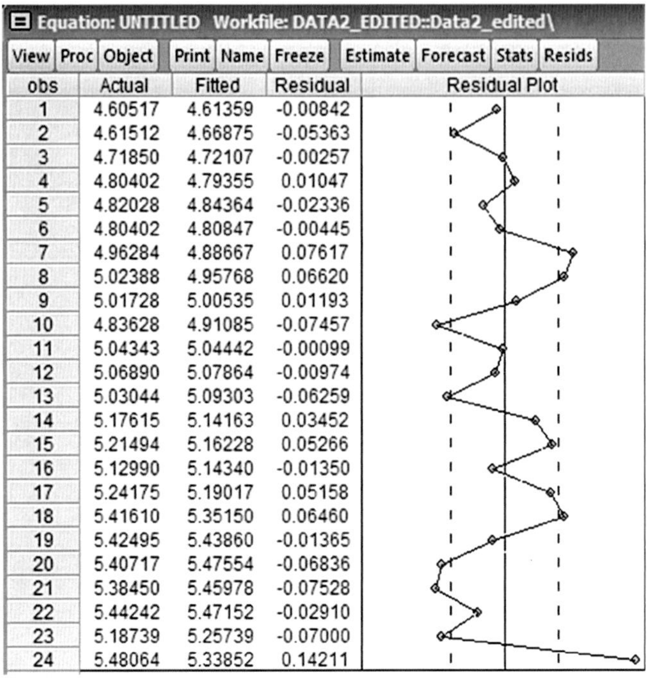

Equation: UNTITLED Workfile: DATA2_EDITED::Data2_edited\

View | Proc | Object | Print | Name | Freeze | Estimate | Forecast | Stats | Resids

obs	Actual	Fitted	Residual	Residual Plot
1	4.60517	4.61359	-0.00842	
2	4.61512	4.66875	-0.05363	
3	4.71850	4.72107	-0.00257	
4	4.80402	4.79355	0.01047	
5	4.82028	4.84364	-0.02336	
6	4.80402	4.80847	-0.00445	
7	4.96284	4.88667	0.07617	
8	5.02388	4.95768	0.06620	
9	5.01728	5.00535	0.01193	
10	4.83628	4.91085	-0.07457	
11	5.04343	5.04442	-0.00099	
12	5.06890	5.07864	-0.00974	
13	5.03044	5.09303	-0.06259	
14	5.17615	5.14163	0.03452	
15	5.21494	5.16228	0.05266	
16	5.12990	5.14340	-0.01350	
17	5.24175	5.19017	0.05158	
18	5.41610	5.35150	0.06460	
19	5.42495	5.43860	-0.01365	
20	5.40717	5.47554	-0.06836	
21	5.38450	5.45978	-0.07528	
22	5.44242	5.47152	-0.02910	
23	5.18739	5.25739	-0.07000	
24	5.48064	5.33852	0.14211	

Fig. 2.18 Actual, fitted and residuals data

According to Fig. 2.18, the residual line (under Residual Plot) has crossed the broken line several times which is considered as a sign of serial correlation problems.

Table 2.2 summarises various residual problems, their causes and remedies. It also suggests formal and informal ways to examine their behaviour. The formal approach involves four major steps:

1. Investigating the Multicollinearity Problem
2. Investigating the Autocorrelation Problem
3. Investigating the Heteroscedasticity Problem
4. Investigating the Normality problem

2.2.5.1 Examine the Multicollinearity Problem in EViews

We are faced with the problem of multicollinearity when we run a multiple regression—that is a regression with more than one explanatory variable. In the case of simple regressions this problem does need to be considered. The formal way involves running a regression between the independent variables without using the dependent variable. For example, run the regression LK C LL (without using LY, the

Table 2.2 A summary of residuals behaviour, causes and remedies

Problem	Common	Causes	Informal	Formal	Elimination
Multicollinearity	In cross section and time series data	By a linear relationship between the explanatory variables	The residual plot shows one or several outliers	– Correlation matrix – Running a regression between the explanatory variables and run the Variable Inflation Factor	– Include further relevant variables in the model – Increase the number of observations
Autocorrelation	Time series data	By a relationship between the value of an observation in time t with its value in time t-1	The residual plot shows positive or negative follow-up among observations	– Durbin-Watson test – LM test – Lagrange multiplier test	Take the first difference of both sides of the equation and run the new equation
Heteroskedasticity	Cross section data	By variance changes of the residual through the time	The residual plot shows a wider variance as observations move from the origin to the right	– White's test – Goldfeld-Quandt test	Run the regression with Logs
Normality	In both types	Misspecifications of the model	The residual histogram shows abnormal distribution	– Jarque-Bera test	Add relevant variables to the model and use bigger size samples

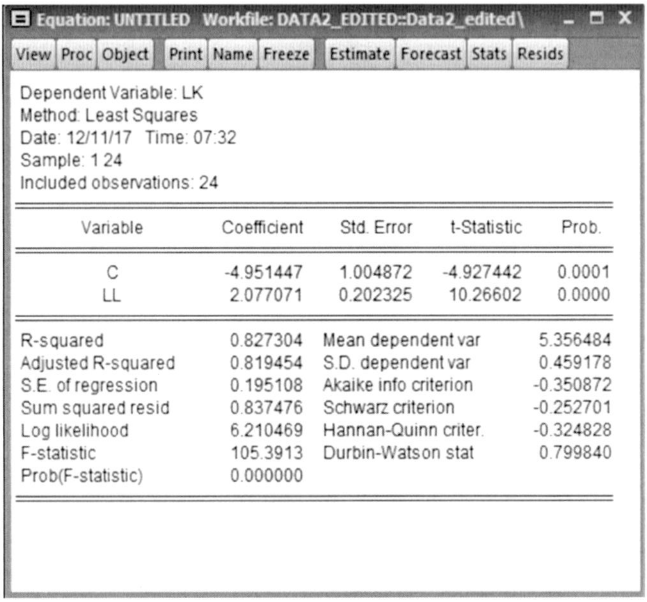

Dependent Variable: LK
Method: Least Squares
Date: 12/11/17 Time: 07:32
Sample: 1 24
Included observations: 24

Variable	Coefficient	Std. Error	t-Statistic	Prob.
C	-4.951447	1.004872	-4.927442	0.0001
LL	2.077071	0.202325	10.26602	0.0000

R-squared	0.827304	Mean dependent var	5.356484
Adjusted R-squared	0.819454	S.D. dependent var	0.459178
S.E. of regression	0.195108	Akaike info criterion	-0.350872
Sum squared resid	0.837476	Schwarz criterion	-0.252701
Log likelihood	6.210469	Hannan-Quinn criter.	-0.324828
F-statistic	105.3913	Durbin-Watson stat	0.799840
Prob(F-statistic)	0.000000		

Fig. 2.19 Testing for Multicollinearity

dependent). Choose one of the independent variables as the dependent and run regression against the rest as shown in Fig. 2.19. The aim of running such a regression is to find the coefficient determination (R^2) which is then used to calculate the Variable Inflation Factor (VIF) as shown below:

$$VIF = \frac{1}{\left(1 - R_j^2\right)} = \frac{1}{1 - 0.82730} = 5.790$$

We then follow the steps below to test for multicollinearity:

1. H_0: There is no problem of multicollinearity
2. H_1: There is problem of multicollinearity.
3. $\alpha = 5\%$ or 0.05
4. VIF = 5.790
5. When the value of VIF is between 5 and 10, we reject the nil hypothesis H_0
6. Based on evidence we conclude that there is a problem of multicollinearity in this series

How can we overcome Multicollinearity?

- Drop one of the collinear variables
- Transform the highly correlated variables into a ratio
- Collect more data e.g. a longer run of data or switch to a higher frequency

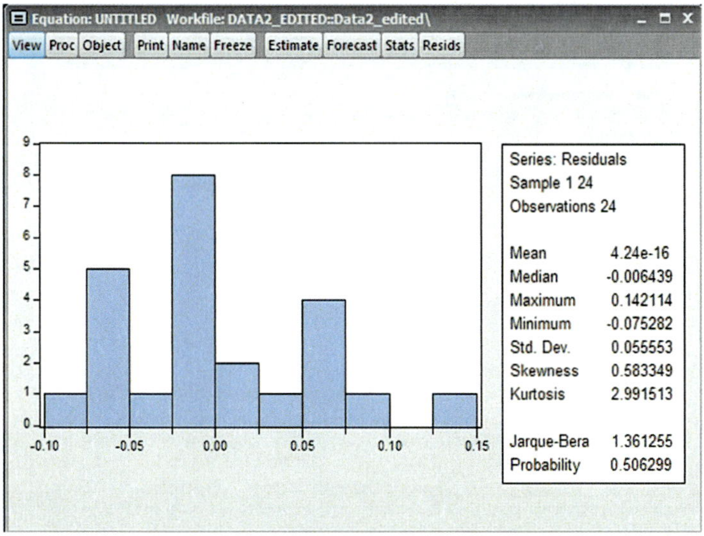

Fig. 2.20 Testing for Normality of the residuals

2.2.5.2 Examine the Normality Problem in EViews

If the residuals are not normally distributed, they should not be used in Z-tests or in any other tests derived from the Normal distribution, such as t-tests (The F-tests and chi-square tests cannot be used as well). Moreover, if the residuals are not normally distributed, then the dependent variable, or at least one explanatory variable, may have the wrong functional form. Another possibility is that one or more important variables are missing, etc. The EViews results of running normality tests on the Cobb-Douglas data (LY C LK LL) is shown in Fig. 2.20.

The steps to run the normality test can be summarized are as follows:

1. H_0: Residuals follow a Normal distribution
2. H_1: Residuals do not follow a Normal distribution
3. $\alpha = 5\%$ or 0.05
4. Jarque-Bera test statistic = 1.361255
5. Prob. = 0.506299 > 0.05 therefore we cannot reject the H_0
6. Based on evidence, the residuals follow a normal distribution.

2.2.5.3 Examine the Heteroscedasticity Problem in EViews

To find out whether the residual variance has remained constant through the whole process we need to run a formal test known as the White Test as seen earlier in this textbook. To interpret the output on Fig. 2.21, we follow the steps below:

Equation: UNTITLED Workfile: DATA2_EDITED::Data2_edited\

View | Proc | Object | Print | Name | Freeze | Estimate | Forecast | Stats | Resids

Heteroskedasticity Test: White

F-statistic	2.864997	Prob. F(5,18)	0.0449
Obs*R-squared	10.63572	Prob. Chi-Square(5)	0.0591
Scaled explained SS	8.108422	Prob. Chi-Square(5)	0.1504

Test Equation:
Dependent Variable: RESID^2
Method: Least Squares
Date: 12/10/17 Time: 18:29
Sample: 1 24
Included observations: 24

Variable	Coefficient	Std. Error	t-Statistic	Prob.
C	-0.948274	1.017984	-0.931522	0.3639
LL^2	-0.094535	0.105155	-0.899006	0.3805
LL*LK	0.051693	0.079661	0.648911	0.5246
LL	0.645639	0.656921	0.982826	0.3387
LK^2	-0.000509	0.016635	-0.030624	0.9759
LK	-0.239810	0.251077	-0.955128	0.3522

R-squared	0.443155	Mean dependent var	0.002958
Adjusted R-squared	0.288476	S.D. dependent var	0.004264
S.E. of regression	0.003596	Akaike info criterion	-8.205473
Sum squared resid	0.000233	Schwarz criterion	-7.910959
Log likelihood	104.4657	Hannan-Quinn criter.	-8.127338
F-statistic	2.864997	Durbin-Watson stat	1.910318
Prob(F-statistic)	0.044905		

Fig. 2.21 The White test statistic in EViews

1. H_0: Residuals are Homoscedastic
2. H_1: Residuals are Heteroskedastic
3. $\alpha = 5\%$ or 0.05
4. The p-value = 0.0591
5. As Prob. = 0.0591 > 0.05 therefore we cannot reject the nil hypothesis and we conclude based on evidence that there is no problem of Heteroscedasticity.

How to overcome the problem of Heteroscedasticity?

If the form (i.e. the cause) of the heteroscedasticity is known, then we can use an estimation method which takes this into account (called generalised least squares, GLS). A simple illustration of GLS is as follows:

Suppose that the error variance is related to another variable z_t by

$$\text{var}(u_t) = \sigma^2 z_t^2$$

To remove the heteroscedasticity, divide the regression equation by z_t

$$\frac{y_t}{z_t} = \beta_1 \frac{1}{z_t} + \beta_2 \frac{x_{2t}}{z_t} + \beta_3 \frac{x_{3t}}{z_t} + v_t$$

where $v_t = \frac{u_t}{z_t}$ is an error term.

Therefore we will have:

$$\text{var}(v_t) = \text{var}\left(\frac{u_t}{z_t}\right) = \frac{\text{var}(u_t)}{z_t^2} = \frac{\sigma^2 z_t^2}{z_t^2} = \sigma^2$$

Which will then mean that the disturbances from the new regression equation will be homoscedastic.

Other solutions include:

1. Transforming the variables into **logs** or reducing by some other measure of "size".
2. Using White's heteroscedasticity consistent standard error estimates. The effect of using White's correction is that, in general, the standard errors for the slope coefficients are increased relative to the usual OLS standard errors.

2.2.5.4 Examine the Autocorrelation Problem in EViews

Autocorrelation happens when there is a correlation between two consecutive observations of the residuals. This is a common problem when time series data is used and a correlogram can be generated to test for autocorrelation. Click:

```
View
  Residual Tests
   Correlogram
     Q-Statistics
```

And click OK. If residuals show no serial correlation, the autocorrelations and partial autocorrelations at all lags should be nearly zero and all Q-Statistics should be insignificant with large p-values. Based on the evidence from Fig. 2.22 the residuals are not auto-correlated.

LM Test: Lagrange Multiplier Test

For autocorrelation, we also use the Durbin-Watson test however, this test cannot be used when we have time lags as independent variables. We must use the serial correlation LM test if we have time lags in the right hand-side of the regression equation.

■■

Serial Correlation using LM Test

After running the main regression select ***View – Residual Tests – Serial Correlation LM*** test to generate the output of Fig. 2.23.

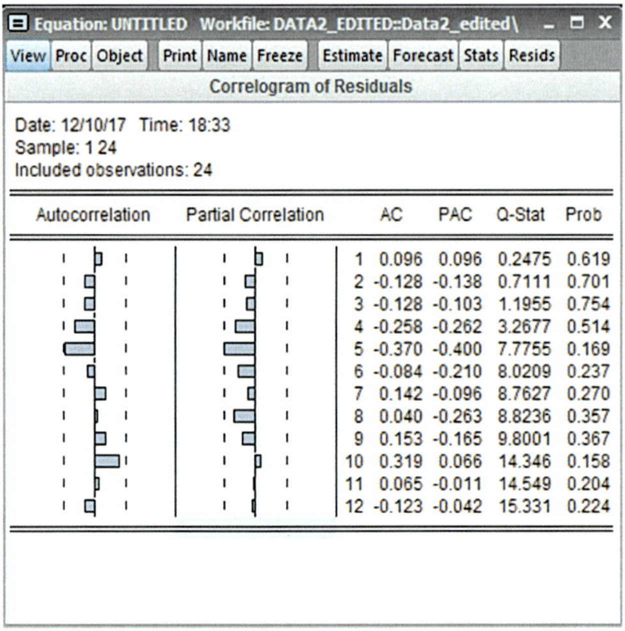

Fig. 2.22 Correlogram of residuals

Note:

The Null Hypothesis and Alternative Hypothesis of the LM Test are completely opposite to the Null Hypothesis and Alternative Hypothesis of the Correlogram Test.

1. H_0: No serial correlation
2. H_a: There is serial correlation
3. $\alpha = 5\%$ or 0.05
4. From Fig. 2.23, the p-value = 0.4997
5. As Prob. = 0.4997 > 0.05, we accept the H_0 and conclude that there is no serial correlation problem.

How can we overcome autocorrelation problems?

Multiply $Y_t = \alpha + \beta X_t + u_t$ by ρ to get $\rho Y_{t-1} = \alpha\rho + \rho\beta X_{t-1} + \rho u_{t-1}$

Subtract ρY_{t-1} from $Y_t = Y_t - \rho Y_{t-1} = \alpha - \alpha\rho + \beta X_t - \rho\beta X_{t-1} + (u_t - \rho u_{t-1})$

The new main equation which is free from autocorrelation is:

$$Y_t - \rho Y_{t-1} = \alpha(1-\rho) + \beta(X_t - \rho X_{t-1}) + (u_t - \rho u_{t-1})$$

If you estimate this new model instead of the main model, the new one will be free of autocorrelation. The last equation is known as AR (1); the first order autocorrelation equation.

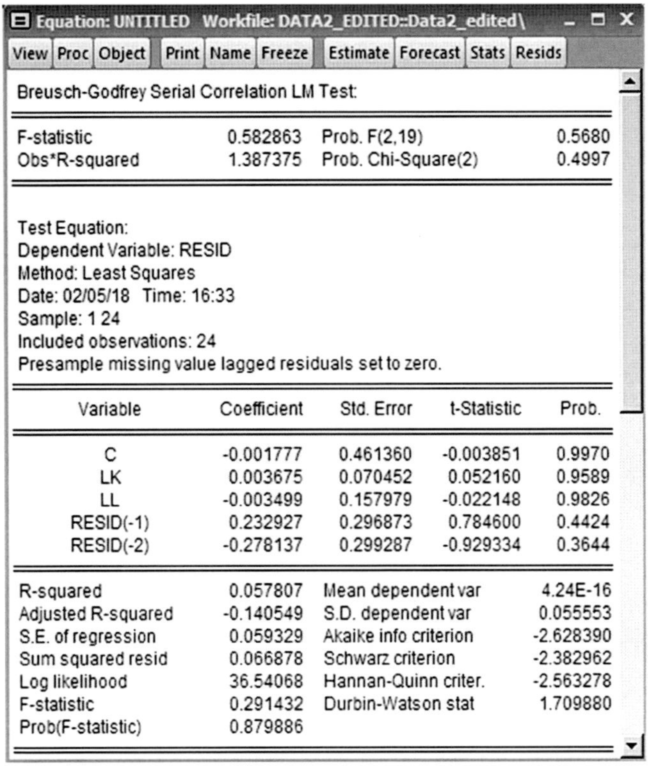

Equation: UNTITLED Workfile: DATA2_EDITED::Data2_edited\

View | Proc | Object | Print | Name | Freeze | Estimate | Forecast | Stats | Resids

Breusch-Godfrey Serial Correlation LM Test:

F-statistic	0.582863	Prob. F(2,19)	0.5680
Obs*R-squared	1.387375	Prob. Chi-Square(2)	0.4997

Test Equation:
Dependent Variable: RESID
Method: Least Squares
Date: 02/05/18 Time: 16:33
Sample: 1 24
Included observations: 24
Presample missing value lagged residuals set to zero.

Variable	Coefficient	Std. Error	t-Statistic	Prob.
C	-0.001777	0.461360	-0.003851	0.9970
LK	0.003675	0.070452	0.052160	0.9589
LL	-0.003499	0.157979	-0.022148	0.9826
RESID(-1)	0.232927	0.296873	0.784600	0.4424
RESID(-2)	-0.278137	0.299287	-0.929334	0.3644

R-squared	0.057807	Mean dependent var	4.24E-16
Adjusted R-squared	-0.140549	S.D. dependent var	0.055553
S.E. of regression	0.059329	Akaike info criterion	-2.628390
Sum squared resid	0.066878	Schwarz criterion	-2.382962
Log likelihood	36.54068	Hannan-Quinn criter.	-2.563278
F-statistic	0.291432	Durbin-Watson stat	1.709880
Prob(F-statistic)	0.879886		

Fig. 2.23 Serial correlation LM test

Final Decision: If a series suffers from autocorrelation, try to take the first order autocorrelation (the first difference) of both sides of the equation; the new equation will be autocorrelation-free.

Chapter 3
Time Series Analysis

Much of the data used and reported in Economics is recorded over time. The term *time series* is given to a sequence of data, (usually inter-correlated), each of which is associated with a moment in time. Examples like daily stock prices, weekly inventory levels or monthly unemployment figures are called discrete series, i.e. readings are taken at set times, usually equally spaced. The form of the data for a time series is, therefore, a single list of readings taken at regular intervals. It is this type of data that will concern us in this and the next chapter.

There are two aspects to the study of time series. Firstly, *the analysis phase* attempts to summarize the properties of a series and to characterize its salient features. Essentially, this involves examination of a variable's past behaviour. Secondly, the *modelling phase* is performed in order to generate future forecasts. This chapter examines the analysis phase. It should be noted that in time series, there is no attempt to relate the variable under study to other variables. This is the goal of regression methods. Rather, in time series analysis, movements in the study variable are 'explained' only in terms of its own past or by its position in relation to time. Forecasts are then made by extrapolation. Graphics are particularly useful in time series studies. They may, for example, highlight regular movements in data and which may assist model specification or selection. Given the excellent graphics capabilities of EViews 10, the package is particularly amenable to time series analysis. The generation of various plots of temporal data over time is assisted if date variables are defined in EViews 10.

In 1936, John M. Keynes wrote his influential book, "The General Theory of Employment, Interest Rates, and Money" in which he developed the theory of money demand, known as the **liquidity preference theory**. His ideas formed the basis for the liquidity preference framework. According to the Keynesian theory, the real money demand (RMD) is a function of the real GDP (RGDP) and the interest rates (INT). The model can be written as:

$$RMD_t = \beta_0 + \beta_1 RGDP_t + \beta_2 INT_t$$

A. Aljandali, M. Tatahi, *Economic and Financial Modelling with EViews*, Statistics and Econometrics for Finance, https://doi.org/10.1007/978-3-319-92985-9_3

Processes to deal with time series data

The processes involve three major steps:

1. Identify whether the series are stationary or not,
2. Run Cointegration test to identify whether there is a long-term relationship between the series or not,
3. Identify the Error Correction Model (ECM).

1. **Identify non-stationary processes**

   There are two methods which we will carry both for each time series data:

Informal method: this involves two steps

(a) Plot the time series
(b) Run the correlogram test.

Formal Method: this involves one step

Run the Augmented Dickey-Fuller (ADF) test.

3.1 Time Series One: The Real Money Demand (RMD)

3.1.1 Informal Method: Plot the Time Series and Generate a Correlogram

To plot the time series follow these steps in EViews:
Go to ***Quick—Graph*** and then write, RMD—Choose ***Line & Symbol*** then click OK which will generate Fig. 3.1

The series appears to be a random walk with a drift: it has moved up and down for some time, and it trended upward latter. It is a non-stationary process. Double click the RMD from the Workfile to get the data of the series RMD.

To generate a correlogram follow these steps in EViews:
Go to ***View—Correlogram***—Select a C***orrelogram of the Level*** and then click OK to generate the output of Fig. 3.2

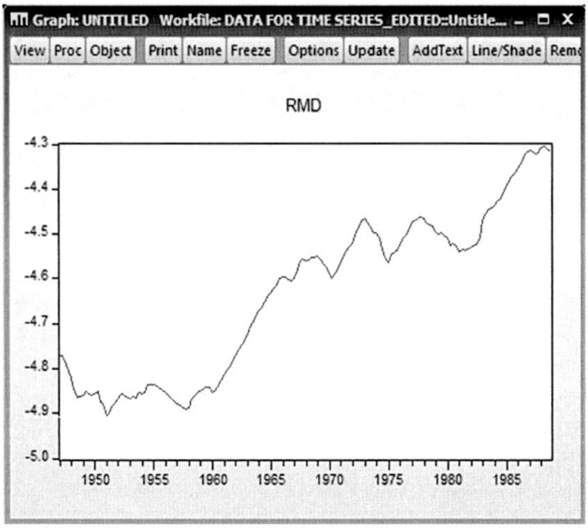

Fig. 3.1 Plot of the RMD series

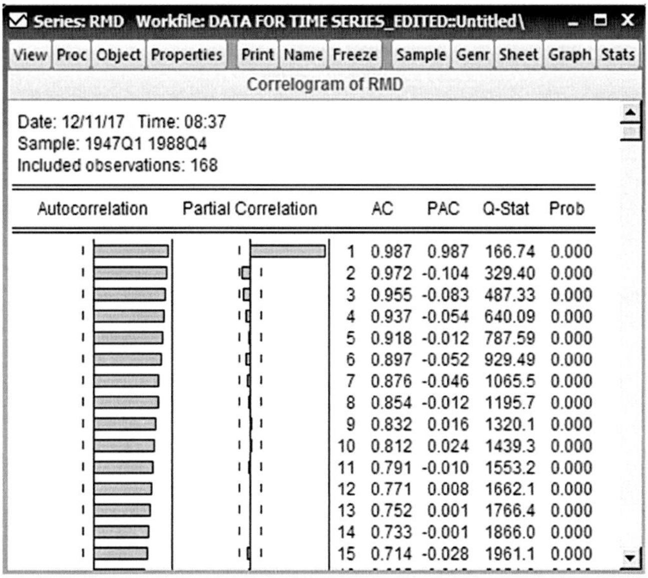

	AC	PAC	Q-Stat	Prob
1	0.987	0.987	166.74	0.000
2	0.972	-0.104	329.40	0.000
3	0.955	-0.083	487.33	0.000
4	0.937	-0.054	640.09	0.000
5	0.918	-0.012	787.59	0.000
6	0.897	-0.052	929.49	0.000
7	0.876	-0.046	1065.5	0.000
8	0.854	-0.012	1195.7	0.000
9	0.832	0.016	1320.1	0.000
10	0.812	0.024	1439.3	0.000
11	0.791	-0.010	1553.2	0.000
12	0.771	0.008	1662.1	0.000
13	0.752	0.001	1766.4	0.000
14	0.733	-0.001	1866.0	0.000
15	0.714	-0.028	1961.1	0.000

Fig. 3.2 Correlogram of RMD series

Note:

The Correlogram Test: the null hypothesis and the alternative hypothesis for the correlogram and for the formal test of the Augmented Dickey-Fuller test are completely opposite. Results of running the Correlogram test are reported in Fig. 3.2

1. H_0: The series has no unit root
2. H_1: The series has a unit root
3. $\alpha = 5\%$ or 0.05

There are no test statistics to be calculated. We look at the "Prob" column in Fig. 3.2

4. The p-values are = 0.000.
5. Since Prob = 0.000 < 0.05, we reject the H_0 and accept the H_1.
6. We conclude that RMD series is not stationary

3.1.2 Formal Method: Run the Augmented Dickey-Fuller (ADF) Test

Before running the ADF test in EViews, we need to decide whether to include a trend component as part of the analysis. If the time series plot exhibits a trend then a trend component should be included in the analysis. If not then do not include the trend but always remember to include an intercept.

Follow these steps in EViews:
Double click on RMD from the Workfile.
From the Window containing RMD, choose ***View—Unit Root test***
Under Test type, select ***Augmented Dickey-Fuller***
Under Test for unit root in, tick ***Level***
Under Include in test equation, choose ***Trend*** and ***Intercept*** if the plot of the series indicates it is trended. Otherwise select only ***Intercept*** and then click OK to generate the output in Fig. 3.3.

1. H_0: series has a unit root and it is not stationary
2. H_1: series has no unit root and it is stationary.
3. $\alpha = 5\%$ or 0.05

ADF test statistic is −3,160,611.

4. The p-values = 0.0961.

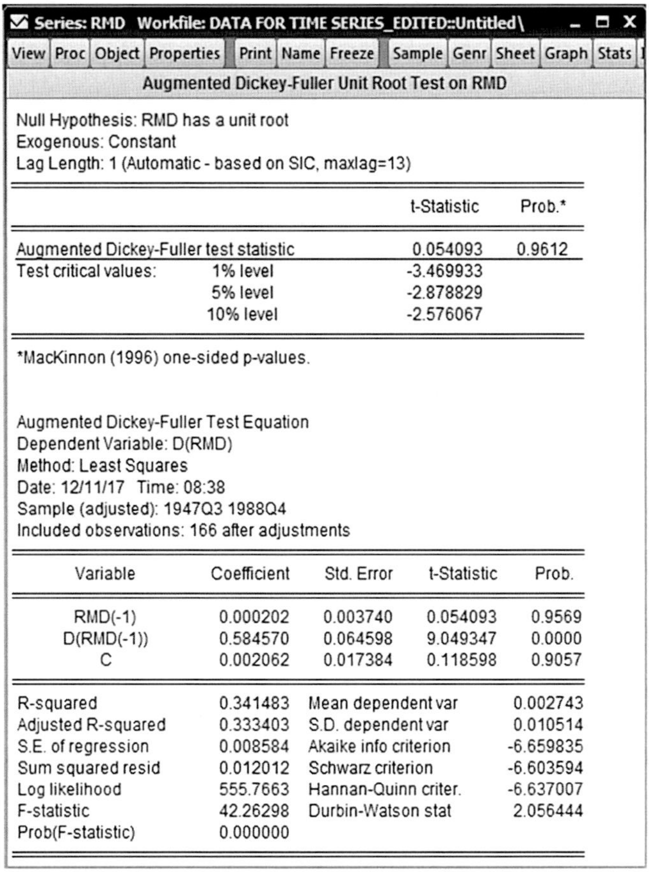

Series: RMD Workfile: DATA FOR TIME SERIES_EDITED::Untitled\

View | Proc | Object | Properties | Print | Name | Freeze | Sample | Genr | Sheet | Graph | Stats

Augmented Dickey-Fuller Unit Root Test on RMD

Null Hypothesis: RMD has a unit root
Exogenous: Constant
Lag Length: 1 (Automatic - based on SIC, maxlag=13)

		t-Statistic	Prob.*
Augmented Dickey-Fuller test statistic		0.054093	0.9612
Test critical values:	1% level	-3.469933	
	5% level	-2.878829	
	10% level	-2.576067	

*MacKinnon (1996) one-sided p-values.

Augmented Dickey-Fuller Test Equation
Dependent Variable: D(RMD)
Method: Least Squares
Date: 12/11/17 Time: 08:38
Sample (adjusted): 1947Q3 1988Q4
Included observations: 166 after adjustments

Variable	Coefficient	Std. Error	t-Statistic	Prob.
RMD(-1)	0.000202	0.003740	0.054093	0.9569
D(RMD(-1))	0.584570	0.064598	9.049347	0.0000
C	0.002062	0.017384	0.118598	0.9057

R-squared	0.341483	Mean dependent var	0.002743
Adjusted R-squared	0.333403	S.D. dependent var	0.010514
S.E. of regression	0.008584	Akaike info criterion	-6.659835
Sum squared resid	0.012012	Schwarz criterion	-6.603594
Log likelihood	555.7663	Hannan-Quinn criter.	-6.637007
F-statistic	42.26298	Durbin-Watson stat	2.056444
Prob(F-statistic)	0.000000		

Fig. 3.3 RMD unit root test

5. Prob = 0.0961 > 0.05 accept the H_0.
6. The series has a unit root and it is not stationary.

If the result found in the formal test is different from the result from the plot and the correlogram test of the series, stick to the result from the formal test. Sometimes the result from the informal test is different from the result from the formal test. We always stick to the result from the formal test.

3.2 Time Series Two: The Real GDP (RGDP)

3.2.1 *Informal Method: Plot the Time Series and Generate a Correlogram*

Go to ***Quick—Graph*** and write RGDP then choose ***Line & Symbol*** and click OK to generate the graph in Fig. 3.4 which appears to be non-stationary. The next step would be to generate a correlogram of the RGDP which can be achieved by going to ***View—Correlogram—Level*** and by clicking OK to generate the output of Fig. 3.5

Note:

The null hypothesis and the alternative hypothesis for the correlogram and for the formal test of the Augmented Dickey-Fuller test are completely opposite.

1. H_0: The series has no unit root
2. H_1: The series has a unit root
3. $\alpha = 5\%$ or 0.05

There are no test statistics to calculate. We look at the "Prob" column.

4. The p-values = 0.000.
5. Prob = 0.000 < 0.05. We reject the H_0 and accept the H_1.
6. The series has a unit root and is not stationary

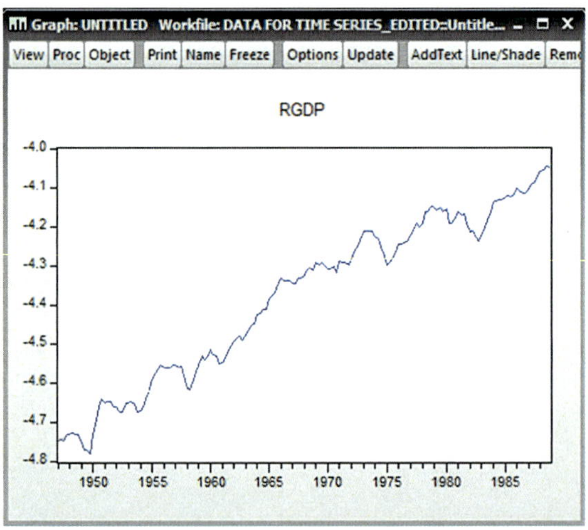

Fig. 3.4 RGDP plot

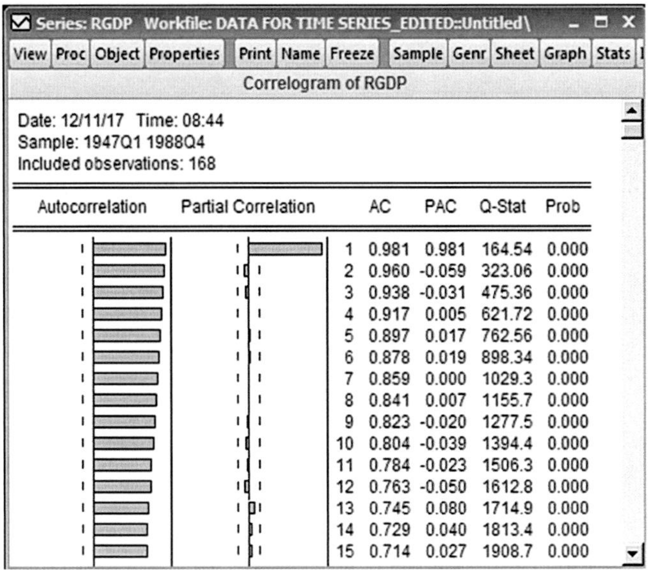

	AC	PAC	Q-Stat	Prob
1	0.981	0.981	164.54	0.000
2	0.960	-0.059	323.06	0.000
3	0.938	-0.031	475.36	0.000
4	0.917	0.005	621.72	0.000
5	0.897	0.017	762.56	0.000
6	0.878	0.019	898.34	0.000
7	0.859	0.000	1029.3	0.000
8	0.841	0.007	1155.7	0.000
9	0.823	-0.020	1277.5	0.000
10	0.804	-0.039	1394.4	0.000
11	0.784	-0.023	1506.3	0.000
12	0.763	-0.050	1612.8	0.000
13	0.745	0.080	1714.9	0.000
14	0.729	0.040	1813.4	0.000
15	0.714	0.027	1908.7	0.000

Fig. 3.5 Corrlogram of RGDP

3.2.2 *Formal Method: Run the Augmented Dickey-Fuller (ADF) Test*

Follow these steps in EViews:
Double click on RGDP from the Workfile.
From the Window containing RGDP, choose ***View—Unit Root test***
Under Test type, select ***Augmented Dickey-Fuller***
Under Test for unit root in, tick ***Level***
Under Include in test equation, choose ***Trend*** and ***Intercept*** if the plot of the series indicates it is trended. Otherwise select only ***Intercept*** and then click OK to generate the output in Fig. 3.6.

1. H_0: series has a unit root and it is not stationary
2. H_1: series has no unit root and it is stationary.
3. $\alpha = 5\%$ or 0.05

ADF test statistic is −3,107,207.

1. The p-values = 0.1081.
2. Prob = 0.1081 > 0.05. We cannot reject the H_0
3. The series has a unit root and it is not stationary.

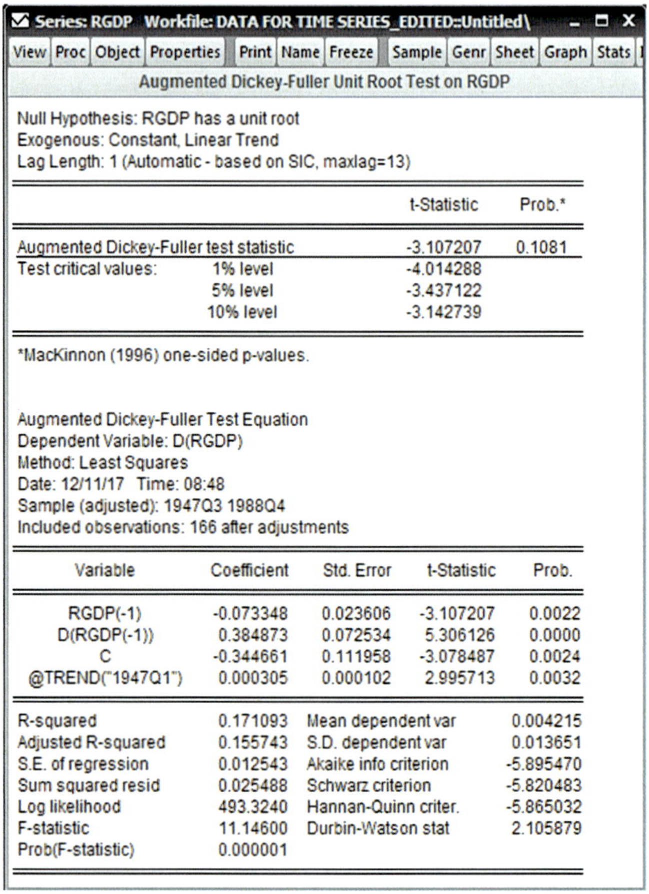

Series: RGDP Workfile: DATA FOR TIME SERIES_EDITED::Untitled\

View | Proc | Object | Properties | Print | Name | Freeze | Sample | Genr | Sheet | Graph | Stats

Augmented Dickey-Fuller Unit Root Test on RGDP

Null Hypothesis: RGDP has a unit root
Exogenous: Constant, Linear Trend
Lag Length: 1 (Automatic - based on SIC, maxlag=13)

		t-Statistic	Prob.*
Augmented Dickey-Fuller test statistic		-3.107207	0.1081
Test critical values:	1% level	-4.014288	
	5% level	-3.437122	
	10% level	-3.142739	

*MacKinnon (1996) one-sided p-values.

Augmented Dickey-Fuller Test Equation
Dependent Variable: D(RGDP)
Method: Least Squares
Date: 12/11/17 Time: 08:48
Sample (adjusted): 1947Q3 1988Q4
Included observations: 166 after adjustments

Variable	Coefficient	Std. Error	t-Statistic	Prob.
RGDP(-1)	-0.073348	0.023606	-3.107207	0.0022
D(RGDP(-1))	0.384873	0.072534	5.306126	0.0000
C	-0.344661	0.111958	-3.078487	0.0024
@TREND("1947Q1")	0.000305	0.000102	2.995713	0.0032

R-squared	0.171093	Mean dependent var	0.004215
Adjusted R-squared	0.155743	S.D. dependent var	0.013651
S.E. of regression	0.012543	Akaike info criterion	-5.895470
Sum squared resid	0.025488	Schwarz criterion	-5.820483
Log likelihood	493.3240	Hannan-Quinn criter.	-5.865032
F-statistic	11.14600	Durbin-Watson stat	2.105879
Prob(F-statistic)	0.000001		

Fig. 3.6 RGDP unit root test

3.3 Time Series Three: Interest Rates (INT)

3.3.1 Informal Method: Plot the Time Series and Generate a Correlogram

Go to ***Quick—Graph*** and write INT then choose ***Line & Symbol*** and click OK to generate the graph of Fig. 3.7 which appears to be non-stationary. The next step would be to generate a correlogram of the INT which can be achieved by going to ***View—Correlogram—Level*** and by clicking OK to generate Fig. 3.8

Fig. 3.7 INT plot

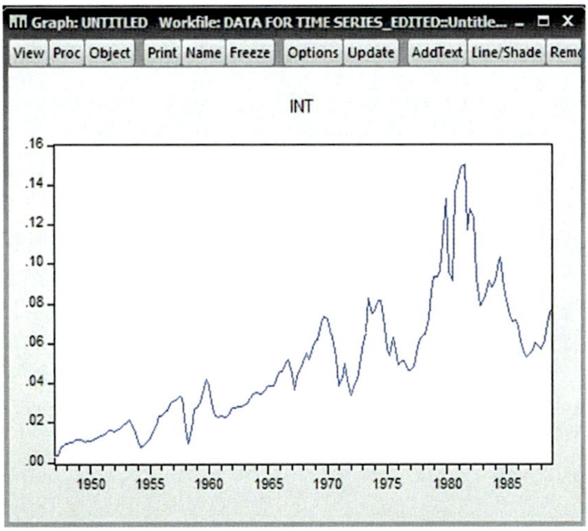

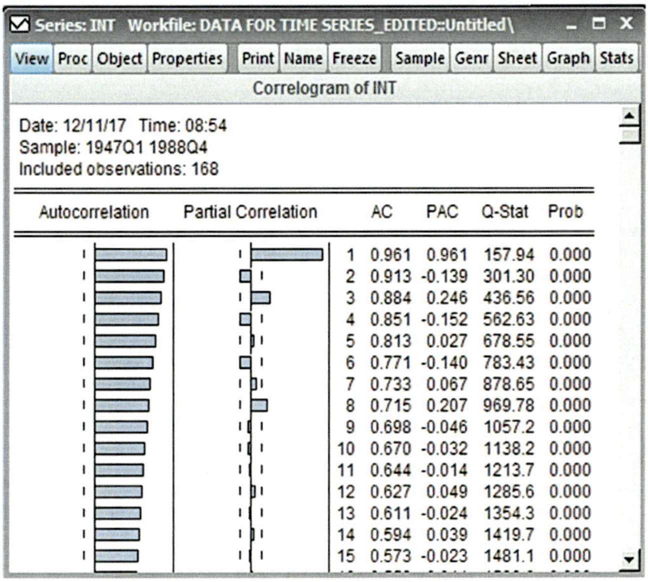

Date: 12/11/17 Time: 08:54
Sample: 1947Q1 1988Q4
Included observations: 168

Autocorrelation	Partial Correlation		AC	PAC	Q-Stat	Prob
		1	0.961	0.961	157.94	0.000
		2	0.913	-0.139	301.30	0.000
		3	0.884	0.246	436.56	0.000
		4	0.851	-0.152	562.63	0.000
		5	0.813	0.027	678.55	0.000
		6	0.771	-0.140	783.43	0.000
		7	0.733	0.067	878.65	0.000
		8	0.715	0.207	969.78	0.000
		9	0.698	-0.046	1057.2	0.000
		10	0.670	-0.032	1138.2	0.000
		11	0.644	-0.014	1213.7	0.000
		12	0.627	0.049	1285.6	0.000
		13	0.611	-0.024	1354.3	0.000
		14	0.594	0.039	1419.7	0.000
		15	0.573	-0.023	1481.1	0.000

Fig. 3.8 Correlogram of INT

1. H_0: The series has no unit root
2. H_1: The series has a unit root
3. $\alpha = 5\%$ or 0.05

There are no test statistics to calculate. We look at the "Prob" column.

1. The p-values = 0.000.
2. Prob = 0.000 < 0.05. We reject the H_0 and accept the H_1.
3. The series has a unit root and is not stationary

3.3.2 Formal Method: Run the Augmented Dickey-Fuller (ADF) Test

The ADF test has been generated for INT series as shown in Fig. 3.9.

1. H_0: series has a unit root and it is not stationary
2. H_1: series has no unit root and it is stationary.
3. $\alpha = 5\%$ or 0.05

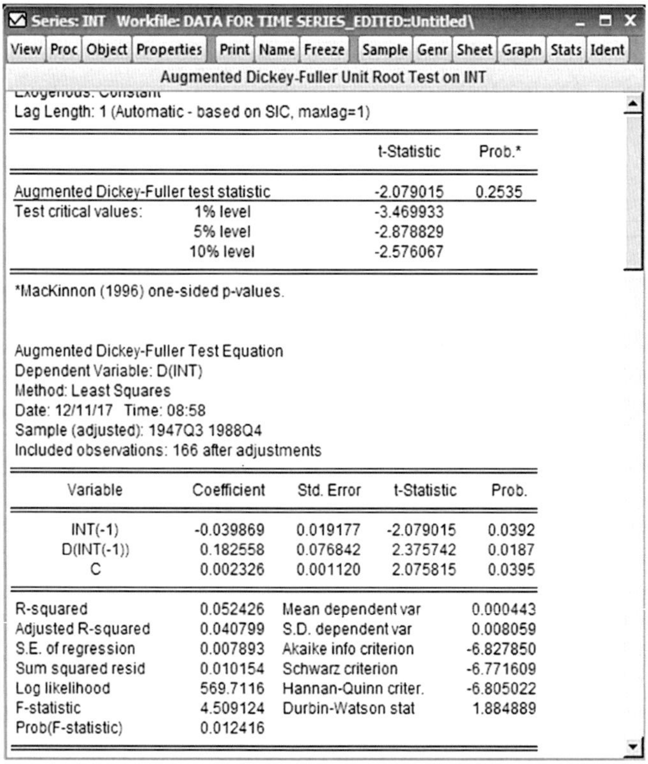

Series: INT Workfile: DATA FOR TIME SERIES_EDITED::Untitled\

View | Proc | Object | Properties | Print | Name | Freeze | Sample | Genr | Sheet | Graph | Stats | Ident

Augmented Dickey-Fuller Unit Root Test on INT

Lag Length: 1 (Automatic - based on SIC, maxlag=1)

		t-Statistic	Prob.*
Augmented Dickey-Fuller test statistic		-2.079015	0.2535
Test critical values:	1% level	-3.469933	
	5% level	-2.878829	
	10% level	-2.576067	

*MacKinnon (1996) one-sided p-values.

Augmented Dickey-Fuller Test Equation
Dependent Variable: D(INT)
Method: Least Squares
Date: 12/11/17 Time: 08:58
Sample (adjusted): 1947Q3 1988Q4
Included observations: 166 after adjustments

Variable	Coefficient	Std. Error	t-Statistic	Prob.
INT(-1)	-0.039869	0.019177	-2.079015	0.0392
D(INT(-1))	0.182558	0.076842	2.375742	0.0187
C	0.002326	0.001120	2.075815	0.0395

R-squared	0.052426	Mean dependent var	0.000443
Adjusted R-squared	0.040799	S.D. dependent var	0.008059
S.E. of regression	0.007893	Akaike info criterion	-6.827850
Sum squared resid	0.010154	Schwarz criterion	-6.771609
Log likelihood	569.7116	Hannan-Quinn criter.	-6.805022
F-statistic	4.509124	Durbin-Watson stat	1.884889
Prob(F-statistic)	0.012416		

Fig. 3.9 INT unit root test

ADF test statistic is −2.079015.

1. The p-values = 0.2532.
2. Prob = 0.2535 > 0.05. We cannot reject the H_0
3. The series has a unit root and it is not stationary.

All series are found to be non-stationary. Making Non-stationary Series Stationary

If a series is not stationary, we have to make the series stationary by taking the first difference. The first difference can be generated as follows:

Follow these steps in EViews:
Under ***Quick—Generate Series*** write:
DRMD = RMD − RMD(−1)
DRGDP = RGDP − RGDP(−1)
DINT = INT − INT(−1)

To make a non-stationary series a stationary one, the first difference of the series, in most cases and in some cases the second or third etc. differences are needed. Each new series, the first difference or the others must be tested informally as well as formally in order to make sure the new series are stationary.

3.4 Time Series Four: The First Difference of the RMD-DRMD

3.4.1 Informal Method: Plot the Time Series and Generate a Correlogram

Go to ***Quick—Graph*** and write DRMD then choose ***Line & Symbol*** and click OK to generate Fig. 3.10 which appears to be stationary. To generate a DRMD correlogram go to ***View—Correlogram—Level*** and click OK to get the output of Fig. 3.11

1. H_0: The series has no unit root
2. H_1: The series has a unit root
3. $\alpha = 5\%$ or 0.05

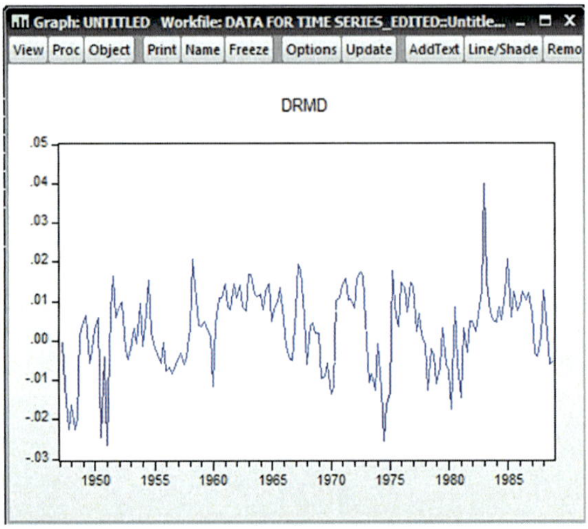

Fig. 3.10 DRMD plot

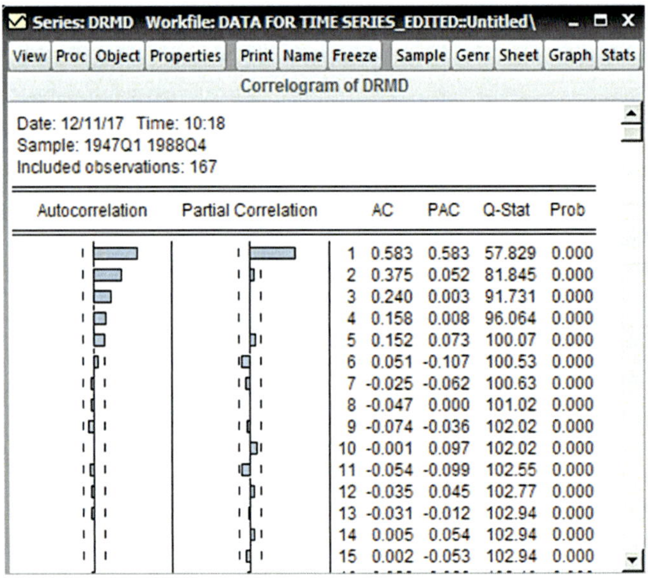

	AC	PAC	Q-Stat	Prob
1	0.583	0.583	57.829	0.000
2	0.375	0.052	81.845	0.000
3	0.240	0.003	91.731	0.000
4	0.158	0.008	96.064	0.000
5	0.152	0.073	100.07	0.000
6	0.051	-0.107	100.53	0.000
7	-0.025	-0.062	100.63	0.000
8	-0.047	0.000	101.02	0.000
9	-0.074	-0.036	102.02	0.000
10	-0.001	0.097	102.02	0.000
11	-0.054	-0.099	102.55	0.000
12	-0.035	0.045	102.77	0.000
13	-0.031	-0.012	102.94	0.000
14	0.005	0.054	102.94	0.000
15	0.002	-0.053	102.94	0.000

Fig. 3.11 Correlogram of DRMD

There are no test statistics to calculate. We look at the Prob column of Fig. 3.11.

4. The p-values = 0.000.
5. Prob = 0.000 < 0.05. We reject the H_0 and accept the H_1.
6. The series has a unit root and is not stationary.

3.4.2 *Formal Method: Run the Augmented Dickey-Fuller (ADF) Test*

Follow these steps in EViews:
Double click the variable, DRMD, from the Workfile.
From the Window containing the data of DRMD, choose ***View-Unit Root test***
From Test type, choose ***Augmented Dickey-Fuller***
From Test for unit root in, choose ***Level***
From Include in test equation, choose ***Trend and Intercept*** if the plot of the series indicates it is trended. Otherwise choose only the ***Intercept***
And then click **OK** to generate the output of Fig. 3.12.

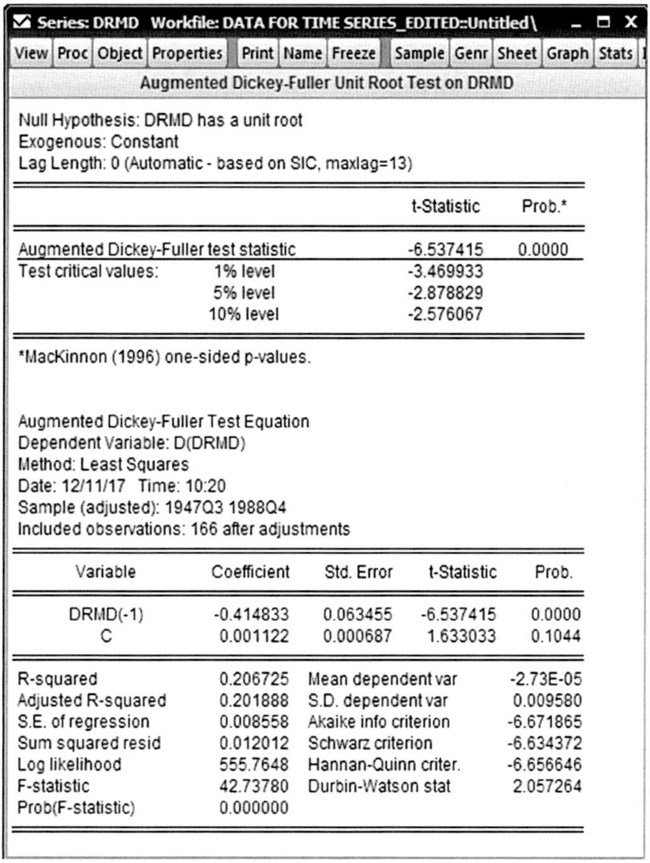

Series: DRMD Workfile: DATA FOR TIME SERIES_EDITED::Untitled\

View | Proc | Object | Properties | Print | Name | Freeze | Sample | Genr | Sheet | Graph | Stats

Augmented Dickey-Fuller Unit Root Test on DRMD

Null Hypothesis: DRMD has a unit root
Exogenous: Constant
Lag Length: 0 (Automatic - based on SIC, maxlag=13)

		t-Statistic	Prob.*
Augmented Dickey-Fuller test statistic		-6.537415	0.0000
Test critical values:	1% level	-3.469933	
	5% level	-2.878829	
	10% level	-2.576067	

*MacKinnon (1996) one-sided p-values.

Augmented Dickey-Fuller Test Equation
Dependent Variable: D(DRMD)
Method: Least Squares
Date: 12/11/17 Time: 10:20
Sample (adjusted): 1947Q3 1988Q4
Included observations: 166 after adjustments

Variable	Coefficient	Std. Error	t-Statistic	Prob.
DRMD(-1)	-0.414833	0.063455	-6.537415	0.0000
C	0.001122	0.000687	1.633033	0.1044

R-squared	0.206725	Mean dependent var	-2.73E-05
Adjusted R-squared	0.201888	S.D. dependent var	0.009580
S.E. of regression	0.008558	Akaike info criterion	-6.671865
Sum squared resid	0.012012	Schwarz criterion	-6.634372
Log likelihood	555.7648	Hannan-Quinn criter.	-6.656646
F-statistic	42.73780	Durbin-Watson stat	2.057264
Prob(F-statistic)	0.000000		

Fig. 3.12 DRMD unit root test

1. H_0: series has a unit root and it is not stationary
2. H_1: series has no unit root and it is stationary.
3. $\alpha = 5\%$ or 0.05

ADF test statistic is −6.537415.

4. The p-values = 0.0000.
5. Prob = 0.0000 < 0.05. We can reject the H_0; we accept the H_1.
6. The series has no unit root and it is stationary

3.5 Time Series Five: The First Difference of the RGDP-DRGDP

3.5.1 *Informal Method: Plot the Time Series and Generate a Correlogram*

Follow these steps in EViews:
Go to ***Quick—Graph*** and write DRGDP then choose ***Line & Symbol*** and click OK to generate Fig. 3.13 which appears to be stationary. To generate a DRGDP correlogram go to ***View—Correlogram—Level*** and click OK to get the output of Fig. 3.14

1. H_0: The series has no unit root
2. H_1: The series has a unit root
3. $\alpha = 5\%$ or 0.05

Fig. 3.13 DRGDP plot

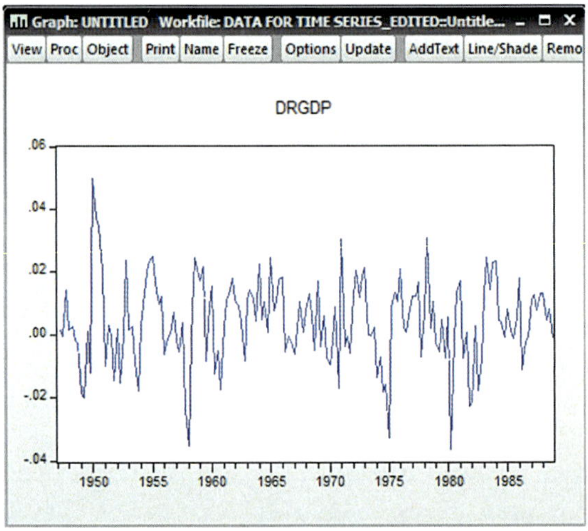

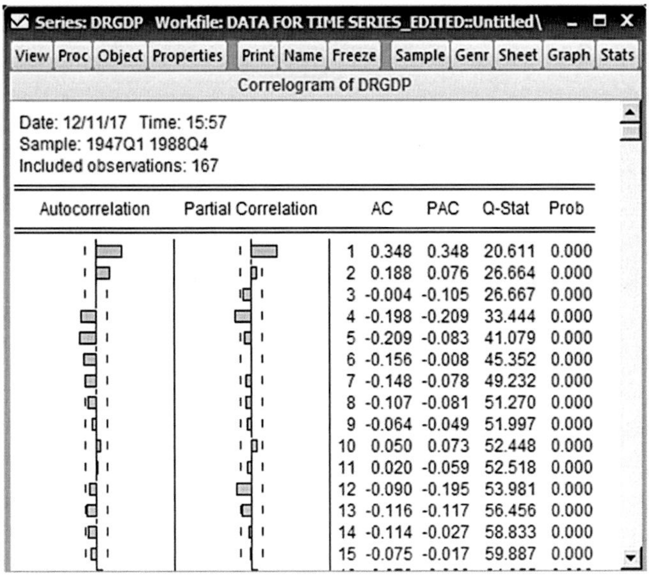
Series: DRGDP Workfile: DATA FOR TIME SERIES_EDITED::Untitled\

View | Proc | Object | Properties | Print | Name | Freeze | Sample | Genr | Sheet | Graph | Stats

Correlogram of DRGDP

Date: 12/11/17 Time: 15:57
Sample: 1947Q1 1988Q4
Included observations: 167

Autocorrelation	Partial Correlation		AC	PAC	Q-Stat	Prob
		1	0.348	0.348	20.611	0.000
		2	0.188	0.076	26.664	0.000
		3	-0.004	-0.105	26.667	0.000
		4	-0.198	-0.209	33.444	0.000
		5	-0.209	-0.083	41.079	0.000
		6	-0.156	-0.008	45.352	0.000
		7	-0.148	-0.078	49.232	0.000
		8	-0.107	-0.081	51.270	0.000
		9	-0.064	-0.049	51.997	0.000
		10	0.050	0.073	52.448	0.000
		11	0.020	-0.059	52.518	0.000
		12	-0.090	-0.195	53.981	0.000
		13	-0.116	-0.117	56.456	0.000
		14	-0.114	-0.027	58.833	0.000
		15	-0.075	-0.017	59.887	0.000

Fig. 3.14 Correlogram of DRGDP

There are no test statistics to calculate. We look at the Prob column.

4. The p-values = 0.000.
5. Prob = 0.000 < 0.05. We reject the H_0; we accept the H_1.
6. The series has a unit root and is not stationary.

3.5.2 *Formal Method: Run the Augmented Dickey-Fuller (ADF) Test*

Follow these steps in EViews:
Double click the variable, DRGDP, from the Workfile.
From the Window containing the data of DRGDP, choose **View-Unit Root test**
From Test type, choose **Augmented Dickey-Fuller**
From Test for unit root in, choose ***Level***
From Include in test equation, choose ***Trend and Intercept*** if the plot of the series indicates it is trended. Otherwise choose only the ***Intercept***
And then click **OK** to generate Fig. 3.15.

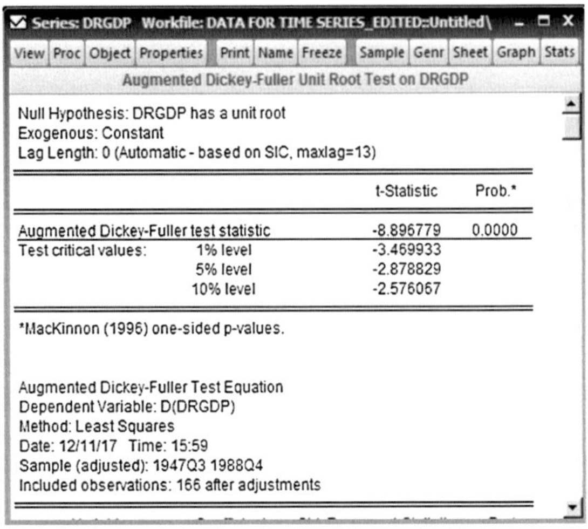

Fig. 3.15 DRGDP unit root test

1. H_0: series has a unit root and it is not stationary
2. H_1: series has no unit root and it is stationary.
3. $\alpha = 5\%$ or 0.05

ADF test statistic is −8.896779.

1. The p-values = 0.0000.
2. Prob = 0.0000 < 0.05. We can reject the H_0; we accept the H_1.
3. The series has no unit root and it is stationary.

3.6 Time Series Six: The First Difference of INT-DINT

3.6.1 *Informal Method: Plot the Time Series and Generate a Correlogram*

Follow these steps in EViews:
Go to ***Quick—Graph*** and write DINT then choose ***Line & Symbol*** and click OK to generate Fig. 3.16 which appears to be stationary. To generate a DINT correlogram go to ***View—Correlogram—Level*** and click OK to get the output of Fig. 3.17

The Correlogram for DINT Series:

1. H_0: The series has no unit root
2. H_1: The series has a unit root
3. $\alpha = 5\%$ or 0.05

Fig. 3.16 DINT plot

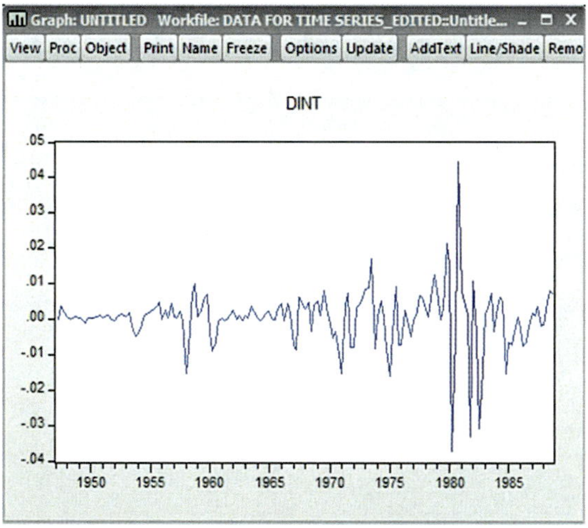

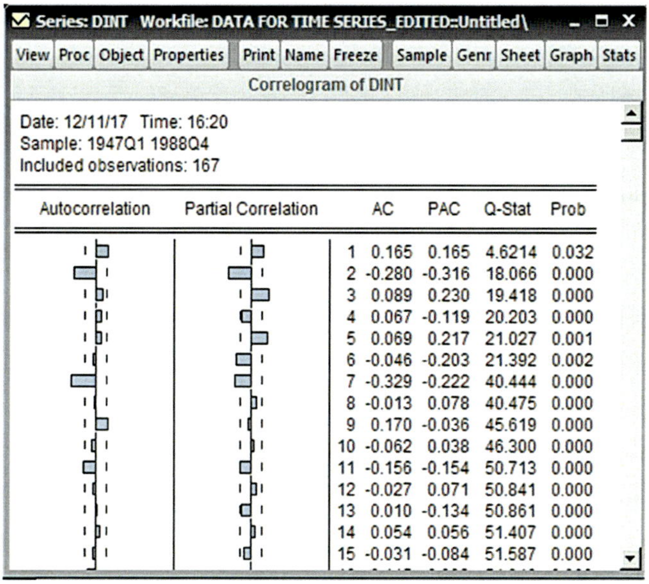

Series: DINT Workfile: DATA FOR TIME SERIES_EDITED::Untitled\

View Proc Object Properties Print Name Freeze Sample Genr Sheet Graph Stats

Correlogram of DINT

Date: 12/11/17 Time: 16:20
Sample: 1947Q1 1988Q4
Included observations: 167

	AC	PAC	Q-Stat	Prob
1	0.165	0.165	4.6214	0.032
2	-0.280	-0.316	18.066	0.000
3	0.089	0.230	19.418	0.000
4	0.067	-0.119	20.203	0.000
5	0.069	0.217	21.027	0.001
6	-0.046	-0.203	21.392	0.002
7	-0.329	-0.222	40.444	0.000
8	-0.013	0.078	40.475	0.000
9	0.170	-0.036	45.619	0.000
10	-0.062	0.038	46.300	0.000
11	-0.156	-0.154	50.713	0.000
12	-0.027	0.071	50.841	0.000
13	0.010	-0.134	50.861	0.000
14	0.054	0.056	51.407	0.000
15	-0.031	-0.084	51.587	0.000

Fig. 3.17 Correlogram of DINT

There are no test statistics to calculate. We look at the "Prob" column in Fig. 3.17.

4. The p-values = 0.000.
5. Prob = 0.000 < 0.05. We reject the H_0; we accept the H_1.
6. The series has a unit root and is not stationary.

3.6.2 *Formal Method: Run the Augmented Dickey-Fuller (ADF) Test*

Follow these steps in EViews:
Double click the variable, DINT, from the Workfile.
From the Window containing the data of DINT, choose View-Unit Root tests—Standard Unit Root test
From 'Test type', choose Augmented Dickey-Fuller
From Test for unit root in, choose level
From Include in test equation, choose Trend and Intercept if the plot of the series indicates it is trended. Otherwise choose only the Intercept
And then click OK to generate Fig. 3.18

Series: DINT Workfile: DATA FOR TIME SERIES_EDITED::Untitled\

View | Proc | Object | Properties | Print | Name | Freeze | Sample | Genr | Sheet | Graph | Stats

Augmented Dickey-Fuller Unit Root Test on DINT

Null Hypothesis: DINT has a unit root
Exogenous: Constant
Lag Length: 1 (Automatic - based on SIC, maxlag=1)

		t-Statistic	Prob.*
Augmented Dickey-Fuller test statistic		-11.38447	0.0000
Test critical values:	1% level	-3.470179	
	5% level	-2.878937	
	10% level	-2.576124	

*MacKinnon (1996) one-sided p-values.

Augmented Dickey-Fuller Test Equation
Dependent Variable: D(DINT)
Method: Least Squares
Date: 12/11/17 Time: 16:22
Sample (adjusted): 1947Q4 1988Q4
Included observations: 165 after adjustments

Variable	Coefficient	Std. Error	t-Statistic	Prob.
DINT(-1)	-1.101321	0.096739	-11.38447	0.0000
D(DINT(-1))	0.318163	0.074791	4.254058	0.0000
C	0.000449	0.000593	0.756917	0.4502

R-squared	0.474970	Mean dependent var	2.18E-05
Adjusted R-squared	0.468488	S.D. dependent var	0.010430
S.E. of regression	0.007604	Akaike info criterion	-6.902299
Sum squared resid	0.009367	Schwarz criterion	-6.845828
Log likelihood	572.4397	Hannan-Quinn criter.	-6.879375
F-statistic	73.27676	Durbin-Watson stat	1.850211
Prob(F-statistic)	0.000000		

Fig. 3.18 DINT unit root test

1. H_0: series has a unit root and it is not stationary
2. H_1: series has no unit root and it is stationary.
3. $\alpha = 5\%$ or 0.05

ADF test statistic is −11.38447.

4. The p-values = 0.0000.
5. Prob = 0.0000 < 0.05. We can reject the H_0; we accept the H_1.
6. The series has no unit root and it is stationary.

All our series are now stationary.

Chapter 4
Time Series Modelling

Time series data consist of readings on a variable taken at equally intervals of time. How would one compute the mean of a time series of a specified length? Calculating the mean of a sequence of observations might appear to be a trivial problem, as we would just sum all readings and divide by their number. However, if the series is steadily increasing overtime, i.e. exhibits a trend and we make decisions based on this mean, we would certainly not, for example, want to use this parameter as a forecast of the future level of the series. We would also not use the overall mean to make inferences (e.g. as the centre of confidence intervals) at time periods at the beginning or end the series. If we regard our gathered series as but one example of all possible series that could be generated by the same mechanism, we are further faced with the problem of estimating the mean for each time period, as we have a sample only of one item. It is similarly impossible to estimate the variance at any one time period.

The observed value of a series at particular time should be viewed as a random value; that is if a new set of data could be obtained under similar conditions, we would not obtain the identical numerical value. Let us measure at equal intervals the thickness of wire made on a continuous extraction machine. Such a list of measurements can be interpreted as a *realization* of wire thickness. If we were repeatedly to stop the process, service the machine and to restart the process to obtain new wires under similar machine conditions, we would be able to obtain new realizations from the same *stochastic* process. These realizations could be used to calculate the mean thickness of the wire after 1, 2 min etc. The term *stochastic* simply means "random" and the term *process* should be interpreted as the mechanism generating data. The problem is that in most situations, we can obtain only one realization. We cannot, for example, stop the economy, go back to some arbitrary point and then restart the economic process to observe a new realization. With a single realization, we cannot estimate with any precision the mean at each time period t and it is impossible to estimate the variance and autocorrelations. Therefore, to estimate the mean, variance and autocorrelation parameters of a stochastic process based on a single realization, the time series analyst must impose restrictions on how the data can be gathered.

A. Aljandali, M. Tatahi, *Economic and Financial Modelling with EViews*, Statistics and Econometrics for Finance, https://doi.org/10.1007/978-3-319-92985-9_4

As discussed in previous chapters, time series data can be used in a regression analysis only when the former is stationary. Based on this very important condition there are three possible time series modelling scenarios which will be discussed in this chapter.

4.1 The Property of Stationarity

A series that measures the cumulative effect of something is called *integrated*. Most of the probability theory of time series is concerned with integrated series that are *stationary*. **Broadly speaking, a time series is said to be stationary if there is no systematic change in mean (no trend) over time, if there is no systematic change in variance and if period variations have been removed.**

4.1.1 Trend Differencing

The assumption of no trend returns us to the problem posed at the start of this chapter. If there is no trend in the series, we might be willing to assume that the mean is constant for each time period and that the observed value at each time period is representative of that mean. The second condition above refers to constant variance. The variance of a series expresses the degree of variation about the assumed constant mean and as such gives a measure of uncertainty around this mean. If the variance is not constant over time, but say increases, it would be incorrect to believe that we can express the uncertainty around a forecasted mean level with a variance based on all the data. Most business and economic time series are *non-stationary*. Time series analysis often requires one to turn a non-stationary series into a stationary one in order to apply various aspects of statistical theory.

The first stage in any time series analysis should be to plot the available observations against time. This is often a very valuable part of any data analysis, since qualitative features such as trend, seasonality and outliers will usually be visible if present in the data. Consider Fig. 4.1, which is a plot of a company's inventory levels over 81 consecutive weeks. A visual inspection clearly evidences that there is a trend in the data. The time series is not stationary. To achieve stationarity, the trend has to be eliminated.

Most economic time series are characterized by movements along a trend time such as in Fig. 4.1. Although there is a general understanding of what a trend is, it is difficult to give a more precise definition of the term trend than "any systematic change in the level of a time series". The difficulty in defining a trend stems from the fact that what looks like a change in the level in a short series of observations may turn out not to be a trend when a longer series becomes available, but rather be part of a cyclical movement.

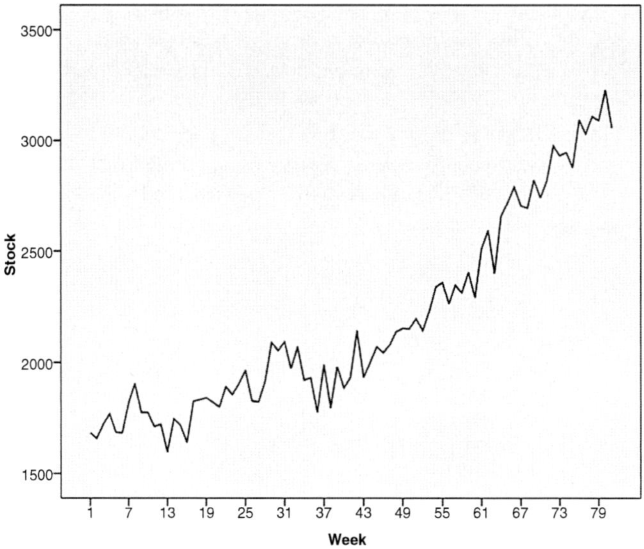

Fig. 4.1 Stock levels over time

Box and Pierce (1970) advocated that an integrated time series can have the trend removed by the method of *differencing*. The method of differencing consists of subtracting the values of the observations from one another in some prescribed time-dependent order. For example, a *first order difference transformation* is defined as the difference between the values of two adjacent observations; second order differencing consists of taking differences of the differenced series; and so on.

Consider the series 1, 3, 5, 7, 9 and 11 which exhibits a constant increase (trend) of two units from one observation to the next. We now take the *first order differences*:

$$3 - 1 = 2$$
$$5 - 3 = 2$$
$$7 - 5 - 2$$
$$9 - 7 = 2$$
$$11 - 9 = 2$$

By taking the first order differences of a series with a linear trend, the trend disappears. Let us apply the method to a series with a non-linear trend: 1, 6, 15, 28, 45, 66 and 91. *The first order differences* are 5, 9, 13, 17, 21 and 25. This differenced series possesses a linear trend with a constant increase of 4. Therefore, by taking the differences of the differences (i.e. *second order differences*), we would obtain a trend-free series. Second order differences, in fact, remove a quadratic trend; third order differencing removes a cubic trend. It is rare for economic time series to involve more than second order differencing. Note that every time that we difference a series, we lose an observation. Due to random fluctuations in the data, such neat results as above cannot always be obtained. However and as said, for many economic time series, first or second order differencing will be sufficient to remove the trend component (called a *detrended* series), so that further analysis can proceed. Note that once the trend has been removed, further differencing will continue to produce a series without a trend. However, each additional differencing results is one additional data point being lost. Therefore, such *overdifferencing* will needlessly complicate the model and should be avoided.

4.1.2 Seasonal Differencing

A lot of economic time series evidence seasonal patterns that make the time series non-stationary. Many monthly or quarterly series will exhibit effects which have a high degree of regularity. The adjustment procedure now to be employed is called *seasonal differencing*, in contrast with consecutive differencing discussed in the last subsection. This involves taking differences among the detrended observations spaced at four-period intervals i.e. if a quarterly pattern is evident, compute the differences between the first quarter value of each successive year and similarly the differences between the second, third and fourth quarters of successive years. Season differencing of order 1 indicates that we are taking the first differences among the same quarters in different years. The seasonal adjustment just described is said to involve a *span* of 4 periods. A span of 4 implies that a lag of 4 periods is used in the seasonal differencing operation.

4.1.3 Homoscedasticity of the Data

The process of differencing attempts to produce stationarity when there is a trend. When the variance of a time series is thought not to be a constant over time, there are several data transformations available. Two of the transformations commonly used are the logarithmic and the square root transform. The logarithmic is particularly effective when (1) the variance of the series is proportional to the mean level of the series or (2) the mean level of the series increases or decreases at a constant percentage. In that logs and roots of negative values are unreal, **such transforms must precede any differencing that may be required**.

4.2 Time Series in Practice

Using time series data for regression depends mainly on whether the time series are stationary or not. Based on this very important condition there are three possible time series modelling scenarios.

1. **All original time series are stationary.**
   Before running regression using the time series, all series need to be checked through necessary tests in order to make sure they are stationary. After this step a cointegration test needs to be carried out in order to find whether the residual of the main regression model is stationary or not. If the residual is stationary the series are cointegrated then this means a long-run relationship exists between the variables examined. The coefficients of the independent variables of this regression will represent the long-run relationship between the dependent and independent variables. We run a hypothesis test for each coefficient in order to make sure they are statistically significant and in addition to a LM test to show the model is not suffering from the autocorrelation problem.
2. **The original time series are not stationary.**
   If the time series of a model are not stationary but the residuals of a given regression model are stationary, a long-run relationship exists between the series and they are cointegrated. If this condition is met the Error Correction Model (ECM) can be calculated to find both short and long-run relationships between the series.
3. **Neither the series nor the residual of the regression (using the original series) are stationary.**
   In such a case, we cannot run the ECM model. We use either a distributed lag (DL) model or an Autoregressive Distributed Lag (ADL).

   Important Notes:

1. For the cointegration test we always run a regression by using the original series, regardless the time series are stationary or not.
2. If the series are not stationary but the cointegration test indicates that they are cointegrated, running a regression between the series will not lead to a situation in which we may lose any information of the time series.

Consider the model below where RMD is the real money demand, RGDP the real GDP and INT the interest rate:

$$RMD_t = \beta_0 + \beta_1 RGDP_t + \beta_2 INT_t + U_t$$

To run regression in EViews
Under ***Quick*** select ***Estimate equation*** and type RMD C RGDP INT then click OK to generate Fig. 4.1.

According to Fig. 4.2, it can be seen that the coefficients of both RGDP and INT are statistically significant, as P-values for both are 0.000 and 0.000, respectively. The R-square is very high (92.78%) and the Durbin-Watson (DW) Stat is very low, 0.0862. As the DW-stat is smaller than the R-square (0.0862 < 0.927801), we may have a case of a spurious (artificial) regression. However, if the residual of this regression is stationary, it will confirm that the series are cointegrated. Thus, there is

Equation: UNTITLED Workfile: DATA FOR TIME SERIES_EDITED:...

View | Proc | Object | Print | Name | Freeze | Estimate | Forecast | Stats | Resids

Dependent Variable: RMD
Method: Least Squares
Date: 12/11/17 Time: 16:32
Sample: 1947Q1 1988Q4
Included observations: 168

Variable	Coefficient	Std. Error	t-Statistic	Prob.
C	-0.351895	0.147514	-2.385501	0.0182
RGDP	0.965542	0.031734	30.42629	0.0000
INT	-1.042427	0.205637	-5.069266	0.0000

R-squared	0.927801	Mean dependent var	-4.635467
Adjusted R-squared	0.926926	S.D. dependent var	0.182267
S.E. of regression	0.049271	Akaike info criterion	-3.165279
Sum squared resid	0.400554	Schwarz criterion	-3.109494
Log likelihood	268.8834	Hannan-Quinn criter.	-3.142639
F-statistic	1060.176	Durbin-Watson stat	0.086269
Prob(F-statistic)	0.000000		

Fig. 4.2 Regression model using the RMD, RGDP and INT variables

a long-run relationship between the series and we are able to calculate an Error Correction Model (ECM).

To save the residuals in EViews
Go to ***Quick—Generate series*** and type U = resid in the process window and then click OK. Double click on U and go to ***View, Graph*** and select the options from Fig. 2.7 to generate a plot of the residuals which shows that the U series is stationary as seen in Fig. 4.3.

To run a correlogram test for the residuals, double click the variable U from the Workfile to get the U series and then click ***View—Correlogram*** and select a ***Correlogram of the Level*** then click OK to generate Fig. 4.4.

1. H_0: The series has no unit root
2. H_1: The series has a unit root
3. $\alpha = 5\%$ or 0.05

There are no test statistics to calculate. We look at the Prob column in Fig. 4.4.

1. The p-values are bigger than 0.05.
2. Prob = 0.000 < 0.05 therefore we reject the H_0 and accept the H_1.
3. The series hast a unit root and it is not stationary

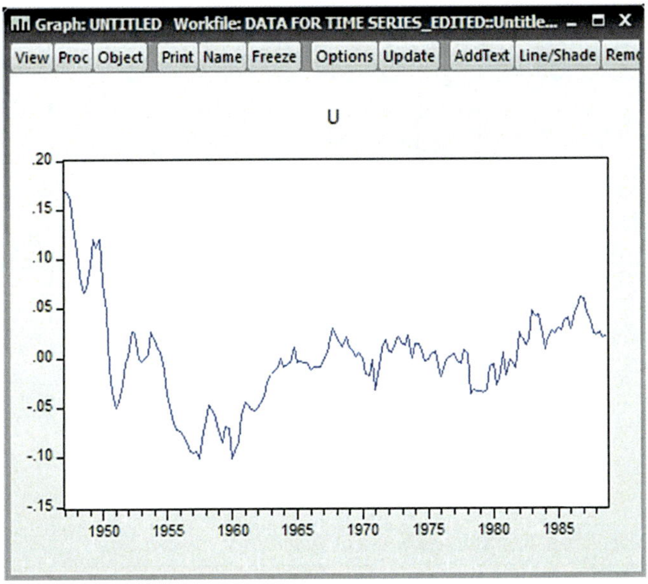

Fig. 4.3 Plot of U residuals

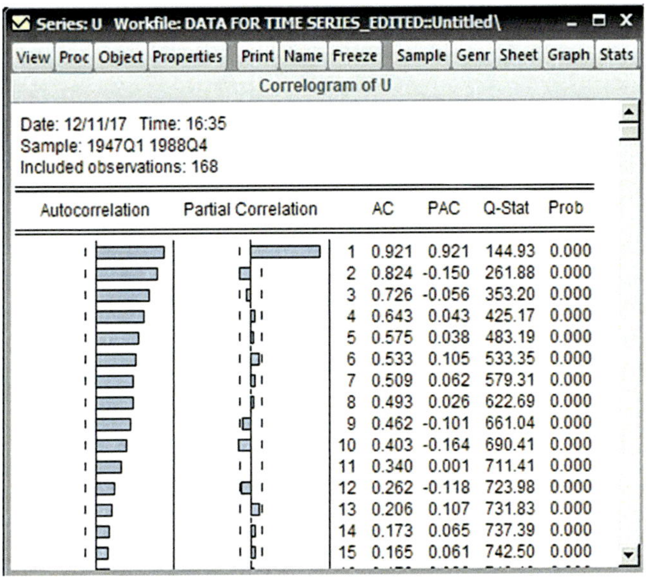
Series: U Workfile: DATA FOR TIME SERIES_EDITED::Untitled\

View Proc Object Properties Print Name Freeze Sample Genr Sheet Graph Stats

Correlogram of U

Date: 12/11/17 Time: 16:35
Sample: 1947Q1 1988Q4
Included observations: 168

	AC	PAC	Q-Stat	Prob
1	0.921	0.921	144.93	0.000
2	0.824	-0.150	261.88	0.000
3	0.726	-0.056	353.20	0.000
4	0.643	0.043	425.17	0.000
5	0.575	0.038	483.19	0.000
6	0.533	0.105	533.35	0.000
7	0.509	0.062	579.31	0.000
8	0.493	0.026	622.69	0.000
9	0.462	-0.101	661.04	0.000
10	0.403	-0.164	690.41	0.000
11	0.340	0.001	711.41	0.000
12	0.262	-0.118	723.98	0.000
13	0.206	0.107	731.83	0.000
14	0.173	0.065	737.39	0.000
15	0.165	0.061	742.50	0.000

Fig. 4.4 Correlogram test for U residuals

Formal method:
Augmented Dickey-Fuller (ADF) Test for the residuals:
Before running the ADF test in EViews, we need to decide whether to include a trend component in the model. If the plot of the series is trended, there is a need to include trend components. Otherwise only Include an intercept.

Follow these steps in EViews:
Double click the variable, U, from the Workfile.

From the Window containing the data of U, choose **View-Unit Root tests—Standard Unit Root test**

From Test type, choose **Augmented Dickey-Fuller**

From Test for unit root in, choose **level**

From Include in test equation, choose "Trend and Intercept" if the plot of the series indicates it is trended otherwise choose only the Intercept

And then click **OK** to generate the output of Fig. 4.5.

1. H_0: series has a unit root and it is not stationary
2. H_1: series has no unit root and it is stationary.
3. $\alpha = 5\%$ or 0.05

According to Fig. 4.4, the ADF test statistic is −3.819725.

1. The p-values = 0.0033.
2. Prob = 0.0033 < 0.05 therefore we reject the H_0 and accept the H_1.
3. The series has no unit root and it is stationary.

Using cointegrated series does not give an idea whether there is <u>**equilibrium**</u> between the series or not. If the series are cointegrated, we use the Error Correction Model (ECM) for three purposes:

1. To identify whether equilibrium between the series exists or not.
2. To find the speed of adjustment from disequilibrium to equilibrium through the coefficient of ECM item. The ECM item has to be negative and it has to be statistically significant. These two properties of ECM coefficient should be held.
3. To ensure short-run relationship among the series can be found through the use of time lags of independent variables in the ECM model.

Steps for implementing an Error correction Model (ECM)
Assume this is the main model:

$$RMD_t = \beta_0 + \beta_1 RGDP_t + \beta_2 INT_t + U_t$$

A one time lag model of this main model can be written as follows:

$$RMD_{t-1} = \beta_0 + \beta_1 RGDP_{t-1} + \beta_2 INT_{t-1} + U_{t-1}$$

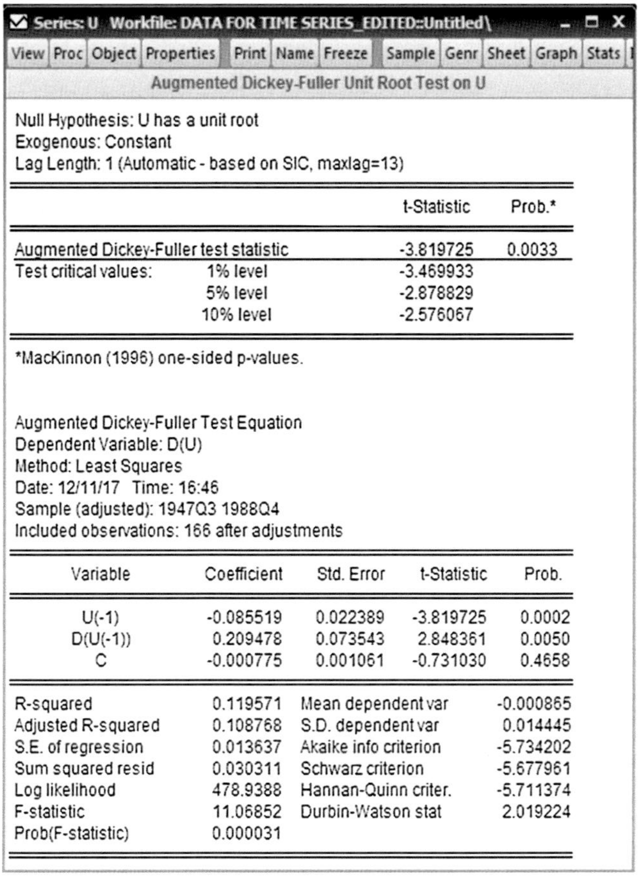

Null Hypothesis: U has a unit root
Exogenous: Constant
Lag Length: 1 (Automatic - based on SIC, maxlag=13)

		t-Statistic	Prob.*
Augmented Dickey-Fuller test statistic		-3.819725	0.0033
Test critical values:	1% level	-3.469933	
	5% level	-2.878829	
	10% level	-2.576067	

*MacKinnon (1996) one-sided p-values.

Augmented Dickey-Fuller Test Equation
Dependent Variable: D(U)
Method: Least Squares
Date: 12/11/17 Time: 16:46
Sample (adjusted): 1947Q3 1988Q4
Included observations: 166 after adjustments

Variable	Coefficient	Std. Error	t-Statistic	Prob.
U(-1)	-0.085519	0.022389	-3.819725	0.0002
D(U(-1))	0.209478	0.073543	2.848361	0.0050
C	-0.000775	0.001061	-0.731030	0.4658

R-squared	0.119571	Mean dependent var	-0.000865
Adjusted R-squared	0.108768	S.D. dependent var	0.014445
S.E. of regression	0.013637	Akaike info criterion	-5.734202
Sum squared resid	0.030311	Schwarz criterion	-5.677961
Log likelihood	478.9388	Hannan-Quinn criter.	-5.711374
F-statistic	11.06852	Durbin-Watson stat	2.019224
Prob(F-statistic)	0.000031		

Fig. 4.5 U residuals unit root test

From this equation we can find the one time lag of the error term, U_{t-1}. Take all items from right hand-side apart from U_{t-1} to the left hand-side of the equation and rewrite the equation as follows:

$$U_{t-1} = RMD_{t-1} - \beta_0 - \beta_1 RGDP_{t-1} - \beta_2 INT_{t-1}$$

We call U_{t-1} the error correction term. As a result of this we can write the last equation using ECM instead of $U_{t-1,}$ in fact we call U_{t-1}, ECM.

$$ECM_{t-1} = RMD_{t-1} - \beta_0 - \beta_1 RGDP_{t-1} - \beta_2 INT_{t-1}$$

Now we can write the main ECM equation as follows:

$$\begin{aligned} DRMD_t = &\beta_0 + \beta_1 DRMD_{t-1} + \beta_2 DRGDP_t + \beta_3 DRGDP_{t-1} + \beta_4 DINT_t \\ &+ \beta_5 DINT_{t-1} + \beta_6 ECM_{t-1} + V_t \end{aligned}$$

B_3 and β_5 are the short-run elasticity of RMD_t with respect to RGDP and INT.

B_2 and β_4 are the long-run elasticity of RMD_t with respect to RGDP and INT.

β_6 is the speed of adjustment of RMD_t to last period's error.

It is expected that β_6 to be smaller than zero ($\beta_4 < 0$).

The term of ECM OR, $[RMD_{t-1} - \beta_0 - \beta_1 \; RGDP_{t-1} - \beta_2 \; INT_{t-1}]$, is a measure of the disequilibrium error in the previous period.

Finally, the long-term effect occurs at a rate dictated by the value on β_6.

- We can thus validly use OLS on the above equation.
- The Granger representation theorem shows that any cointegrating relationship can be expressed as an equilibrium correction model.
- The test statistic for the above model will be focused on the coefficient, β_6.
- The null hypothesis is: $H_0 : \widehat{\beta}_6 = 0$ the estimated coefficient on the error term is equal to zero, such that the two series are not cointegrated and that the residuals from the cointegrating regression are classified I(1).
- The alternative hypothesis is: $H_1 : \widehat{\beta}_6 < 0$ and the two series are cointegrated, with residuals which are classified as I(0).
- Under the null hypothesis, the test-statistic $\dfrac{\widehat{\beta}_6}{SE(\widehat{\beta}_6)}$ follows a non-standard distribution
- When the null hypothesis is rejected and the alternative is accepted, we can come to the conclusion that there is a long-run relationship between the series.
- If the null hypothesis cannot be rejected, we can conclude that no cointegration exists between the series. More specifically, the above test indicates that it is not possible to find an error correction model; therefore, a long run relationship between pairs cannot be established.
- If we can reject the null hypothesis and three is a long-run relationship between the series, the long-run equilibrium is:

$$RMD_t = \beta_0 + \beta_1 RGDP_t + \beta_2 INT_t$$

The analyst is faced with three possibilities

1. The independent variables, $DRGDP_t$ and $DINT_t$, may have only contemporaneous effects, where they affect DRMD immediately, but that effect does not persist into the future. In this example, this occurs when $\widehat{\beta}_6$ is equal to 0.
2. The independent variables may have a contemporaneous effect as well as an equilibrium component that persists across future time periods and decaying at some rate. This is a situation in which all coefficients: β_2, β_3, β_4, β_5 and $\widehat{\beta}_6$ are statistically significant.

3. The independent variables may have no contemporaneous effect, but instead have an equilibrium effect, where the causal effect on DRMD only occurs across future time points. In that context β_3 and β_5 are equal to 0.

The power of error correcting models is that they allow us to estimate and test for all three types of effects. The ability to estimate such effects ought to encourage the development of theories that incorporate these questions. For applied analysts, this means greater attention should be paid to whether their dynamic theory implies effects that are contemporaneous, equilibrium or both. ECMs present a number of advantages. Firstly, if an ECM is correctly specified, it can be assumed that its disequilibrium error term is a stationary variable; providing that the number of observations is sufficiently large, i.e. it is asymptotically efficient. Moreover, an ECM can be estimated using classical regression methods and it can be relied upon as feasible stationary variables, despite the fact that we are dealing with large samples. Secondly, ECMs lend themselves from general to specific modelling.

Using EViews for implementing the Error Correction Model (ECM)

The ECM equation can be written as:

$$DRMD_t = \beta_0 + \beta_1 DRMD_{t-1} + \beta_2 DRGDP_t + \beta_3 DRGDP_{t-1} + \beta_4 DINT_t + \beta_5 DINT_{t-1} + \beta_6 ECM_{t-1} + V_t$$

Where D stands for the first difference of the series. We need to generate The ECM first, therefore we have to run the regression of the main equation to get the residual series. So we run the main model:

$$RMD_t = \beta_0 + \beta_1 RGDP_t + \beta_2 INT_t + U_t$$

In EViews do as follows:

Go to ***Quick*** choose ***Estimate equation*** and run the regression as follows

RMD C RGDP INT

Click **OK** to generate Fig. 4.6

To generate the ECM data

Go to ***Quick*** again and select ***Generate Series*** then write:

ECM = RESID

Click **OK**. Now we have the term ECM, which represents U_t, the residual variable of the main regression model.

To run the ECM model

Go to ***Quick*** choose ***Estimate equation*** and run the regression model as follows

DRMD C DRMD(−1) DRGDP DRGDP(−1) DINT DINT(−1) ECM (−1) whose output is reported in Fig. 4.7.

Equation: UNTITLED Workfile: DATA FOR TIME SERIES_EDITED::...

View | Proc | Object | Print | Name | Freeze | Estimate | Forecast | Stats | Resids

Dependent Variable: RMD
Method: Least Squares
Date: 12/11/17 Time: 17:34
Sample: 1947Q1 1988Q4
Included observations: 168

Variable	Coefficient	Std. Error	t-Statistic	Prob.
C	-0.351895	0.147514	-2.385501	0.0182
RGDP	0.965542	0.031734	30.42629	0.0000
INT	-1.042427	0.205637	-5.069266	0.0000

R-squared	0.927801	Mean dependent var	-4.635467
Adjusted R-squared	0.926926	S.D. dependent var	0.182267
S.E. of regression	0.049271	Akaike info criterion	-3.165279
Sum squared resid	0.400554	Schwarz criterion	-3.109494
Log likelihood	268.8834	Hannan-Quinn criter.	-3.142639
F-statistic	1060.176	Durbin-Watson stat	0.086269
Prob(F-statistic)	0.000000		

Fig. 4.6 Regression output using RMD, RGDP and INT

Equation: UNTITLED Workfile: DATA FOR TIME SERIES_EDITED::...

View | Proc | Object | Print | Name | Freeze | Estimate | Forecast | Stats | Resids

Dependent Variable: DRMD
Method: Least Squares
Date: 12/11/17 Time: 17:37
Sample (adjusted): 1947Q3 1988Q4
Included observations: 166 after adjustments

Variable	Coefficient	Std. Error	t-Statistic	Prob.
C	0.001168	0.000620	1.883321	0.0615
DRMD(-1)	0.528092	0.058336	9.052610	0.0000
DRGDP	0.129273	0.049152	2.630079	0.0094
DRGDP(-1)	-0.049828	0.048224	-1.033272	0.3030
DINT	-0.212607	0.076788	-2.768748	0.0063
DINT(-1)	-0.431866	0.077732	-5.555847	0.0000
ECM(-1)	-0.039013	0.012344	-3.160519	0.0019

R-squared	0.527991	Mean dependent var	0.002743
Adjusted R-squared	0.510180	S.D. dependent var	0.010514
S.E. of regression	0.007359	Akaike info criterion	-6.944635
Sum squared resid	0.008610	Schwarz criterion	-6.813406
Log likelihood	583.4047	Hannan-Quinn criter.	-6.891368
F-statistic	29.64302	Durbin-Watson stat	2.212199
Prob(F-statistic)	0.000000		

Fig. 4.7 DRMD's ECM model

The regression equation can be written as follows:

$$
\begin{aligned}
DRMD_t = {} & 0.001168 + 0.528092\,DRMD_{t-1} + 0129273\,DRGDP_t \\
& -0.049828\,DRGDP_{t-1} - 0.212607\,DINT_t - 0.431866\,DINT_{t-1} \\
& -0.039013\,ECM_{t-1}
\end{aligned}
$$

The next step would be to test whether the gradients β_2, β_3, β_4, β_5 and $\widehat{\beta}_6$ are statistically significant

Test for β_2:

1. H_0: $\beta_2 = 0$
2. H_1: $\beta_2 \neq 0$
3. $\alpha = 5\%$ or 0.05
4. $t_{\beta_2} = \frac{Coefficient\ \ \beta_2 - \beta_2(\beta_2\ \ in\ THEH_0)}{S.E.(\beta_2)} = \frac{0.129273-0}{0.049152} = 2.630079$
5. P-value of 0.0094 < 0.05 therefore we reject the H_0
6. We conclude that this coefficient is statistically significant. It means there is a long-term relationship between the real money demand and the aggregate income.

Test for β_3:

1. H_0: $\beta_3 = 0$
2. H_1: $\beta_3 \neq 0$
3. $\alpha = 5\%$ or 0.05
4. $t_{\beta_2} = \frac{Coefficient\ \ \beta_3 - \beta_3(\beta_3\ \ in\ \ THEH_0)}{S.E.(\beta_3)} = \frac{-0.049828-0}{0.048224} = -1.033272$
5. P-value of 0.3030 > 0.05 therefore cannot reject the H_0
6. We conclude that this coefficient is not statistically significant. It means there is no short-term relationship between the real money demand and the aggregate income.

The aggregate income representing by the real GDP in this context can only affect the real money demand in the long-run not in the short-run.

Test for β_4:

1. H_0: $\beta_4 = 0$
2. H_1: $\beta_4 \neq 0$
3. $\alpha = 5\%$ or 0.05
4. $t_{\beta_3} = \frac{Coefficient\ \ \beta_4 - \beta_4(\beta_4\ \ in\ \ THEH_0)}{S.E.(\beta_4)} = \frac{-0.212607-0}{0.076788} = -2.768748$
5. P-value of 0.0063 < 0.05 therefore reject the H_0
6. We conclude that this coefficient is statistically significant. It means there is a long-term relationship between the real money demand and the interest rates. The sign of the coefficient (−) proves theoretically that there is a negative relationship between the real money demand and the interest rates.

Test for β_5:

1. H_0: $\beta_5 = 0$
2. H_1: $\beta_5 \neq 0$
3. $\alpha = 5\%$ or 0.05
4. $t_{\beta_3} = \frac{Coefficient\, \beta_5 - \beta_5 (\beta_5\, in\, THE H_0)}{S.E.(\beta_5)} = \frac{-0.431866-0}{0.077732} = -5.555847$
5. P-value of $0.0000 < 0.05$ therefore reject the H_0.
6. We conclude that this coefficient is statistically significant. It means there is a short-term relationship between the real money demand and the interest rates. The sign of the coefficient (−) proves theoretically that there is a negative relationship between the real money demand and the interest rates. The interest rates can affect the real money demand in both short and long-run.

Test for β_6:

1. H_0: $\beta_6 = 0$
2. H_1: $\beta_6 \neq 0$
3. $\alpha = 5\%$ or 0.05
4. $t_{\beta_4} = \frac{Coefficient\, \beta_6 - \beta_6 (\beta_6\, in\, THE H_0)}{S.E.(\beta_6)} = \frac{-0.039013-0}{0.012344} = -3.160519$
5. P-value of $0.0019 < 0.05$ therefore reject the H_0
6. We conclude that this coefficient is statistically significant. It means there is a long-term or equilibrium between the real money demand and real GDP in one hand, and the interest rates on the other hand. The sign of the coefficient (−) proves theoretically that this is the case and such an equilibrium exists. The coefficient amount, −0.039013, represents the speed of adjustment from disequilibrium to the equilibrium.

As the coefficient for ECM is statistically significant, the two below models for long-run relationship and the equilibrium between series are identical.

$$DRMD_t = \beta_0 + \beta_1 DRMD_{t-1} + \beta_2 DRGDP_t + \beta_3 DRGDP_{t-1} + \beta_4 DINT_t + \beta_5 DINT_{t-1} + \beta_6 ECM_{t-1} + V_t$$

Or

$$RMD_t = \beta_0 + \beta_1 RGDP_t + \beta_2 INT_t$$

Residual Tests

The Serial Correlation LM Test

For this test we cannot use Durbin-Watson test because, when we use lags as independent variables, this test is not a valid test.

1. H_0: There is no problem of serial correlation
2. H_1: There is a problem of serial correlation

Equation: UNTITLED Workfile: DATA FOR TIME SERIES_EDITED::...

View | Proc | Object | Print | Name | Freeze | Estimate | Forecast | Stats | Resids

Breusch-Godfrey Serial Correlation LM Test:

F-statistic	2.115929	Prob. F(2,157)	0.1239
Obs*R-squared	4.357008	Prob. Chi-Square(2)	0.1132

Test Equation:
Dependent Variable: RESID
Method: Least Squares
Date: 12/12/17 Time: 12:13
Sample: 1947Q3 1988Q4
Included observations: 166
Presample missing value lagged residuals set to zero.

Variable	Coefficient	Std. Error	t-Statistic	Prob.
C	-0.000229	0.000636	-0.359963	0.7194
DRMD(-1)	0.134238	0.102478	1.309911	0.1921
DRGDP	-0.018150	0.049675	-0.365370	0.7153
DRGDP(-1)	-0.016802	0.048733	-0.344769	0.7307
DINT	-0.011992	0.076876	-0.155987	0.8762
DINT(-1)	0.032272	0.079268	0.407118	0.6845
ECM(-1)	0.006657	0.013449	0.494973	0.6213
RESID(-1)	-0.248475	0.137943	-1.801283	0.0736
RESID(-2)	-0.008114	0.098171	-0.082655	0.9342

R-squared	0.026247	Mean dependent var	-8.83E-19
Adjusted R-squared	-0.023371	S.D. dependent var	0.007224
S.E. of regression	0.007308	Akaike info criterion	-6.947136
Sum squared resid	0.008384	Schwarz criterion	-6.778414
Log likelihood	585.6123	Hannan-Quinn criter.	-6.878651
F-statistic	0.528982	Durbin-Watson stat	2.013567
Prob(F-statistic)	0.833412		

Fig. 4.8 Breusch-Godfrey serial correlation LM test

3. $\alpha = 5\%$ or 0.05
4. $T^*R^2 = 4.357008$ with P-value = 0.1132

Testing for serial correlation in EViews (Fig. 4.8), we are not able to reject the null hypothesis as 0.1239 and 0.1132 > 0.05. We conclude that he test is not significant. The evidence shows that there is no problem of serial correlation.

Chapter 5
Further Properties of Time Series

As part of the study of time series, it is important to distinguish between the terms **stochastic** and **deterministic** when considering time series. Stochastic comes from the Greek word $\sigma\tau o\chi o\varsigma$ pronounced "stokhos" and which means "a target". If you throw a dart at the bulls-eye on a target many times, you will probably hit the bulls-eye only a few times. At other times, the dart will miss the target and be spread randomly about that point. Stochastic processes contain such random errors.

For example, you may wish to forecast inflation (Y_t) in terms of interest rates, unemployment etc. Although the explanatory variables will be logically and sensibly chosen, it will be impossible to forecast inflation exactly. There is bound to be some random variability inherent to inflation that cannot be fully explained no matter how many explanatory variables are considered. On the other hand, deterministic processes are held to be free from error. For example, Ohm's Law states that in an electric circuit, the current C is inversely proportional to the voltage V i.e. $C = \frac{1}{k}V$ where $\frac{1}{k}$ is the constant of proportionality. Other physical laws such as Newton's laws are also regarded as deterministic. Of course, if $\frac{1}{k}$ does contain an error of measurement, this deterministic relationship becomes a stochastic one.

5.1 Stochastic and Deterministic Trends

Consider the following time series model:

$$Y_t = \beta_1 + \beta_2 t + \beta_3 Y_{t-1} + e_t$$

in which e_t is a white noise error term with constant variance σ^2 and time t is measured chronologically. There are the following possibilities:

(A) $\beta_1 \neq 0, \beta_2 \neq 0$ and $\beta_3 = 0$ this implies that:

$$Y_t = \beta_1 + \beta_2 t + e_t, \tag{5.1}$$

A. Aljandali, M. Tatahi, *Economic and Financial Modelling with EViews*, Statistics and Econometrics for Finance, https://doi.org/10.1007/978-3-319-92985-9_5

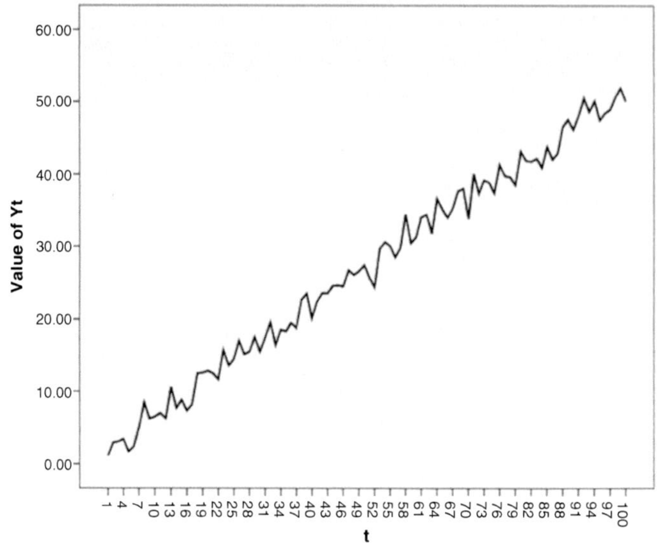

Fig. 5.1 A time series with a deterministic trend

which is called a **deterministic trend** model. This model assumes that the trend is a predictable (error free) function of time with Y_t trending upwards or downwards according to the sign of β_2. A deterministic trend could be a function of t^2, t^3 or any other function of time.

Figure 5.1 shows a plot of $Y_t = 2 + 0.5\ t + e_t$ for $1 \leq t \leq 100$, where e_t are normally distributed with a mean of zero and variance of 1.5. The data values are seen to fluctuate around the deterministic trend of 2 + 0.5 t with no obvious tendency for the amplitude of the fluctuations to increase or decrease. For this reason, a time series Y_t with a deterministic trend is said to be a **trend stationary process (TS)**.

Note that the steady increase in the mean over time in Fig. 5.1 renders that time series non-stationary. (**Remember for a series to be stationary, it should have three conditions met—constant mean, constant variance and constant covariance for all t**). Replacing t by t − 1 in Eq. (5.1) reveals that $Y_t - Y_{t-1} = \Delta Y_t = \beta_2 + e_t - e_{t-1} = \beta_2 + \Delta e_t$. This latter equation has "highly undesirable properties" (Brooks 2004: 372). It is, therefore, most certainly **not recommended to use differencing to remove a deterministic trend**. The proper way to eliminate a deterministic trend is to regress Y_t on t and the residuals obtained have the linear trend removed. The undesirability of the above equation is explained later on in this chapter.

Also, note that a deterministic trend implies that as t tends to infinity so too does Y_t—an unlikely property of an economic time series. Also note that Brooks (*op cit*, p. 375) is of the opinion that stochastic trends (see below) apply to far more real time series in Finance and Economics than do deterministic trends.

(B) $\beta_1 \neq 0, \beta_2 = 0$ and $\beta_3 = 1$

$$Y_t = \beta_1 + Y_{t-1} + e_t, \tag{5.2}$$

This generates the model which is called the **random walk model with drift**. β_1 is called the **drift term**. The above can be written as

$$\Delta Yt = \beta_1 + e_t. \tag{5.3}$$

The random walk with drift model in Eq. (5.2) is non-stationary (the coefficient of Y_{t-1} should be less than 1 for stationarity), but the time series is made stationary via first differences as per Eq. (5.3). Equation (5.3) infers that Y_t will exhibit a positive ($\beta_1 > 0$) or negative ($\beta_1 < 0$) trend. Y_t is said to **drift** upward or downward depending on whether β_1 is positive or negative. Such a trend is called a **stochastic trend**. Unlike the plot in Fig. 5.1, time series with stochastic trends tend to show greater amplitudes and deviations from the trend die out much more slowly. The way to remove stochastic trends is via differencing. Once a stochastic trend has been removed the time series is said to be **a difference stationary process (DSP)**.

If $\beta_1 = 0$ in Eq. (5.2), then the time series is said to be a **random walk without drift**. In this situation, $\Delta Y_t = e_t$. Note that both random walk models (with and without drift) are AR(1) processes. The random walk without drift model is again a non-stationary stochastic process, but first differencing makes the series stationary, as was the case with the equivalent model with drift. The random walk without drift model is therefore also a DSP after differencing. It is often said that asset prices such as stock prices or exchange rates follow random walk models (Gujarati and Porter 2009).

(C) $\beta_1 \neq 0, \beta_2 \neq 0$ and $\beta_3 = 1$
This generates the model $Y_t = \beta_1 + \beta_2 t + Y_{t-1} + e_t$ which is a **random walk with drift and deterministic trend**. Taking first differences, $\Delta Yt = \beta_1 + \beta_2 t + e_t$. This shows that the first differences possess a deterministic trend. Therefore, the first differences are non-stationary as explained earlier, which also means that Y_t is a non-stationary time series. This series is not a DSP after differencing.

(D) $\beta_1 \neq 0, \beta_2 \neq 0$ and $\beta_3 < 1$
This generates the model $Y_t = \beta_1 + \beta_2 t + \beta_3 Y_{t-1} + e_t$ which is a **deterministic trend with a stationary AR(1) component** (due to the fact that $\beta_3 < 1$). Consequently, this model is difference stationary about a deterministic trend.

5.2 The Lag Operator and Invertibility

Before reading this section you are advised to work through the Binomial Theorem in Appendix 5.1 at the end of the chapter.

The lag operator L is defined as $LY_t = Y_{t-1}$ and more generally as $L^i Y_t = Y_{t-i}$.

- The lag of a constant is a constant i.e. $Lc = c$.
- The distributive law of lag operators states that $(L^i + L^j)Y_t = L^i Y_t + L^j Y_t = Y_{t-i} + Y_{t-j}$.
- The associative law of lag operators states that $L^i L^j Y_t = L^i(L^j Y_t) = L^i Y_{t-j} = Y_{t-i-j}$.
- $L^0 Y_t$ is defined to equal Y_t.

Lag operators offer a concise way of writing out the general ARIMA model. ARIMA models are discussed in further detail Chap. 7. Consider the autoregressive part:

$$Y_t = \mu + \varphi_1 Y_{t-1} + \ldots + \varphi_p Y_{t-p} + e_t \text{ which implies that}$$

$$\left(1 - \varphi_1 \mathrm{L} - \varphi_2 L^2 - \ldots - \varphi_p L^p\right) Y_t = \mu + e_t \text{ or more compactly as}$$

$$\mathrm{A(L)} Y_t = \mu + e_t \text{ where } \mathrm{A(L)} = 1 - \varphi_1 \mathrm{L} - \varphi_2 L^2 - \ldots - \varphi_p L^p.$$

A(L) is called a **polynomial in the lag operator of order p**, just as a quadratic is a polynomial in a particular variable, of order 2. Recall that the ARMA(p,q) model is:

$$Y_t = \mu + \varphi_1 Y_{t-1} + \ldots + \varphi_p Y_{t-p} + e_t + \theta_1 e_{t-1} + \ldots + \theta_q e_{t-q}$$

where we have written plus (+) in front of the theta (θ) coefficients instead of the conventional minus sign (this makes no difference), then:

$$Y_t - \varphi_1 Y_{t-1} + \ldots - \varphi_p Y_{t-p} = \mu + e_t + \theta_1 e_{t-1} + \ldots + \theta_q e_{t-q}$$

$$\left(1 - \varphi_1 \mathrm{L} - \varphi_2 L^2 - \ldots - \varphi_p L^p\right) Y_t = \mu + \left(1 + \theta_1 \mathrm{L} + \theta_2 L^2 \ldots + \theta_q L^q\right) e_t$$

$$\boxed{\mathrm{A(L)} Y_t = \mu + \mathrm{B(L)} e_t}$$

in which $\mathrm{A(L)} = 1 - \varphi_1 \mathrm{L} - \varphi_2 L^2 - \ldots - \varphi_p L^p$ is a polynomial of order p and $\mathrm{B(L)} = 1 + \theta_1 \mathrm{L} + \theta_2 L^2 \ldots + \theta_q L^q$ is a polynomial of order q. For simplicity, ignore the intercept term μ. Using the lag operator, we can now prove one of the most fundamental properties of the Box-Jenkins class of models:

(A) The above representation implies that any ARMA(p,q) process may be expressed as a pure MA process of infinite order, $Y_t = A^{-1}(L)B(L)e_t$.
(B) The above representation also implies that any ARMA(p,q) process may be written as a pure AR process of infinite order, $e_t = B^{-1}(L)A(L)Y_t$.
Condition (B) above is called the condition of *invertibility*.

We shall illustrate condition (A) above by considering an AR(1) process with intercept:

$$Y_t = \beta_1 + \beta_2 Y_{t-1} + e_t.$$

Using the preceding notation, this may be written as:

$$Y_t - \beta_2 Y_{t-1} = (1 - \beta_2 L)Y_t = \beta_1 + e_t \text{ and}$$
$$Y_t = (1 - \beta_2 L)^{-1}(\beta_1 + e_t).$$

Using the Binomial Theorem in Appendix 5.1 to expand $(1 - \beta_2 L)^{-1}$:

$$Y_t = \left(1 + (-1)(-\beta_2 L) + \frac{(-1)(-2)}{2!}(-\beta_2 L)^2 + \frac{(-1)(-2)(-3)}{3!}(-\beta_2 L)^3 + \ldots\right)$$
$$\times(\beta_1 + e_t)Y_t = \left(1 + \beta_2 L + \beta_2^2 L^2 + \beta_2^3 L^3 + \ldots\right)\beta_1 + \left(1 + \beta_2 L + \beta_2^2 L^2 + \beta_2^3 L^3 + \ldots\right)e_t.$$

We can drop the lag operator, L, from the left hand bracket above because any lag of a constant β_1 is equal to β_1. Therefore, for this AR(1) process, we derive:

$$Y_t = \left(1 + \beta_2 + \beta_2^2 + \beta_2^3 + \ldots\right)\beta_1 + \left(e_t + \beta_2 e_{t-1} + \beta_2^2 e_{t-2} + \beta_2^3 e_{t-3} + \ldots\right) \quad (5.4)$$

Equation (5.4) illustrates that an AR(1) process can be represented as a pure MA process of infinite order, as was stated on the previous page.

The expression in the left hand bracket of Eq. (5.4) is an infinite geometric series with common ratio equal to β_2. **Provided that $|\beta_2| < 1$**, the sum of this infinite geometric series converges to $\frac{1}{1-\beta_2}$. (Recall that the sum of an infinite geometric series with first term = a, and fractional common ration = r is given by $\frac{a}{1-r}$). Hence, from Eq. (5.4), the expected (or mean) value of Y_t is $E(Y_t) = \frac{\beta_1}{1-\beta_2}$, since in the right hand bracket of this equation, $E(e_t) = E(e_{t-1}) = E(e_{t-2}) = \ldots = 0$, because the white noise error terms have zero mean.

Given that $\text{var}(e_t) = \text{var}(e_{t-1}) = \text{var}(e_{t-2}) = \text{var}(e_{t-3})\ldots = \sigma^2$ i.e. the error terms exhibit constant variance (they are homoscedastic), then from Eq. (5.4) we derive that:

$$\text{var}(Y_t) = \sigma^2\left(1 + \beta_2^2 + \beta_2^4 + \beta_2^6 + \ldots\right) \quad (5.5)$$

(You need to recall that varA = 0 and that var(AX) = A^2var(X) if A is a constant. There is also zero covariance between the errors since they are uncorrelated). The term in brackets on the right hand side of Eq. (5.5) is an infinite geometric series with common ratio equal to β_2^2 and **provided that $|\beta_2| < 1$**, the sum of this geometric series is $\frac{1}{1-\beta_2^2}$, whereby $\text{var}(Y_t) = \frac{\sigma^2}{1-\beta_2^2}$.

Therefore, an AR(1) process has a constant mean and a constant variance, both independent of time, if and only if $|\beta_2| < 1$. The covariance of Y_t may also be shown to be constant for all t if $|\beta_2| < 1$. The condition $|\beta_2| < 1$ is, therefore, by definition the requirement for an AR(1) process to be stationary (constant mean, variance and covariance). Also note from Eq. (5.4) that an AR(1) process can be written as an infinite moving average process MA(∞).

Let us revisit the property of **invertibility**—condition (B). This condition implies that an MA(1) process, for example, is capable of being represented by a pure AR process of infinite order.

Consider the MA(1) process $Y_t = e_t - \gamma e_{t-1}$. Therefore, $Y_t = (1 - \gamma L)e_t$ or $(1 - \gamma L)^{-1}Y_t = e_t$. Using the Binomial Theorem on $(1 - \gamma L)^{-1}$:

$$\left(1 + \gamma L + \gamma^2 L^2 + +\gamma^3 L^3 + \ldots\right)Y_t = e_t$$
$$Y_t + \gamma Y_{t-1} + \gamma^2 Y_{t-2} + \gamma^3 Y_{t-3} + \ldots = e_t.$$

This shows that an MA(1) process can be expressed as an infinite AR process. For convergence, we require that $|\gamma| < 1$, which is the condition for the above AR process (or the MA(1) process) to be stationary. This condition also shows that the AR representation of the MA(1) process has coefficients that are declining in magnitude over time—a desirable property. In this case, we say that the MA (1) time series is **invertible**. If $|\gamma| \geq 1$, then the time series is not invertible. The reason that we require an MA process to be invertible is that the autocorrelation function (ACF) and the partial autocorrelation function (PACF) require the time series to be well approximated by an autoregressive model so that the autocorrelations can be computed. SPSS gives a warning if an MA process is not invertible.

5.3 The Characteristic Equation and Stationarity

Before reading this section, you are advised to read the material on quadratics in Appendix 5.2 at the end of this chapter.

We have seen that an AR(1) process $Y_t = \beta_1 + \beta_2 Y_{t-1} + e_t$ may be written as $A(L)Y_t = \beta_1 + e_t$, where A(L) is the order one polynomial in the lag operator, $A(L) = (1 - \beta_2 L)$. Consider an AR(2) process:

$$Y_t = \beta_1 + \beta_2 Y_{t-1} + \beta_3 Y_{t-2} + e_t$$
$$Y_t - \beta_2 Y_{t-1} - \beta_3 Y_{t-2} = \beta_1 + e_t$$
$$\left(1 - \beta_2 L - \beta_3 L^2\right)Y_t = \beta_1 + e_t$$
$$\text{or } A(L)Y_t = \beta_1 + e_t,$$

where $A(L) = 1 - \beta_2 L - \beta_3 L^2$ is an order two polynomial (quadratic) in the lag operator. In these and all other cases, the A(L) polynomial that precedes Y_t is called the **characteristic equation**. Suppose we have an AR(p) process, then the associated characteristic equation is:

$$A(L) = 1 - \beta_2 L - \beta_3 L^2 - \beta_4 L^3 - \ldots - \beta_{p+1} L^p.$$

It is possible to show that a stochastic process will be stationary only if ALL of the roots of A(L) = 0 are all greater than unity in absolute value. The roots are said to be "*outside the unit circle*". Otherwise, the process is non-stationary. (Thomas 1997: 416)

Reconsider the AR(1) process $Y_t = \beta_1 + \beta_2 Y_{t-1} + e_t$ for which the characteristic equation is $A(L) = (1 - \beta_2 L)$. Solving $A(L) = 0$, we obtain $L = 1/\beta_2$. For stationarity, we require the absolute value of the root L to be greater than unity, i.e. $|1/\beta_2| > 1$ which implies that $|\beta_2| < 1$ which is the stationarity condition obtained in the previous section.

Consider the AR(2) process $Y_t = \beta_1 + \beta_2 Y_{t-1} + \beta_3 Y_{t-2} + e_t$ for which $A(L) = 1 - \beta_2 L - \beta_3 L^2$. Solving for $A(L) = 0$, we have:

$$\beta_3 L^2 + \beta_2 L - 1 = 0$$

whose roots are given by $L = \dfrac{-\beta_2 \pm \sqrt{\beta_2^2 + 4\beta_3}}{2\beta_3}$.

We require $|L| > 1$ for stationarity, so if for example β_3 is positive, we require the value of the numerator above to be greater than the value of the denominator to achieve $|L| > 1$:

$$-\beta_2 \pm \sqrt{\beta_2^2 + 4\beta_3} > 2\beta_3$$

$$\pm\sqrt{\beta_2^2 + 4\beta_3} > 2\beta_3 + \beta_2 \text{ and squaring both sides}$$

$$\beta_2^2 + 4\beta_3 > 4\beta_3^2 + \beta_2^2 + 4\beta_2\beta_3 \text{ and cancel the } \beta_2^2 \text{ terms}$$

$$4\beta_3 > 4\beta_3^2 + 4\beta_2\beta_3 \text{ and } \textit{divide} \text{ both sides by } 4\beta_3$$

$$1 > \beta_3 + \beta_2 \text{ or}$$

$$\boxed{\beta_3 + \beta_2 < 1}$$

For example, consider the AR(2) process $Y_t = 1.6 - 0.4Y_{t-1} + 0.5Y_{t-2} + e_t$ for which the characteristic equation is $A(L) = (1 + 0.4L - 0.5L^2)$ and solving $A(L) = 0$, we have:

$$L = \frac{-0.4 \pm \sqrt{0.16 + 2}}{-1},$$

which yields the roots $L_1 = 1.8697$ and $L_2 = -1.0697$. Both roots are in excess of unity in absolute value so this AR(2) process is stationary. Note that we have obeyed the stationarity condition $\beta_3 + \beta_2 < 1$. Note from the results in Appendix 5.2 that the sum of these two real roots is 0.8 and their product is -2 as is required by theory.

Of course, an AR(2) process may have a characteristic equation with complex roots, for example, $Y_t = 1.6 + 0.4Y_{t-1} - 0.5Y_{t-2} + e_t$. The characteristic equation is $A(L) = 1 - 0.4\,L + 0.5L^2$ and the negative discriminant immediately informs us that A(L) has two complex roots. The complex roots are given by:

$$L = \frac{0.4 \pm \sqrt{0.16 - 2}}{1},$$

which yields $L_1 = 0.4 + 1.3565i$ and $L_2 = 0.4 - 1.3565i$ (using $\sqrt{1.84} = 1.3565$). When the roots of the characteristic equation are complex, stationarity requires that their moduli are greater than unity. Here, both complex roots have the same modulus, namely by Pythagoras $\sqrt{0.4^2 + 1.3565^2} = 1.414$, which is greater than unity as required for stationarity.

Consider an AR(3) process $Y_t = 2 + 3.9Y_{t-1} + 0.6Y_{t-2} - 0.8Y_{t-3} + e_t$ which has $A(L) = 1 - 3.9L - 0.6L^2 + 0.8L^3 = 0$. This cubic (fortunately) factorises into $(1 - 0.4L)(1 + 0.5L)(1 - 4L) = 0$ with roots $L_1 = 2.5$, $L_2 = -2$ and $L_3 = 0.25$. Since one of these roots is less than unity in absolute value, this AR(3) process is not stationary.

Consider the random walk models with and without drift, respectively $Y_t = \beta_1 + Y_{t-1} + e_t$ and $Y_t = Y_{t-1} + e_t$. Both have characteristic equations of $A(L) = 1 - L$ and for $A(L) = 0$, $L = 1$. This shows that both random walks are not stationary since the root of the characteristic equation is not greater than unity. When $L = 1$, the time series is said to have **a unit root** and the time series is not convergent (stationary).

If you return to the beginning of this chapter for a moment, you will recall that we examined the equation $\Delta Y_t = \beta_2 + \Delta e_t$ in the context of deterministic trends. Now the above equation in differences implies that $(1 - L)Y_t = \beta_2 + (1 - L)e_t$. Therefore, $Y_t = (1 - L)^{-1}\beta_2 + e_t$ and using the Binomial Theorem, $Y_t = (1 + L + L^2 + L^3 + \ldots)\beta_2 + e_t$. Recalling that the lag of a constant equals that constant itself, this result implies that Y_t is becoming infinite as the lag process proceeds; β_2 terms are being continuously added. This is why Brooks (*op cit*) stated that the above equation had "highly undesirable properties".

5.4 Unit Root Tests

Consider the AR(1) process $Y_t = \beta_1\, Y_{t-1} + e_t$. If $\beta_1 = 1$, we have the random walk without drift model $(1 - L)Y_t = e_t$ and since $L = 1$, the model is not stationary. Indeed, it has a unit root. If we subtract Y_{t-1} from both sides, we obtain an equivalent form of this model, $\Delta Y_t = \gamma Y_{t-1} + e_t$ where $\gamma = \beta_1 - 1$. If we wanted to test H_0: this model has a unit root, we would be testing H_0: $\beta_1 = 1$ or equivalently H_0: $\gamma = 0$. The alternative hypothesis is H_1: $\beta_1 < 1$ (which implies that $\gamma < 0$) i.e. there is not a unit root. **Consequently, this statistical test is one-tailed.** From earlier in this chapter and given that we are focussed on stochastic models, the technically correct formulation is:

H_0 : the time series is not difference stationary because it has a unit root.

Consider the random walk with drift model, $Y_t = \beta_0 + \beta_1 Y_{t-1} + e_t$ which is AR (1). Again subtracting Y_{t-1} from both sides, $\Delta Y_t = \beta_0 + \gamma Y_{t-1} + e_t$ with $\gamma = \beta_1 - 1$. Hence, if we wanted to test H_0: this second model has a unit root, we would again be testing H_0: $\beta_1 = 1$ or equivalently H_0: $\gamma = 0$.

Finally, consider an AR(1) model with drift and deterministic trend, $Y_t = \beta_1 + \beta_2 t + \beta_3 Y_{t-1} + e_t$. If $\beta_3 = 1$, we have the random walk with drift model which is non-stationary since it has a unit root. If we subtract Y_{t-1} from both sides, we derive $\Delta Y_t = \beta_1 + \beta_2 t + \gamma Y_{t-1} + e_t$, where $\gamma = \beta_3 - 1$. If we wanted to test H_0: this third model has a unit root, we would be testing H_0: $\beta_3 = 1$ or equivalently H_0: $\gamma = 0$.

Dickey and Fuller (1979) considered all three of these different equations for the possibility of a unit root:

$$\Delta Y_t = \gamma Y_{t-1} + e_t \tag{5.6}$$

$$\Delta Y_t = \beta_0 + \gamma Y_{t-1} + e_t \tag{5.7}$$

$$\Delta Y_t = \beta_1 + \beta_2 t + \gamma Y_{t-1} + e_t \tag{5.8}$$

The difference between these three models concerns the presence of the deterministic elements β_0 (and β_1) and $\beta_2 t$. The parameter of interest in these three equations is γ and if $\gamma = 0$ the time series has a unit root. The DF test involves estimating one (or more) of Eqs. (5.6), (5.7), and (5.8) using least squares and a test statistic for γ is computed. The test statistic has a different notation according to which of the above three models the researcher is considering. Also, the critical values are different according to which of the above three models is examined. For model (5.6), no constant or time trend, the test statistic is denoted by τ (the Greek letter tau). For model (5.7), a drift (intercept) term and no time trend, the test statistic is denoted by τ_μ. For model (5.8), a drift term and time trend, the test statistic is denoted by τ_t. For example and from statistical tables, if the sample size is $n = 100$, the critical 5% values are $\tau = -1.95$, $\tau_\mu = -2.89$ and $\tau_t = -3.45$ respectively for the three models. Remember that the DF is one-tailed on the negative side, so only one critical value is required at the selected significance level. Clearly, the Dickey-Fuller table values are built into EViews.

So far, I have assumed that the time series under investigation can be modelled by a first order AR process. However, the series might well be the result of a higher order AR process. A solution to this is to generalise the AR(1) model $Y_t = \beta_0 + \beta_1 Y_{t-1} + e_t$ and this gives rise to what is known as the **Augmented Dickey-Fuller (ADF) test**. As an illustration, consider the third order AR process:

$$Y_t = \beta_0 + \beta_1 Y_{t-1} + \beta_2 Y_{t-2} + \beta_3 Y_{t-3} + e_t \tag{5.9}$$

which has characteristic equation:

$$\left(1 - \beta_1 L - \beta_2 L^2 - \beta_3 L^3\right) = 0 \tag{5.10}$$

We are going to **reparameterise** Eq. (5.9). Recall that when we studied cointegration, reparameterisation simply means expressing an equation in another, equivalent algebraic form. Subtract Y_{t-1} from both sides of Eq. (5.9) and rearrange terms:

$$\Delta Y_t = \beta_0 + (\beta_1 + \beta_2 + \beta_3 - 1)Y_{t-1} - (\beta_2 + \beta_3)Y_{t-1} + \beta_2 Y_{t-2} + \beta_3 Y_{t-3} + e_t \tag{5.11}$$

Add and subtract $\beta_3 Y_{t-2}$ from the right side of Eq. (5.11),

$$\begin{aligned}
\Delta Y_t &= \beta_0 + (\beta_1 + \beta_2 + \beta_3 - 1)Y_{t-1} - (\beta_2 + \beta_3)Y_{t-1} + \boldsymbol{\beta_3 Y_{t-2}} + \beta_2 Y_{t-2} + \beta_3 Y_{t-3} - \boldsymbol{\beta_3 Y_{t-2}} + e_t \\
\Delta Y_t &= \beta_0 + (\beta_1 + \beta_2 + \beta_3 - 1)Y_{t-1} - (\beta_2 + \beta_3)Y_{t-1} + (\beta_2 + \beta_3)Y_{t-2} - \beta_3(Y_{t-2} - Y_{t-3}) + e_t \\
\Delta Y_t &= \beta_0 + (\beta_1 + \beta_2 + \beta_3 - 1)Y_{t-1} - (\beta_2 + \beta_3)(Y_{t-1} - Y_{t-2}) - \beta_3 \Delta Y_{t-2} + e_t \\
\Delta Y_t &= \beta_0 + (\beta_1 + \beta_2 + \beta_3 - 1)Y_{t-1} - (\beta_2 + \beta_3)\Delta Y_{t-1} - \beta_3 \Delta Y_{t-2} + e_t \\
\text{or } \Delta Y_t &= \beta_0 + \beta^* Y_{t-1} + \beta_1^* \Delta Y_{t-1} + \beta_2^* \Delta Y_{t-2} + e_t
\end{aligned} \tag{5.12}$$

where $\beta^* = \beta_1 + \beta_2 + \beta_3 - 1$, $\beta_1^* = -(\beta_2 + \beta_3)$ and $\beta_2^* = -\beta_3$.

We know that Eq. (5.9) will be non-stationary if any of the roots of the characteristic Eq. (5.10) are less than or equal to unity in absolute value. If a unit root exists i.e. $L = 1$ then it must be possible to write the characteristic equation in the form $(1 - L)(1 + \lambda L + \mu L^2) = 0$. Therefore if there is a unit root, Eq. (5.9) could be written as:

$$\left(1 + \lambda L + \mu L^2\right)(1 - L)Y_t = \beta_0 + e_t.$$

However, $(1 - L)Y_t = Y_t - Y_{t-1} = \Delta Y_t$, so **if there is a unit root** then we can write Eq. (5.9) as:

$$\left(1 + \lambda L + \mu L^2\right)\Delta Y_t = \beta_0 + e_t \tag{5.13}$$

Using the facts that $\lambda L(\Delta Y_t) = \lambda \Delta Y_{t-1}$ and that $\mu L^2(\Delta Y_t) = \mu \Delta Y_{t-2}$, we may rewrite Eq. (5.13) as:

$$\Delta Y_t = \beta_0 - \lambda \Delta Y_{t-1} - \mu \Delta Y_{t-2} + e_t \tag{5.14}$$

The key point is that Eq. (5.14) is a reparameterisation of the AR(3) process **if we assume that there is a unit root**. Hence, if we compare Eqs. (5.14) and (5.12), we will see that when there is a unit root, then the coefficient (β^*) of Y_{t-1} in Eq. (5.12) must be zero. Therefore, to examine difference stationarity in an AR(3) process, the

ADF test requires us to test $H_0 : \beta^* = 0$ (not difference stationary) versus $H_1 : \beta^* < 0$ (is difference stationary). In the reparameterised Eq. (5.12).

Equation (5.11) is simply a reparameterisation of the AR(3) process in Eq. (5.9). An AR(2) process would have one differenced term ΔY_{t-1} in Eq. (5.12) and have shown that an AR(3) process has two such differenced terms. to select a model with intercept (like Eq. (5.7) on p17), with an intercept and trend (like Eq. (5.8)) or neither an intercept nor trend (like Eq. (5.6)). A graph may well be able to assist here. It is a question of whether a series is stochastic, deterministic or a mixture of both. There is evidence that short run forecasts from all of these types of models have very close forecasting performance. When in doubt, include an intercept and disregard the possibility of a deterministic trend.

This result may be generalised for a pth order autoregressive process:

$$Y_t=\beta_0+\beta_1 Y_{t-1}+\beta_2 Y_{t-2}+\ldots+\beta_{p-2} Y_{t-p+2}+\beta_{1p-1} Y_{t-p+1}+\beta_p Y_{t-p}+e_t,$$

which can be reparameterised into:

$$\Delta Y_t=\beta_0+\beta^* Y_{t-1}+\beta_1^* \Delta Y_{t-1}+\beta_2^* \Delta Y_{t-2}+\ldots+\beta_{p-1}^* \Delta Y_{t-p+1}+e_t, \tag{5.15}$$

The great advantage of the reparameterised Eq. (5.15) is that testing a pth order process for stationarity is equivalent to testing whether or not $\beta^* = 0$. If we can reject $H_0 = \beta^* = 0$, then the implication is that we have a stationary pth order process. The coefficients in Eq. (5.15) are determined by least squares. Note that in this test, the notation ADF(p) is used, where p is the number of differenced terms included on the right hand side of Eq. (5.15).

The above now permits to explain EViews output in respect of unit roots in more detail.

Equation (5.15) will take a slightly different form if a deterministic trend is included.

Three other points. There may be more than one unit root. However, it only takes one unit root for the time series not to be difference stationary. Secondly, the DF and ADF processes apply only to AR processes. What if the time series is purely MA or mixed AR/M(A)? The beauty is that if you remember the invertibility condition, MA terms can be expressed in AR form e.g. an MA(1) process can be expressed as an infinite AR process. Thirdly, the DF and ADF tests assume that the residuals are uncorrelated and have constant variance. The Phillips-Perron test was derived to relax these two assumptions.

Appendix 5.1: The Binomial Theorem

It may be verified that $(x + a)^1 = (x + a)$, $(x + a)^2 = x^2 + 2ax + a^2$, $(x + a)^3 = x^3 + 3ax^2 + 3a^2x + a^3$, $(x + a)^4 = x^4 + 4ax^3 + 6a^2x^2 + 4a^3x + a^4$ etc. The powers of x descend by one and the powers of a ascend by one as one moves from term to term. The numerical coefficients are given by Pascal's triangle:

$$
\begin{array}{ccccccccc}
 & & & & 1 & & & & \\
 & & & 1 & & 1 & & & \\
 & & 1 & & 2 & & 1 & & \\
 & 1 & & 3 & & 3 & & 1 & \\
1 & & 4 & & 6 & & 4 & & 1
\end{array}
$$

Simply sum two consecutive digits in any row and write the total in between them a row lower. Hence the next row would be 1, 5, 10, 10, 5, 1 whereby:

$$(x + a)^5 = x^5 + 5ax^4 + 10a^2x^3 + 10a^3x^2 + 5a^4x + a^5.$$

Pascal's triangle only works for $(x + a)^n$ where n is an integer. For negative and fractional values of the power n we turn to the Binomial Theorem which states that **for all n**:

$$(x + a)^n = x^n + nax^{n-1} + \frac{n(n-1)}{2!}a^2x^{n-2} + \frac{n(n-1)(n-2)}{3!}a^3x^{n-3} + \frac{n(n-1)(n-2)(n-3)}{4!}a^4x^{n-4} + \ldots$$

where ! is the factorial notation such that, for example, $4! = 4 \times 3 \times 2 \times 1 = 24$. You may wish to verify the result for $(x + a)^5$ via the Binomial Theorem. If the power n is an integer, then the number of terms involved in the expansion is (n + 1), so an expansion of the quadratic $(x + a)^2$ involves three terms, expanding a cubic involves four terms etc. However, if n is a fraction or negative then the expansion has an infinite number of terms.

As an example, consider the expansion of $(1 + x)^{-1}$, by replacing x by 1, a by x and n by −1 in the binomial expansion on the previous page:

$$(1 + x)^{-1} = 1^{-1} + (-1)x.1^{-2} + \frac{(-1)(-2)}{2!}x^2.1^{-3}$$
$$+ \frac{(-1)(-2)(-3)}{3!}x^3.1^{-4} + \frac{(-1)(-2)(-3)(-4)}{4!}x^4.1^{-5} + \ldots (1 + x)^{-1} = 1 - x$$
$$+x^2 - x^3 + x^4 + \ldots$$

This series expansion is **convergent** if and only if $|x| < 1$. The same rule of convergence applies to the series expansion for $(1 - x)^n$. By convergent, I mean that the sum of the series approaches a finite limit. For example, the geometric series 1 $+\frac{1}{2}+\frac{1}{4}+\frac{1}{8}+\ldots$ converges towards a limit of 2 as you add more and more terms. Conversely the geometric series 1 + 4 + 16 + 64 + ... is **divergent** because it becomes larger and larger as you add more terms. The latter series does not approach a finite limit; rather the sum becomes infinite.

Given the above rule for convergence, we can insert x = 0.01 into the above series expansion to establish that:

$$(1.01)^{-1} = 1 - 0.01 + (0.01)^2 - (0.01)^3 + (0.01)^4 + \ldots$$

and summing the five terms on the right hand side, we obtain that $(1.01)^{-1}$ is approximately equal to 0.99009901. In fact, $(1.01)^{-1} = 0.990099009$. Note that it would be invalid to insert x = 3 in the above series expansion in order to approximate $(4)^{-1}$, because x = 3 is outside the range of convergence. The series would diverge away from the correct answer that $(4)^{-1} = 0.25$.

Appendix 5.2: The Quadratic Equation

The equation $y = ax^2 + bx + c$ in which a, b and c are constants and $a \neq 0$ is called a quadratic. When we solve a quadratic equals zero we are said to be establishing **the roots** of that quadratic (i.e. where it cross the x-axis). For example, to solve $x^2 - x - 12 = 0$ we use the quadratic's factors i.e. $(x - 4)(x + 3) = 0$ and the roots are x = 4 and x = −3. The quadratic $x^2 - 10x + 25 = 0$ has **repeated roots**, since $(x - 5)(x - 5) = 0$ and x = 5.

Sometimes the roots of a quadratic may be **a complex number**. A complex number is of the form c + id where c and d are positive or negative constants and is defined to equal $\sqrt{-1}$. For example, $4 + 7i = 4 + 7\sqrt{-1}$ is a complex number in which 4 is called the **real (Re) part** and 7i is called the **imaginary (Im) part**. The real part of a complex number can be zero, for example $\sqrt{-16} = 4i$ since $4i \times 4i = 16 \times i^2 = -16$. Similarly, $\sqrt{-289} = 17i$. A complex number may be represented on what is called an **Argand diagram** in which the horizontal axis is the Re part and the vertical axis is the Im part. Figure 5.2 graphs the complex number 2 + 3i. The line joining the origin to this point is called a vector. The length (or modulus) of this vector is found by Pythagoras to be $\sqrt{2^2 + 3^2} = 3.606$. Similarly, the length of the vector representing 2 − 3i would also be 3.606.

Consider solving the quadratic $ax^2 + bx + c = 0$. Divide throughout by a to derive $x^2 + \frac{b}{a}x + \frac{c}{a} = 0$. Add $\frac{b^2}{4a^2}$ to both sides:

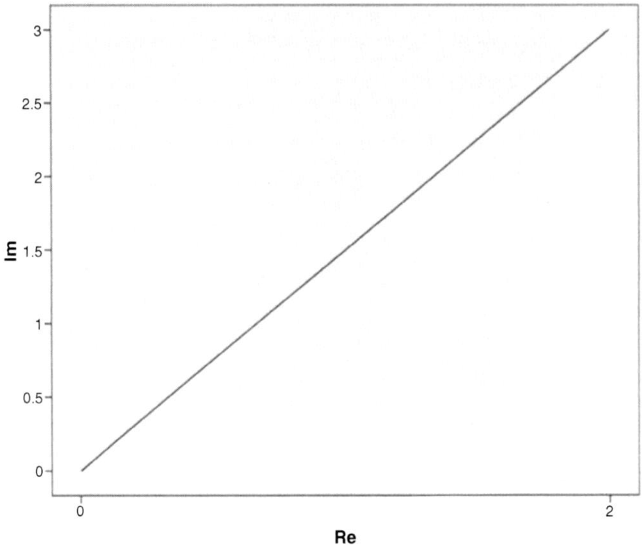

Fig. 5.2 An Argand plot of the complex number 2 + 3i

$$X^2 + \frac{b}{a}x + \frac{b^2}{4a^2} = \frac{b^2}{4a^2} - \frac{c}{a}$$

$$\left(x - \tfrac{b}{2a}\right)^2 = \frac{b^2 - 4ac}{4a^2}$$

$$x - \frac{b}{2a} = \frac{\pm\sqrt{b^2 - 4ac}}{2a}$$

(since a square root can be + or −, e.g. $\sqrt{144} = \pm 12$) and we establish that:

$$\boxed{x = \frac{-b \pm \sqrt{b^2 - 4ac}}{2a}.}$$

This formula gives the roots of a quadratic. For example, if we are to solve $x^2 - 13x + 42 = 0$, we have that a = 1, b = −13 and c = 42, whereby:

$$X = \frac{13 \pm \sqrt{169 - 168}}{2} = \frac{13 \pm 1}{2}$$

and the roots are x = 7 and x = 6. These are real roots as opposed to complex roots. Hence when factorised, the quadratic must be (x − 6)(x − 7) = 0. Note that a quadratic can only have either two real roots, two complex roots or a repeated root; it cannot have one real root and one complex one. For a general polynomial of order n, complex roots always appear in pairs i.e. there is an even number of them.

The quantity $b^2 - 4ac$ in the above formula for the roots of a quadratic is called the **discriminant**. If the discriminate is negative, the quadratic must have complex roots (since we are taking the square root of a negative number). If the discriminant equals zero, then the quadratic has repeated roots. Consider $x^2 - 4x + 28 = 0$. The discriminant, $16 - 4(28)$, is clearly negative and

$$X = \frac{4 \pm \sqrt{16 - 112}}{2} = \frac{4 \pm \sqrt{-96}}{2} = \frac{4 \pm \sqrt{(16)(-6)}}{2} = 2 \pm 2\sqrt{-6}$$
$$= 2 \pm 4.898i \quad \text{since } \sqrt{6} = 2.449.$$

This quadratic has a pair of complex roots $x = 2 + 4.898i$ and $x = 2 - 4.898i$.

There is a relationship between the coefficients of a quadratic and the roots of the quadratic. Suppose a quadratic has roots δ and γ. Then solving the quadratic equals zero:

$$ax^2 + bx + c = 0 \text{ implies that (divide throughout by a)}$$
$$x^2 + \frac{b}{a}x + \frac{c}{a} = 0$$

but we know the roots are δ and γ, so $(x - \delta)(x - \gamma) = x^2 - (\delta + \gamma)x + \delta\gamma = 0$.

The only way that the last two lines can be equal to zero is if:

$$\text{Sum of the roots, } \delta + \gamma = -\frac{b}{a}$$
$$\text{and Product of the roots, } \delta\gamma = \frac{c}{a}$$

Reconsider the roots of $x^2 - 4x + 28 = 0$, which we have seen are $x = 2 + 4.898i$ and $x = 2 - 4.898i$. Now for this quadratic, $a = 1$, $b = -4$ and $c = 28$. Therefore the sum of the roots should be equal to $-\frac{b}{a} = 4$. The sum of the roots is $(2 + 4.898i) + (2 - 4.898i)$ which is indeed 4. The product of the roots should be equal to $\frac{c}{a} = 28$. As a check, the product of the roots is $(2 + 4.898i)(2 - 4.898i) = 4 + 0i + (4.898i)(-4.898i) = 4 - 24i^2 = 4 + 24 = 28$, as required, since $i^2 = -1$.

Chapter 6
Economic Forecasting Using Regression

An essential part of managing any organisation be it governmental, commercial, industrial or social, is planning for the future by generating adequate forecasts of factors that are central to that organisation's successful operation. Methods of forecasting fall into two groups; qualitative and quantitative. Among the former fall expert judgment and intuitive approaches. Such methods are particularly used by management when conditions in the past are unlikely to hold in the future. In all branches of scientific enquiry, no statistical tool has received the attention given to *regression analysis* during the last 30 years. A frequent misconception among non-statistical users of regression is that the technique's prime objective is to generate forecasts. Besides being a forecasting tool, regression analysis attempts to shed light on the mechanisms that relate variables. Knowledge of such mechanisms would, in some circumstances, permit a degree of control.

For example, knowledge of how certain factors contribute towards the production of defective industrial items might assist in reducing the defective rate. Knowledge of the factors that drive changes in share prices would help in portfolio selection. Regression is a tool for forecasting and explanation. (Time series analysis, a technique which was discussed in Chaps. 3 and 4, is solely a forecasting tool.) Of all regression models, the linear model is the most widely applied. The objective of forecasting is to provide quantitative estimate(s) of the possibility of the future course of the object of interest. This objective can be carried out based on past and current information. Economic Forecasting can be useful for several economic areas such as:

1. Operational planning, inventory management, production planning and sales
2. GDP, unemployment, consumption, investment and interest rates forecast
3. Financial asset management like asset returns, exchange rates and commodity prices
4. Business and government budgeting revenue forecasts
5. Demographic changes like fertility and mortality rates
6. Crisis management related to the issues like: probabilities of default, currency devaluations, military coups, and so forth.

A. Aljandali, M. Tatahi, *Economic and Financial Modelling with EViews*, Statistics and Econometrics for Finance, https://doi.org/10.1007/978-3-319-92985-9_6

In this book, we will consider two prominent methods of forecasting:

1. Regression models, which are covered in this chapter
2. The **autoregressive integrated moving average** (**ARIMA**) models, popularized by statisticians Box and Jenkins and known as the **Box–Jenkins (BJ) methodology**. The ARIMA forecasting methods will be discussed in Chapter 7.

6.1 Forecasting with Regression Models

For business, government and many other economic agents, forecasting is probably the most important purpose of estimating regression models. To keep things simple, we will first consider the following bivariate regression:

$$PCE_t = B_1 + B_2 PDI_t + u_t$$

PCE is per capita personal consumption expenditure, *PDI* is per capita personal disposable (after-tax) income, and *u* is the error term. This regression is a *consumption function*. The slope coefficient in this regression represents the **marginal propensity to consume** (**MPC**). It is the consumption expenditure for an additional dollar's increase in income. We use US data of these variables collected between 1967Q1 and 1996Q3. To estimate this consumption function, we need to follow time series analysis approaches to make sure the series are stationary and are cointegrated.

We use the observations from1967Q1–1995Q2 and save the last five observations, called the hold back period, to evaluate the performance of the estimated model when we run the main regression (a regression of stationary series).

We first plot the data to get an idea of the nature of the relationship between the two variables. According to Fig. 6.1, there is almost a linear relationship between PCE and PDI.

To estimate the above model we need to check whether the variables are stationary. In this process we need to consider the cointegration between series and establish whether a long-run relationship exists among the series or not.

6.2 Step One: Checking the Stationarity of the Series

Both variables are converted to logs as follows:

LPCE = @LOG(PCE)
LPDI = @LOG(PDI)

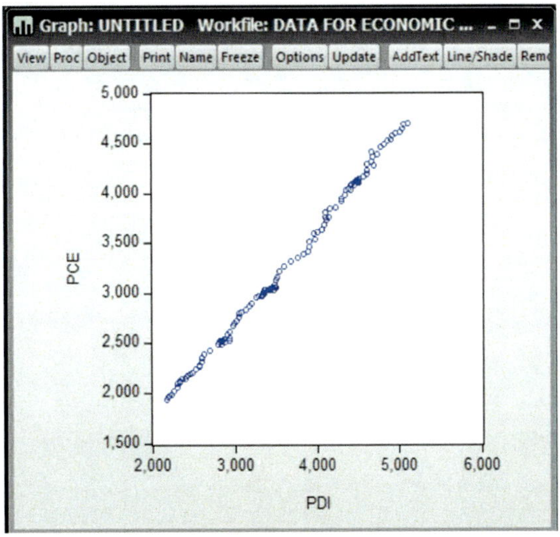

Fig. 6.1 Data plot of PCE and PDI

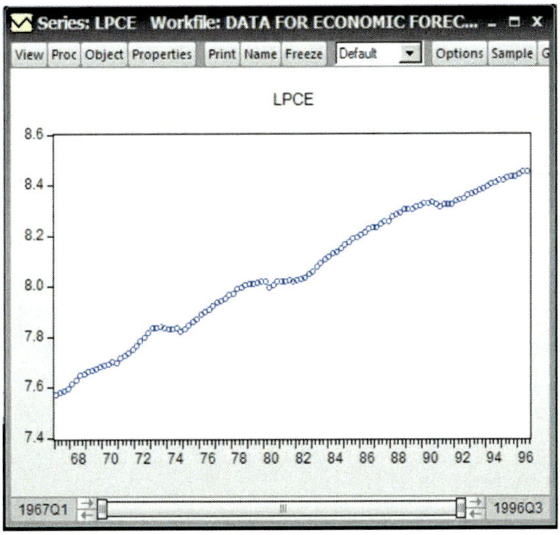

Fig. 6.2 Plot of the Natural Log of PCE (LPCE)

Figure 6.2 shows that LPCE is not stationary and has a deterministic upward trend.

Correlogram of LPCE

1. H_0: The series has no unit root
2. H_1: The series has a unit root
3. $\alpha = 5\%$ or 0.05

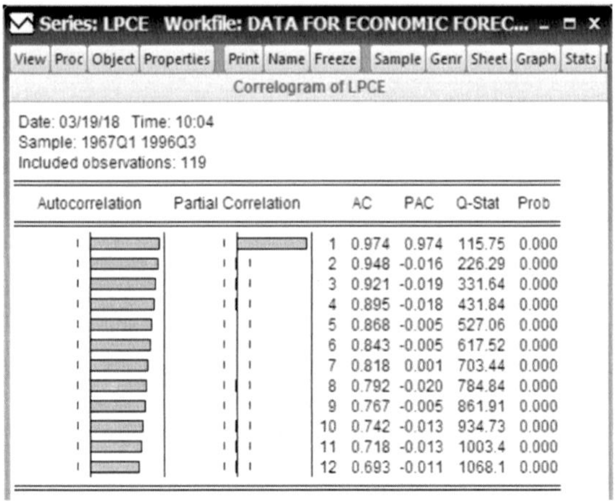

Series: LPCE Workfile: DATA FOR ECONOMIC FOREC...

View | Proc | Object | Properties | Print | Name | Freeze | Sample | Genr | Sheet | Graph | Stats

Correlogram of LPCE

Date: 03/19/18 Time: 10:04
Sample: 1967Q1 1996Q3
Included observations: 119

Autocorrelation	Partial Correlation		AC	PAC	Q-Stat	Prob
		1	0.974	0.974	115.75	0.000
		2	0.948	-0.016	226.29	0.000
		3	0.921	-0.019	331.64	0.000
		4	0.895	-0.018	431.84	0.000
		5	0.868	-0.005	527.06	0.000
		6	0.843	-0.005	617.52	0.000
		7	0.818	0.001	703.44	0.000
		8	0.792	-0.020	784.84	0.000
		9	0.767	-0.005	861.91	0.000
		10	0.742	-0.013	934.73	0.000
		11	0.718	-0.013	1003.4	0.000
		12	0.693	-0.011	1068.1	0.000

Fig. 6.3 Correlogram of LPCE

There are no test statistics to calculate. We look at the Prob column in Fig. 6.3.

4. The p-values are 0.000
5. Prob = 0.000 < 0.05 therefore we reject the H_0
6. We conclude that LPCE is not stationary

Formal Approach

1. H_0: series has a unit root and it is not stationary
2. H_1: series has no unit root and it is stationary.
3. $\alpha = 5\%$ or 0.05

According to Fig. 6.4, ADF test statistic is −1.518590.

1. The p-values = 0.5208
2. Prob = 0.5208 > 0.05 therefore we cannot reject the H_0
3. The series has a unit root and it is not stationary.

Figure 6.5 shows LPCE is not stationary and has a deterministic upward trend. The graph shows LPDI is not stationary and has a deterministic upward trend.

Correlogram of LPDI

1. H_0: LPDI has no unit root
2. H_1: LPDI has a unit root
3. $\alpha = 5\%$ or 0.05

There are no test statistics to calculate. We look at the Prob column in Fig. 6.6.

1. The p-values are 0.000
2. Prob = 0.000 < 0.05 therefore we reject the H_0
3. LPDI is not stationary

Series: LPCE Workfile: DATA FOR ECONOMIC FOREC...

View | Proc | Object | Properties | Print | Name | Freeze | Sample | Genr | Sheet | Graph | Stats

Augmented Dickey-Fuller Unit Root Test on LPCE

Null Hypothesis: LPCE has a unit root
Exogenous: Constant
Lag Length: 1 (Automatic - based on SIC, maxlag=12)

		t-Statistic	Prob.*
Augmented Dickey-Fuller test statistic		-1.518590	0.5208
Test critical values:	1% level	-3.487046	
	5% level	-2.886290	
	10% level	-2.580046	

*MacKinnon (1996) one-sided p-values.

Augmented Dickey-Fuller Test Equation
Dependent Variable: D(LPCE)
Method: Least Squares
Date: 03/19/18 Time: 10:05
Sample (adjusted): 1967Q3 1996Q3
Included observations: 117 after adjustments

Variable	Coefficient	Std. Error	t-Statistic	Prob.
LPCE(-1)	-0.003991	0.002628	-1.518590	0.1316
D(LPCE(-1))	0.221487	0.090964	2.434881	0.0164
C	0.037914	0.021281	1.781625	0.0775

R-squared	0.076673	Mean dependent var	0.007448
Adjusted R-squared	0.060474	S.D. dependent var	0.007311
S.E. of regression	0.007086	Akaike info criterion	-7.035948
Sum squared resid	0.005725	Schwarz criterion	-6.965123
Log likelihood	414.6030	Hannan-Quinn criter.	-7.007194
F-statistic	4.733247	Durbin-Watson stat	2.049983
Prob(F-statistic)	0.010599		

Fig. 6.4 LPCE Unit root test

Formal Approach

1. H_0: series has a unit root and is not stationary
2. H_1: series has no unit root and is stationary.
3. $\alpha = 5\%$ or 0.05

According to Fig. 6.7, ADF test statistic is -1.581621.

4. The p-values = 0.4888.
5. Prob = 0.4888 > 0.05. We cannot reject the H_0; the test is not significant.
6. The series has a unit root and it is not stationary.

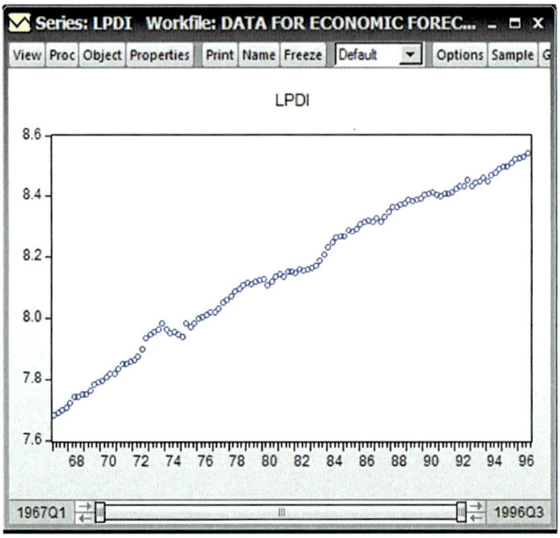

Fig. 6.5 Plot of the Natural Log PDI-LPI

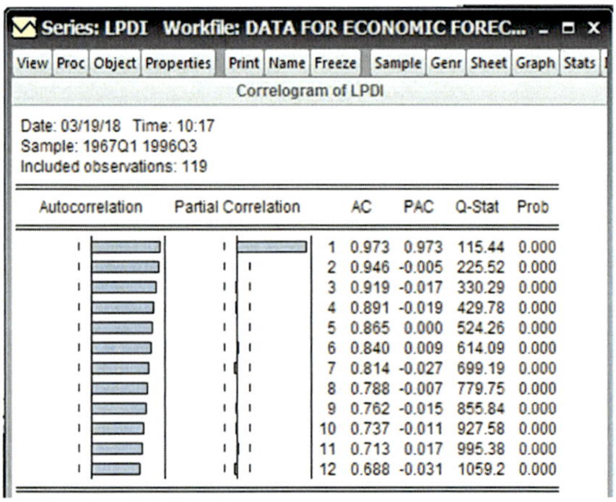

Correlogram of LPDI

Date: 03/19/18 Time: 10:17
Sample: 1967Q1 1996Q3
Included observations: 119

Autocorrelation	Partial Correlation		AC	PAC	Q-Stat	Prob
		1	0.973	0.973	115.44	0.000
		2	0.946	-0.005	225.52	0.000
		3	0.919	-0.017	330.29	0.000
		4	0.891	-0.019	429.78	0.000
		5	0.865	0.000	524.26	0.000
		6	0.840	0.009	614.09	0.000
		7	0.814	-0.027	699.19	0.000
		8	0.788	-0.007	779.75	0.000
		9	0.762	-0.015	855.84	0.000
		10	0.737	-0.011	927.58	0.000
		11	0.713	0.017	995.38	0.000
		12	0.688	-0.031	1059.2	0.000

Fig. 6.6 Correlogram of LPDI

6.3 Step Two: Making Series Stationary

The first difference of the above series can be found as follows:

DLPCE = LPCE − LPCE(−1)
DLPDI = LPDI − LPDI(−1)

According to Fig. 6.8, The DLPCE series is now stationary.

Correlogram of DLPCE

1. H_0: The series has no unit root
2. H_1: The series has a unit root
3. $\alpha = 5\%$ or 0.05

Series: LPDI Workfile: DATA FOR ECONOMIC FOREC...

View | Proc | Object | Properties | Print | Name | Freeze | Sample | Genr | Sheet | Graph | Stats

Augmented Dickey-Fuller Unit Root Test on LPDI

Null Hypothesis: LPDI has a unit root
Exogenous: Constant
Lag Length: 0 (Automatic - based on SIC, maxlag=12)

		t-Statistic	Prob.*
Augmented Dickey-Fuller test statistic		-1.581621	0.4888
Test critical values:	1% level	-3.486551	
	5% level	-2.886074	
	10% level	-2.579931	

*MacKinnon (1996) one-sided p-values.

Augmented Dickey-Fuller Test Equation
Dependent Variable: D(LPDI)
Method: Least Squares
Date: 03/19/18 Time: 10:18
Sample (adjusted): 1967Q2 1996Q3
Included observations: 118 after adjustments

Variable	Coefficient	Std. Error	t-Statistic	Prob.
LPDI(-1)	-0.005974	0.003777	-1.581621	0.1165
C	0.055951	0.030796	1.816814	0.0718

R-squared	0.021110	Mean dependent var	0.007265
Adjusted R-squared	0.012671	S.D. dependent var	0.010027
S.E. of regression	0.009963	Akaike info criterion	-6.363069
Sum squared resid	0.011514	Schwarz criterion	-6.316108
Log likelihood	377.4211	Hannan-Quinn criter.	-6.344001
F-statistic	2.501526	Durbin-Watson stat	2.177505
Prob(F-statistic)	0.116459		

Fig. 6.7 LPDI unit root test

There are no test statistics to calculate. We look at the Prob column in Fig. 6.9.

4. The p-values are bigger than 0.05.
5. As p-values are bigger than 0.05, we cannot reject the H_0
6. The series is stationary

Formal Approach

1. H_0: series has a unit root and is not stationary
2. H_1: series has no unit root and is stationary.
3. $\alpha = 5\%$ or 0.05

According to Fig. 6.10, ADF test statistic is -8.381064.

4. The p-values = 0.0000.
5. Prob = 0.0000 < 0.05 we reject the H_0
6. The series has no unit root and it is stationary (Fig. 6.11).

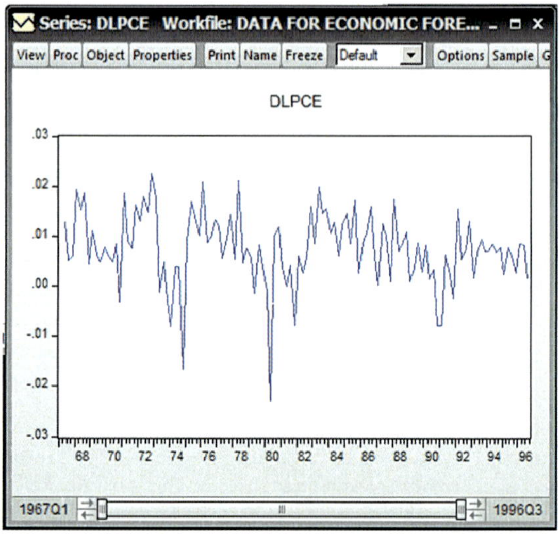

Fig. 6.8 Plot of the first difference (DLPCE)

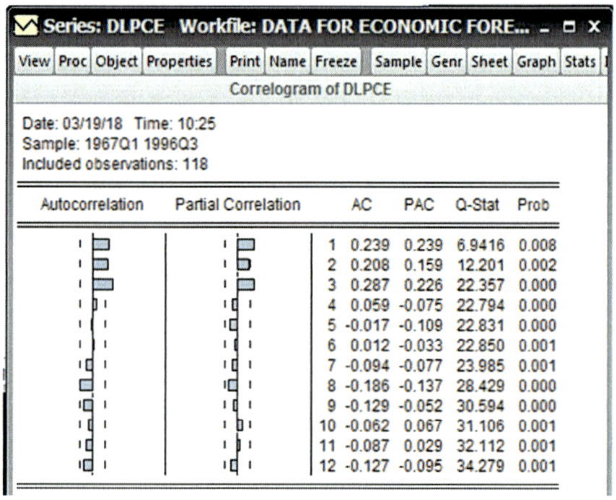

Series: DLPCE Workfile: DATA FOR ECONOMIC FORE...

Correlogram of DLPCE

Date: 03/19/18 Time: 10:25
Sample: 1967Q1 1996Q3
Included observations: 118

	AC	PAC	Q-Stat	Prob
1	0.239	0.239	6.9416	0.008
2	0.208	0.159	12.201	0.002
3	0.287	0.226	22.357	0.000
4	0.059	-0.075	22.794	0.000
5	-0.017	-0.109	22.831	0.000
6	0.012	-0.033	22.850	0.001
7	-0.094	-0.077	23.985	0.001
8	-0.186	-0.137	28.429	0.000
9	-0.129	-0.052	30.594	0.000
10	-0.062	0.067	31.106	0.001
11	-0.087	0.029	32.112	0.001
12	-0.127	-0.095	34.279	0.001

Fig. 6.9 Correlogram of DLPCE

Correlogram of DLPDI

1. H_0: DLPDI has no unit root
2. H_1: DLPDI has a unit root
3. $\alpha = 5\%$ or 0.05

There are no test statistics to calculate. We look at the Prob column in Fig. 6.12.

4. The p-values are bigger than 0.05.
5. As p-values are bigger than 0.05, we cannot reject the H_0.
6. DLPDI is stationary

Series: DLPCE Workfile: DATA FOR ECONOMIC FORE...

View | Proc | Object | Properties | Print | Name | Freeze | Sample | Genr | Sheet | Graph | Stats

Augmented Dickey-Fuller Unit Root Test on DLPCE

Null Hypothesis: DLPCE has a unit root
Exogenous: Constant
Lag Length: 0 (Automatic - based on SIC, maxlag=12)

		t-Statistic	Prob.*
Augmented Dickey-Fuller test statistic		-8.381064	0.0000
Test critical values:	1% level	-3.487046	
	5% level	-2.886290	
	10% level	-2.580046	

*MacKinnon (1996) one-sided p-values.

Augmented Dickey-Fuller Test Equation
Dependent Variable: D(DLPCE)
Method: Least Squares
Date: 03/19/18 Time: 10:26
Sample (adjusted): 1967Q3 1996Q3
Included observations: 117 after adjustments

Variable	Coefficient	Std. Error	t-Statistic	Prob.
DLPCE(-1)	-0.759025	0.090564	-8.381064	0.0000
C	0.005629	0.000949	5.930373	0.0000

R-squared	0.379191	Mean dependent var	-9.79E-05
Adjusted R-squared	0.373793	S.D. dependent var	0.009006
S.E. of regression	0.007127	Akaike info criterion	-7.033015
Sum squared resid	0.005841	Schwarz criterion	-6.985798
Log likelihood	413.4314	Hannan-Quinn criter.	-7.013846
F-statistic	70.24223	Durbin-Watson stat	2.061287
Prob(F-statistic)	0.000000		

Fig. 6.10 DLPCE unit root test

Fig. 6.11 Plot of the first difference (DLPDI)

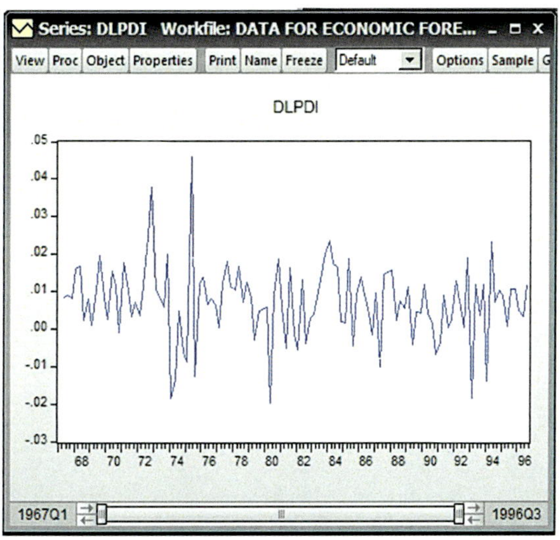

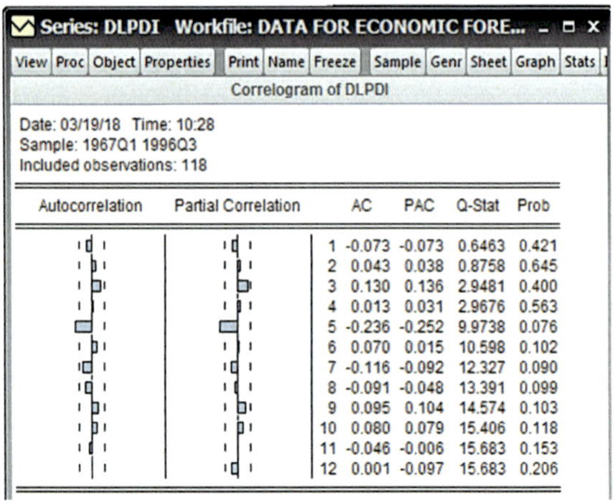

Series: DLPDI Workfile: DATA FOR ECONOMIC FORE...

View Proc Object Properties Print Name Freeze Sample Genr Sheet Graph Stats

Correlogram of DLPDI

Date: 03/19/18 Time: 10:28
Sample: 1967Q1 1996Q3
Included observations: 118

Autocorrelation	Partial Correlation		AC	PAC	Q-Stat	Prob
		1	-0.073	-0.073	0.6463	0.421
		2	0.043	0.038	0.8758	0.645
		3	0.130	0.136	2.9481	0.400
		4	0.013	0.031	2.9676	0.563
		5	-0.236	-0.252	9.9738	0.076
		6	0.070	0.015	10.598	0.102
		7	-0.116	-0.092	12.327	0.090
		8	-0.091	-0.048	13.391	0.099
		9	0.095	0.104	14.574	0.103
		10	0.080	0.079	15.406	0.118
		11	-0.046	-0.006	15.683	0.153
		12	0.001	-0.097	15.683	0.206

Fig. 6.12 Correlogram of DLPDI

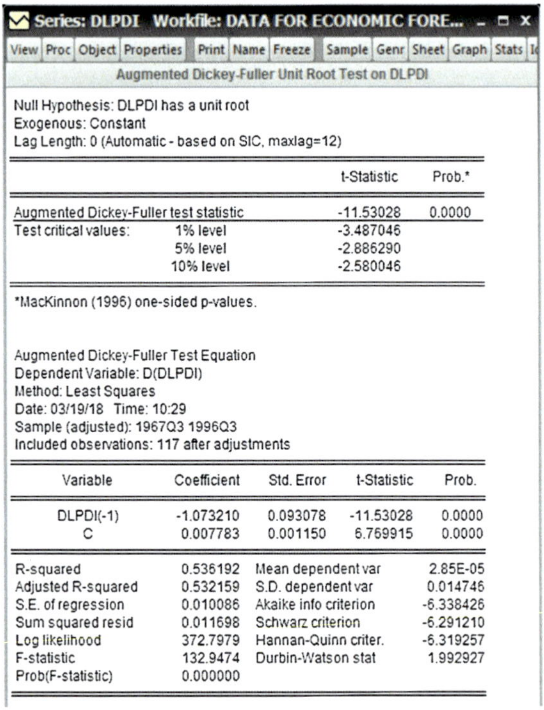

Series: DLPDI Workfile: DATA FOR ECONOMIC FORE...

View Proc Object Properties Print Name Freeze Sample Genr Sheet Graph Stats

Augmented Dickey-Fuller Unit Root Test on DLPDI

Null Hypothesis: DLPDI has a unit root
Exogenous: Constant
Lag Length: 0 (Automatic - based on SIC, maxlag=12)

		t-Statistic	Prob.*
Augmented Dickey-Fuller test statistic		-11.53028	0.0000
Test critical values:	1% level	-3.487046	
	5% level	-2.886290	
	10% level	-2.580046	

*MacKinnon (1996) one-sided p-values.

Augmented Dickey-Fuller Test Equation
Dependent Variable: D(DLPDI)
Method: Least Squares
Date: 03/19/18 Time: 10:29
Sample (adjusted): 1967Q3 1996Q3
Included observations: 117 after adjustments

Variable	Coefficient	Std. Error	t-Statistic	Prob.
DLPDI(-1)	-1.073210	0.093078	-11.53028	0.0000
C	0.007783	0.001150	6.769915	0.0000

R-squared	0.536192	Mean dependent var	2.85E-05
Adjusted R-squared	0.532159	S.D. dependent var	0.014746
S.E. of regression	0.010086	Akaike info criterion	-6.338426
Sum squared resid	0.011698	Schwarz criterion	-6.291210
Log likelihood	372.7979	Hannan-Quinn criter.	-6.319257
F-statistic	132.9474	Durbin-Watson stat	1.992927
Prob(F-statistic)	0.000000		

Fig. 6.13 DLPDI unit root test

Formal Approach

1. H_0: series has a unit root and is not stationary
2. H_1: series has no unit root and is stationary.
3. $\alpha = 5\%$ or 0.05

According to Fig. 6.13, ADF test statistic is −11.5328.

4. The p-values = 0.0000
5. Prob = 0.0000 < 0.05. We can reject the H_0; the test is significant.
6. DLPDI has no unit root and it is stationary.

6.4 Step Three: The Cointegration Test

We use the original data and run the regression as shown in Fig. 6.14:

We go to **Quick-Generate series** and we write: u = resid and then **click OK** to generate the plot of Fig. 6.15.

According to Fig. 6.15, the residuals are stationary.

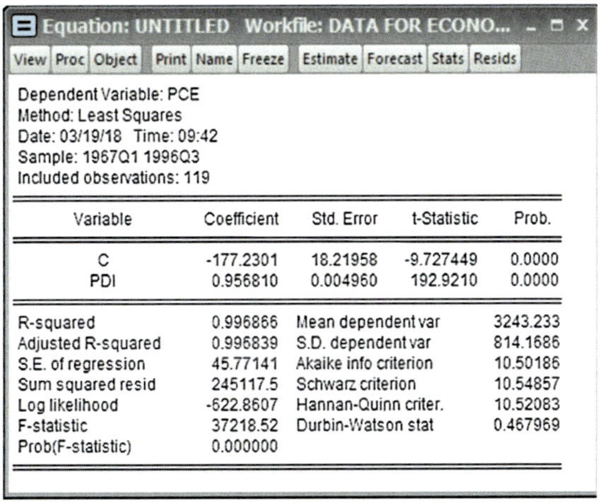
Equation: UNTITLED Workfile: DATA FOR ECONO...
View | Proc | Object | Print | Name | Freeze | Estimate | Forecast | Stats | Resids

Dependent Variable: PCE
Method: Least Squares
Date: 03/19/18 Time: 09:42
Sample: 1967Q1 1996Q3
Included observations: 119

Variable	Coefficient	Std. Error	t-Statistic	Prob.
C	-177.2301	18.21958	-9.727449	0.0000
PDI	0.956810	0.004960	192.9210	0.0000

R-squared	0.996866	Mean dependent var	3243.233
Adjusted R-squared	0.996839	S.D. dependent var	814.1686
S.E. of regression	45.77141	Akaike info criterion	10.50186
Sum squared resid	245117.5	Schwarz criterion	10.54857
Log likelihood	-622.8607	Hannan-Quinn criter.	10.52083
F-statistic	37218.52	Durbin-Watson stat	0.467969
Prob(F-statistic)	0.000000		

Fig. 6.14 PCE PDI Regression output

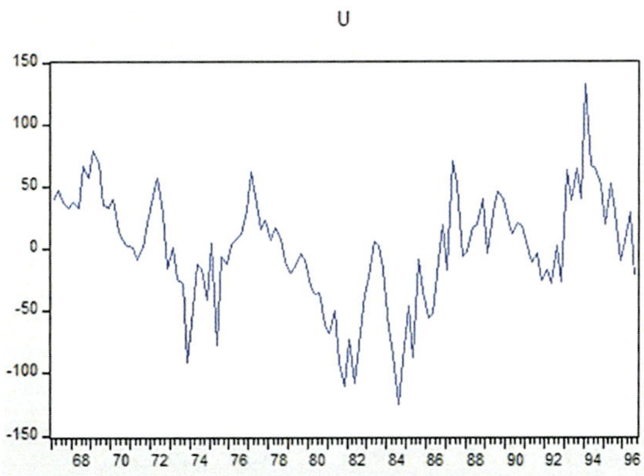

Fig. 6.15 Graph of the residuals

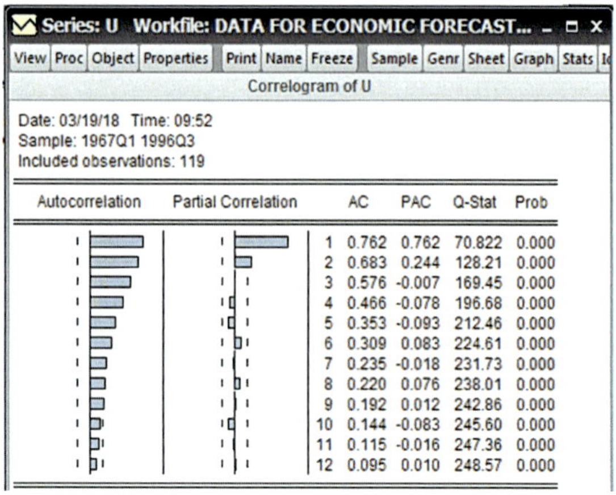

	AC	PAC	Q-Stat	Prob
1	0.762	0.762	70.822	0.000
2	0.683	0.244	128.21	0.000
3	0.576	-0.007	169.45	0.000
4	0.466	-0.078	196.68	0.000
5	0.353	-0.093	212.46	0.000
6	0.309	0.083	224.61	0.000
7	0.235	-0.018	231.73	0.000
8	0.220	0.076	238.01	0.000
9	0.192	0.012	242.86	0.000
10	0.144	-0.083	245.60	0.000
11	0.115	-0.016	247.36	0.000
12	0.095	0.010	248.57	0.000

Fig. 6.16 Correlogram of the residuals

Correlogram of the residuals:

1. H_0: The residuals have no unit root
2. H_1: The residuals have a unit root
3. $\alpha = 5\%$ or 0.05

There are no test statistics to calculate. We look at the Prob column in Fig. 6.16.

4. The p-values are smaller than 0.05
5. As p-values are <0.05, we reject the H_0
6. The residuals are not stationary.

The unit root test of the residual

1. H_0: The residuals have a unit root and are not stationary
2. H_1: The residuals have no unit root and are stationary.
3. $\alpha = 5\%$ or 0.05

According to Fig. 6.17, the ADF test statistic $= -2.935775$.

4. The p-values $= 0.0443$
5. Prob $= 0.0443 < 0.05$ therefore we reject the H_0
6. The residuals have no unit root and are stationary.

We have found the series to be cointegrated. It means that there is a long-run relationship between the first differences of these two series.

Fig. 6.17 Residuals unit root test

Series: U Workfile: DATA FOR ECONOMIC FORECAST...

View | Proc | Object | Properties | Print | Name | Freeze | Sample | Genr | Sheet | Graph | Stats

Augmented Dickey-Fuller Unit Root Test on U

Null Hypothesis: U has a unit root
Exogenous: Constant
Lag Length: 1 (Automatic - based on SIC, maxlag=12)

		t-Statistic	Prob.*
Augmented Dickey-Fuller test statistic		-2.935775	0.0443
Test critical values:	1% level	-3.487046	
	5% level	-2.886290	
	10% level	-2.580046	

*MacKinnon (1996) one-sided p-values.

Augmented Dickey-Fuller Test Equation
Dependent Variable: D(U)
Method: Least Squares
Date: 03/19/18 Time: 09:54
Sample (adjusted): 1967Q3 1996Q3
Included observations: 117 after adjustments

Variable	Coefficient	Std. Error	t-Statistic	Prob.
U(-1)	-0.182085	0.062023	-2.935775	0.0040
D(U(-1))	-0.250174	0.091320	-2.739546	0.0071
C	-0.640830	2.661250	-0.240800	0.8101

R-squared	0.175810	Mean dependent var	-0.590730
Adjusted R-squared	0.161350	S.D. dependent var	31.43295
S.E. of regression	28.78561	Akaike info criterion	9.582935
Sum squared resid	94461.70	Schwarz criterion	9.653760
Log likelihood	-557.6017	Hannan-Quinn criter.	9.611689
F-statistic	12.15880	Durbin-Watson stat	1.981301
Prob(F-statistic)	0.000016		

6.5 Step Four: Model Forecasting

We run a regression between these two stationary series with a sample of 43 observations for each series. We will be using the last four observations for the forecasting processes. As we lost one observation, when we made the first difference of the series, the amount of observations have now reduced to 43 observations.

Go to **Quick-Estimate Equation**-change the sample from 1967Q1 1996Q3 to 1967Q1 1995Q2 and then click **OK** (Fig. 6.18).

This is the final regression output (Fig. 6.19):

We can use this output to forecast the future value(s) of personal consumption expenditure.

Suppose we want to find out ***E*(*PCE*1996Q1|*PDI*1996Q1)**, that is the population or true mean personal consumption expenditure value in 1996Q1 given the value of total disposable income (X) for 1996Q1, which is 0.004915 (note that our sample regression is based on the period 1967Q1–1995Q2).

At this stage we need to learn some special terms used in forecasting such as:

1. *ex post* (after the fact) and *ex ante*

Fig. 6.18 Equation specification

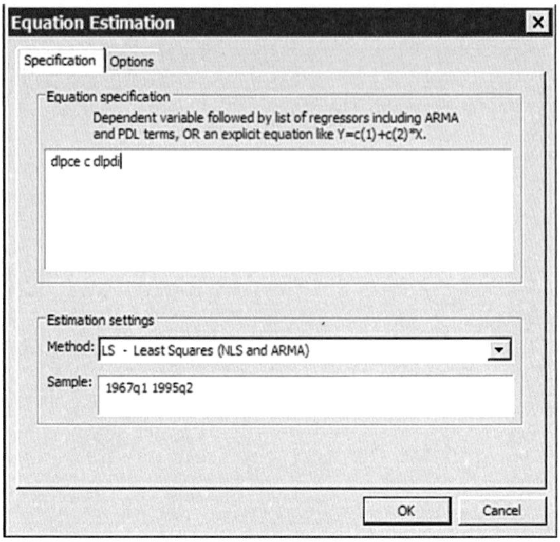

Fig. 6.19 DLPCE Regression output

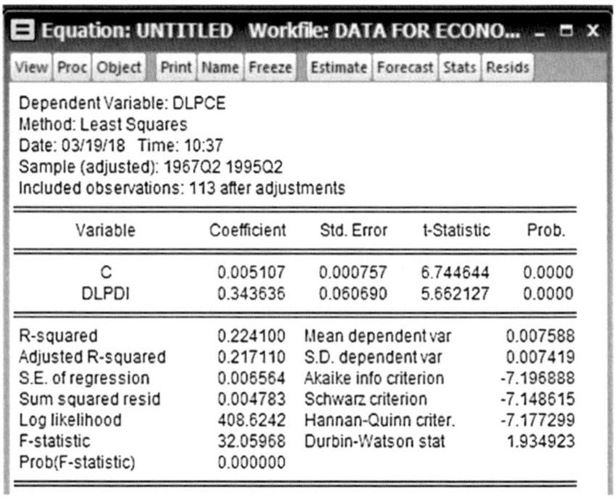

Dependent Variable: DLPCE
Method: Least Squares
Date: 03/19/18 Time: 10:37
Sample (adjusted): 1967Q2 1995Q2
Included observations: 113 after adjustments

Variable	Coefficient	Std. Error	t-Statistic	Prob.
C	0.005107	0.000757	6.744644	0.0000
DLPDI	0.343636	0.060690	5.662127	0.0000

R-squared	0.224100	Mean dependent var	0.007588
Adjusted R-squared	0.217110	S.D. dependent var	0.007419
S.E. of regression	0.006564	Akaike info criterion	-7.196888
Sum squared resid	0.004783	Schwarz criterion	-7.148615
Log likelihood	408.6242	Hannan-Quinn criter.	-7.177299
F-statistic	32.05968	Durbin-Watson stat	1.934923
Prob(F-statistic)	0.000000		

1. **Point forecasts** and **interval forecasts**:
   In point forecasts we provide a **single value** for each forecast period.
   In interval forecasts we obtain **a range, or an interval**, that will include the realized value with some probability. In other words, the interval forecast provides **a margin of uncertainty** about the point forecast.

2. ***Ex post*** **and** ***ex ante*** **forecasts**:

Estimation period	Ex-post forecast Period	Ex-ante forecast Period
1967Q1-1995Q2	1995Q2-1996Q3	1996Q3 forward

(a) In the estimation period we have data on all the variables in the model.
(b) In the *ex post* forecast period we also know the values of the dependent variable and independent variables (this is the holdover period). We can use these values to get some idea about the performance of the fitted model.
(c) In the *ex ante* forecast we estimate the values of the dependent variable beyond the estimation period but we may not know the values of the regressors with certainty, in which case we may have to estimate these values before we can forecast.

3. **Conditional and unconditional forecasts**:

   In conditional forecasts, we forecast the variable of interest *conditional* on the assumed values of the regressors. This type of conditional forecasting is also known as **scenario analysis** or **contingency analysis**.

   In unconditional forecasts, we know the values of the regressors with certainty instead of picking some arbitrary values of them, as in conditional forecasting.

Now we estimate the point forecast of consumption expenditure for 1996Q1.

The value of personal disposable income *DLPDI* for 1996Q1 is 0.004915; this can be found by looking at the data set of this series.

Now it can be shown that the best mean prediction of *Y* (1996Q1) given the *X* value is:

$$\widehat{DLPCE}_t = \beta_0 + \beta_1 DLPDI_t + \epsilon_t$$
$$\widehat{DLPCE}_t = 0.005107 + 0.343636 DLPDI_t$$
$$\widehat{DLPCE}_t = 0.005107 + 0.343636\,(0.004915)$$
$$\widehat{DLPCE}_t = 0.006796$$

This the mean predicted value of personal consumption expenditure in 1996Q1, 0.006796, given the value of DLPDI, 0.004915. From the data it can be seen that the actual value of DLPCE for 1996Q1 was 0.008569. So the actual value was greater than the estimated value by 0.001773. We can call this the **forecast error**.

We do not expect the estimated regression line to forecast the actual values of the regressors and without some error.

Now it can be shown that if the error term in the above model is normally distributed, the estimated DLPCE for the time, 1996Q1, is normally distributed with a mean equal to $\beta_0 + \beta_1 DLPDI_{1996Q1}$ and a variance of

$$var\left(\widehat{DLPCE}_{1996Q1}\right) = \sigma^2\left[\frac{1}{n} + \frac{\left(DLPDI_{1996Q1} - \overline{DLPDI}\right)^2}{\sum\left(DLPDI_i - \overline{DLPDI}\right)^2}\right]$$

where $\overline{DLPDI}$ is the sample mean of the *DLPDI* values in our sample period of 1967Q1–1995Q2, σ^2 is the variance of the error term ε and n is the sample size.

Since we do not observe the true variance of ε we estimate it from the sample as:

$$\widehat{\sigma^2} = \frac{\sum e_t^2}{(n-1)}$$

Using this information, and given the *DLPDI* value for 1996Q1, we can establish, say, a 95%confidence interval for true $E(DLPCE_{1996Q1})$ as follows:

Let DLPCE to be Y:

$$\Pr\left[\widehat{Y}_{1996Q1} - t_{\frac{\alpha}{2}} se\left(\widehat{Y}_{1996Q1}\right) \leq E\left(\widehat{Y}_{1996Q1}\right) \leq \widehat{Y}_{1996Q1} + t_{\frac{\alpha}{2}} se\left(\widehat{Y}_{1996Q1}\right)\right]$$

where $se(Y_{1996Q1})$ is the standard error obtained from above, and where $\alpha = 5\%$.

Notice that in establishing this confidence interval, we are using the *t* distribution rather than the normal distribution because we are estimating the true error variance.

The confidence intervals found in this way is known as a **confidence band**.

We can use a software package such as *Stata* or *EViews* to calculate the above confidence interval.

Using *EViews*, we obtain the confidence band for our example as follows:

From the regression output, go to Proc and choose Forecast (Fig. 6.20).

Before hitting OK, change the forecast sample from 1996Q3 to 1995Q2 as shown in Fig. 6.21.

Clicking OK generates the output of Fig. 6.22.

The solid line in this figure is the estimated regression line (curve) and the two broken lines show the 95% confidence band for it. The complementary table gives some measures of the quality of the forecast, namely, the **root mean square**, **mean absolute error**, **mean absolute percentage error** and the **Theil Inequality Coefficient**. The **Theil Inequality Coefficient** should lie between 0 and 1. A value closer to zero indicates the model is better. These forecasting performance measures are useful if we are comparing two or more methods of forecasting, as we will discuss shortly.

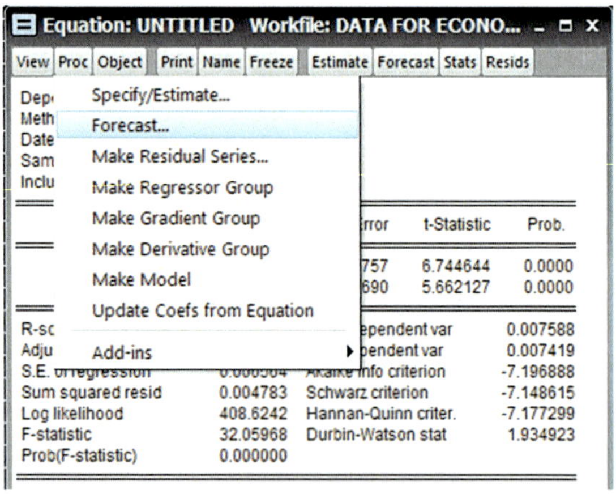

Fig. 6.20 The Forecast option

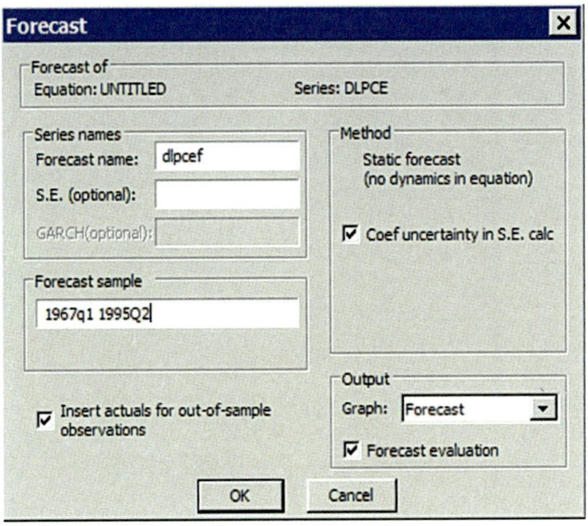

Fig. 6.21 Forecast characteristics

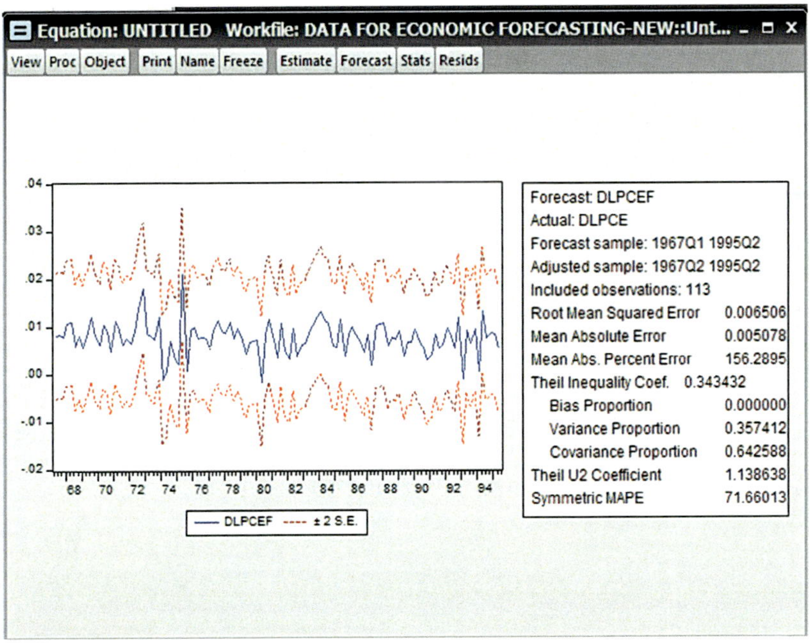

Fig. 6.22 Plot of real vs. forecasted data

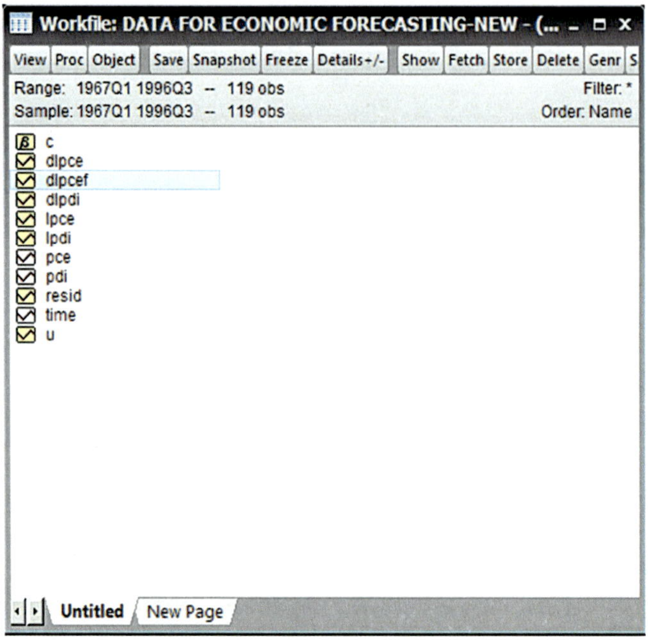

Fig. 6.23 Data sets

6.6 Step Five: Making a Joint Graph of the Dependent Variable and Its Forecast

EViews automatically generate DLPCEF in the workfile as shown in Fig. 6.23.

Now we can make a joint graph of dlpce and DLPCEF by following steps in Fig. 6.24:

EViews opens the data set of these two variables. Now go to **View** and choose **the Graph-Line & Symbol** then click **OK** (Fig. 6.25).

The right part of this joined graph shows in several occasions the forecasts have moved to different directions, opposite to the movement of DLNPCE.

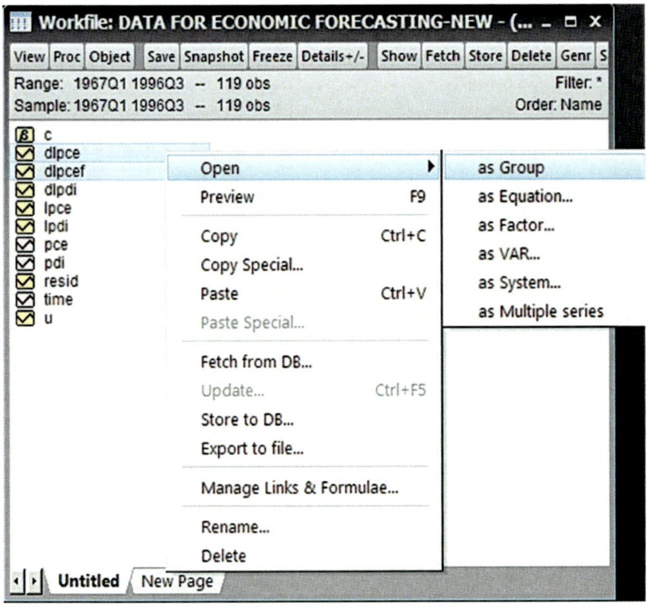

Fig. 6.24 Generate forecast plots

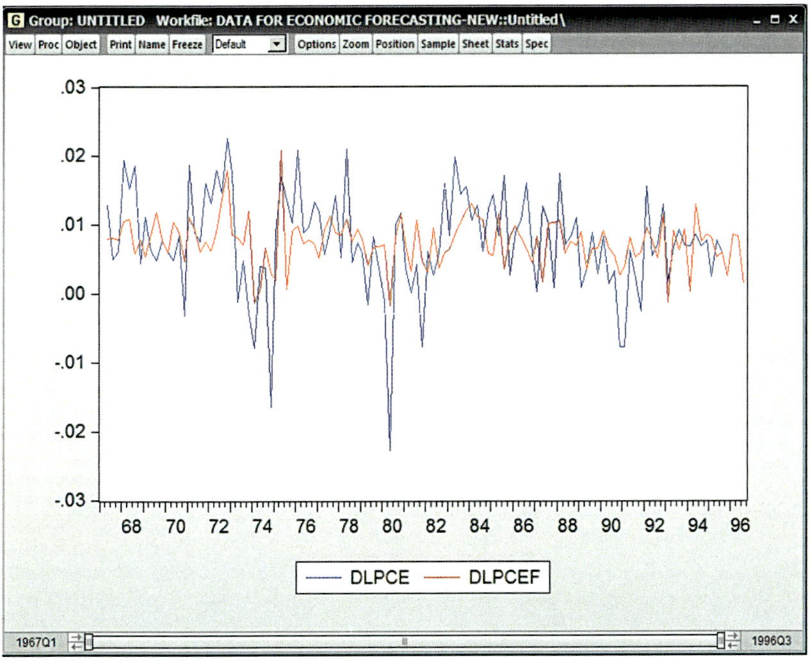

Fig. 6.25 Multiple charts on EViews

6.7 Step Six: Adding Autocorrelation of the Error Term

EViews can estimate the original model by allowing for autocorrelation in the error term. For example, if we assume that the error term follows the first-order autoregressive scheme [AR(1)] discussed in time series analysis as:

$\varepsilon_t = \rho\varepsilon_{t-1} + u_t; -1 \leq \rho \leq 1$, where ρ is the coefficient of (first-order) autocorrelation and u_t is the white noise error term.

In EViews we go to Quick-Estimate Equation-and we write these:

DLPCE C DLPDI AR(1)

Change the sample to 1967Q1–1995Q2.

If we hit OK, we will get this output (Fig. 6.26):

Compared with the last regression output (without AR(1)) we see that the marginal propensity to consume has changed slightly, and its standard error is now less. From this table we also see that the coefficient of the first-order autocorrelation is about 0.042845. Now we repeat the forecast process like earlier. From the regression output go to Proc and choose Forecast as shown in Fig. 6.27.

Change the Forecast sample to 1995Q2 in Fig. 6.28. Hit OK to generate the output of Fig. 6.29.

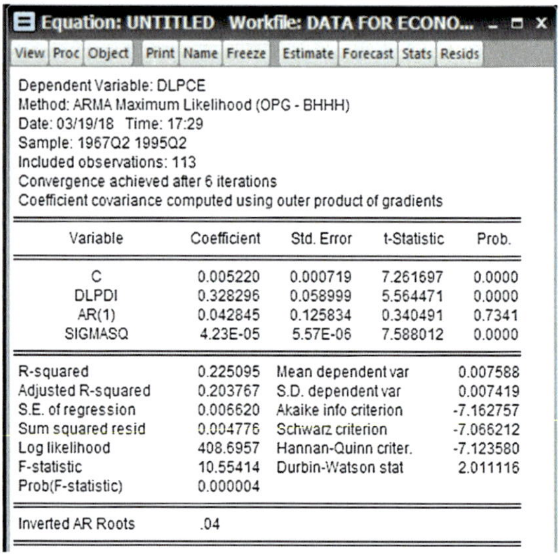

Equation: UNTITLED Workfile: DATA FOR ECONO...

View | Proc | Object | Print | Name | Freeze | Estimate | Forecast | Stats | Resids

Dependent Variable: DLPCE
Method: ARMA Maximum Likelihood (OPG - BHHH)
Date: 03/19/18 Time: 17:29
Sample: 1967Q2 1995Q2
Included observations: 113
Convergence achieved after 6 iterations
Coefficient covariance computed using outer product of gradients

Variable	Coefficient	Std. Error	t-Statistic	Prob.
C	0.005220	0.000719	7.261697	0.0000
DLPDI	0.328296	0.058999	5.564471	0.0000
AR(1)	0.042845	0.125834	0.340491	0.7341
SIGMASQ	4.23E-05	5.57E-06	7.588012	0.0000

R-squared	0.225095	Mean dependent var	0.007588
Adjusted R-squared	0.203767	S.D. dependent var	0.007419
S.E. of regression	0.006620	Akaike info criterion	-7.162757
Sum squared resid	0.004776	Schwarz criterion	-7.066212
Log likelihood	408.6957	Hannan-Quinn criter.	-7.123580
F-statistic	10.55414	Durbin-Watson stat	2.011116
Prob(F-statistic)	0.000004		

Inverted AR Roots .04

Fig. 6.26 ARIMA modelling

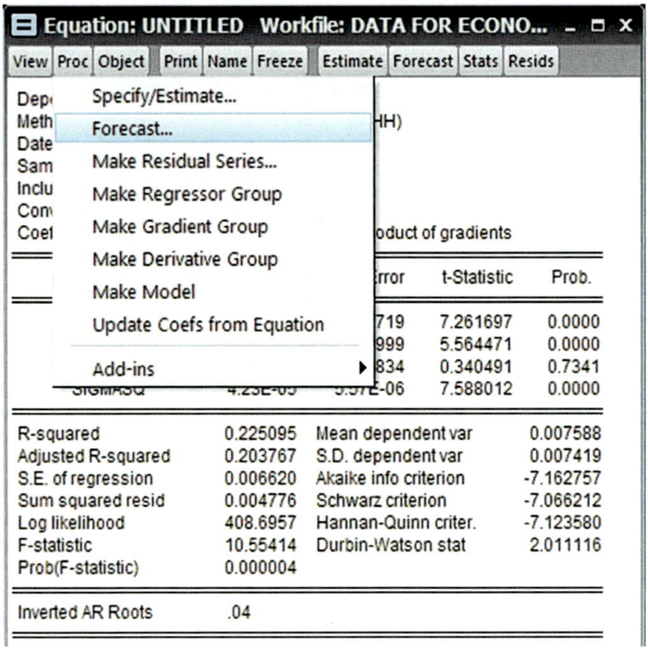

Fig. 6.27 Forecast option

Fig. 6.28 Dynamic forecast

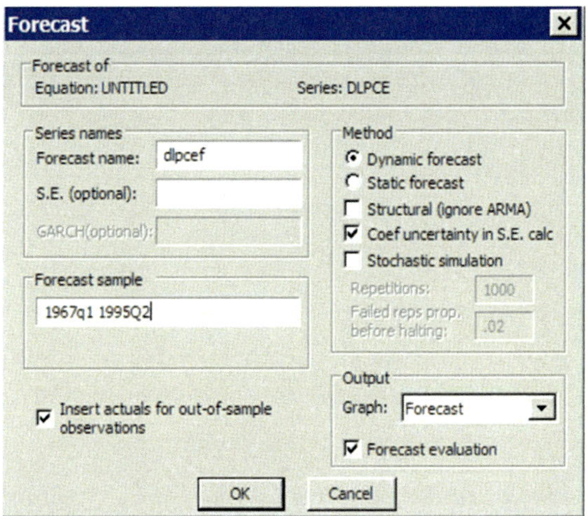

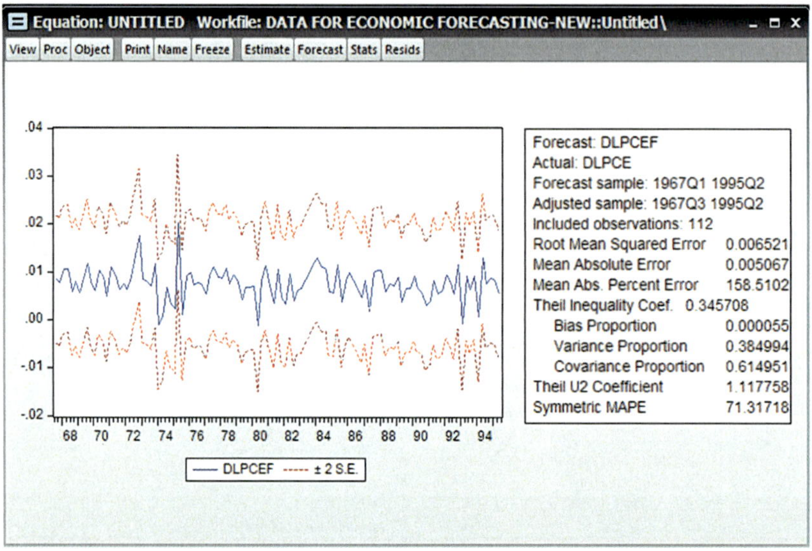

Fig. 6.29 Forecast vs. real data

We have obtained the 95% confidence band for the estimated regression line. If you compare this output with the similar output above you can make a decision which model is better. In order to do this we compare the Theil Inequality Coefficient of these two models. The one with less value indicates a better model:

1. The Theil Inequality Coefficient of the earlier model is: 0.343432.
2. The Theil Inequality Coefficient of the latter model is: 0.345708.

The earlier model is better than the latter model.

Chapter 7
Economic Forecasting using ARIMA Modelling

The Box-Jenkins approach to time series modelling consists of extracting predictable movements (or patterns) from the observed data through a series of iterations. The univariate Box-Jenkins method is purely a forecasting tool; no explanation is offered in that there are no regressor-type variables. The Box-Jenkins approach follows a three phase procedure:

- **Model identification:** a particular category of Box-Jenkins (B-J) model is identified by using various statistics computed from an analysis of the historical data.
- **Model estimation and verification:** once identified, the "best model" is estimated such that the fitted values come as close as possible to capturing the pattern exhibited by the actual data.
- **Forecasting:** the final model is used to forecast the time series and to develop confidence intervals that measure the uncertainty associated with the forecast.

7.1 The Box-Jenkins Methodology

The purpose of BJ methodology for forecasting is to analyse the probabilistic, or stochastic, properties of economic time series on their own. The process does not follow traditional regression modelling, in which the dependent variable, for example, Y_t is explained by other explanatory variables like: X_1, X_2, X_3, ..., X_k. The BJ methodology allows the dependent variable, Y_t, to be explained by the past, or lagged, values of Y_t itself and the current and lagged values of the residual, u_t. The u_t is assumed to be an uncorrelated random error term with zero mean and constant variance or in general a ***white noise error term***. The BJ methodology has several ways of forecasting a time series. We will look at different BJ approaches in general with examples. The BJ methodology is based on the *assumption that the time series under study is stationary*.

A. Aljandali, M. Tatahi, *Economic and Financial Modelling with EViews*, Statistics and Econometrics for Finance, https://doi.org/10.1007/978-3-319-92985-9_7

Model One: The Autoregressive (AR) Model

Consider the following model:

$$Y_t + \beta_0 + \beta_1 Y_{t-1} + \beta_2 Y_{t-2} + \cdots + \beta_p Y_{t-p} + U_t$$

We assume the u_t is a white noise error term. The model is called an autoregressive model of order *p*, **AR (*p*)**. The model involves regressing *Y* at time *t* on its values lagged *p* periods into the past. The value of *p* will be determined by using some criterion, such as the Akaike information criterion.

Model Two: The Moving Average (MA) Model

This model has an equation as follows:

$$Y_t + C_0 + C_1 u_t + C_2 u_{t-1} + \cdots + C_q u_{t-q}$$

The dependent variable, Y_t is a weighted, or moving, average of the current and past white noise error terms. This model is known as the **MA(*q*)** model, the value of *q* being determined empirically.

Model Three: The Autoregressive Moving Average (ARMA) Model

When we combine the AR and MA models, we can form a new model which we call the **ARMA (*p*, *q*)** model, with *p* autoregressive terms and *q* moving average terms. Again, the values of *p* and *q* are determined empirically.

Model Four: The Autoregressive Integrated Moving Average (ARIMA) Model

The BJ methodology is based on the assumption that the underlying time series is ***stationary*** or can be made stationary by differencing it one or more times. Using this approach is known as the **ARIMA (*p*, *d*, *q*) model**, where *d* denotes the number of times a time series has to be differenced to make it stationary. If a time series is already stationary, then an ARIMA (*p*, *d*, *q*) becomes an ARMA (*p*, *q*) model.

$$\begin{aligned} Y_t = \beta_0 + \beta_1 Y_{t-1} + \beta_2 Y_{t-2} + \cdots + \beta_p Y_{t-p} + C_1 u_{t-1} + C_2 u_{t-2} + \ldots \\ + C_q u_{q-1} + v_t \end{aligned}$$

7.2 The ARIMA Model

BJ models are known as Auto Regressive Integrated Moving Average (ARIMA). The methods used to solve the parameters of ARIMA models require quite a lot of computation, so for practical use, software is needed. The methods used in identifying, estimating and diagnosing ARIMA models are quite evolved.

The ARIMA procedure is carried out on stationary data. The notation Z_t is used for the stationary data at time t, whereas Y_t is the non-stationary datum value at that time. The ARIMA process considers linear models of the form:

$$Z_t = \mu + \varnothing_1 Z_{t-1} + \varnothing_2 Z_{t-2} + \ldots - \theta_1 e_{t-1} - \theta_2 e_{t-2} - \ldots + e_t$$

where Z_t, Z_{t-1} are stationary data points; e_t, e_{t-1} are present and past forecast errors and μ, $\varnothing_1$,$\varnothing_2$..., θ_1,θ_2...are parameters of the model to be estimated.

If a successful model involved only $\varnothing_1$i.e. was of the form:

$$Z_t = \mu + \varnothing_1 Z_{t-1} + e_t$$

The series is said to be governed by a first order autoregressive process, written AR(1). $\varnothing_1$is called *the autoregressive parameter* and the model above, describes the effect of a unit change in Z_{t-1} on Z_t. Similarly the model:

$$Z_t = \mu + \varnothing_1 Z_{t-1} + \varnothing_2 Z_{t-2} + \ldots + \varnothing_p Z_{t-p} + e_t$$

is called a p-order autoregressive process, written as AR(p). The sum of the coefficients $\varnothing_i$, i= 1, 2, ...,p of an autoregressive process must always be less than unity.

If a successful model only involved θ_1 i.e. was of the form:

$$Z_t = \mu - \theta_1 e_{t-1} + e_t$$

Then the time series is said to be governed by a first order moving average process, written as MA(1). $\theta_1$is called *the moving average parameter*. Similarly, the model:

$$Z_t = \mu - \theta_1 e_{t-1} - \theta_2 e_{t-2} - \cdots - \quad \theta_q e_{t-q} + e_t$$

is called a q-order moving average model written as MA(q).

Models involving both autoregressive and moving average processes are called *mixed models*. If a mixed model contained an autoregressive process of order 1 and a moving average process of order 2, the n the model is written as ARIMA(1,2) and would be of the form:

$$Z_t = \mu + \varnothing_1 Z_{t-1} - \theta_1 e_{t-1} - \theta_2 e_{t-2} + e_t$$

When differencing has been used to generate stationarity, the model is said to be *integrated* and is written as ARIMA (p,d,q). The middle parameter d is simply the number of times that the series had to be differenced before stationarity was achieved. If the (stationary) Z_t in the above equation had to be differenced twice before stationarity was achieved, then that model would be written as ARIMA (1,2,2).

7.3 Autocorrelations

To identify the model that best describes the time series under consideration, two sets of statistics are used: autocorrelations (AC) and partial autocorrelations (PAC). Both measure how much interdependence there is among the observations and take values that range between ± 1, depending on the pattern of the relationship. If, for example, values of the time series that are above the mean value of the series are immediately followed by values that are below the mean, then both the AC and PAC will be negative. This is said to be *negative autocorrelation.*

7.3.1 Autocorrelation Functions

AC's provide us with a numerical measure of the relationship of specific values of a time series to other values in the time series. That is, they measure the relationship of a variable to itself over time. AC's are normally computed for different time lags. For example, given n readings $Z_1, Z_2, \ldots, Z_n$, we can form n − 1 pairs of observations (Z_1, Z_2), ((Z_2, Z_3) … ((Z_{n-1}, Z_n). Regarding the first observation in each pair as one variable and the second observation as the second variable, we can compute the Pearsonian correlation coefficient at, in the example of the data in Fig. 7.1, a lag of 1 week. This measures correlation between successive readings and is called the *first order autocorrelation coefficient.* Similarly, we could compute the correlation between observations at a distance k apart, which is called the *kth. order autocorrelation coefficient.*

For example, consider the data:

$$51, 52, 54, 60, 55, 61, 62, 66, 60, 62, 66\ldots$$

The first order autocorrelation coefficient is calculated using the standard formula for the Pearsonian coefficient, involving the pairs:

$$(51, 52)\ (52, 54)\ (54, 60)\ (60, 55)\ldots$$

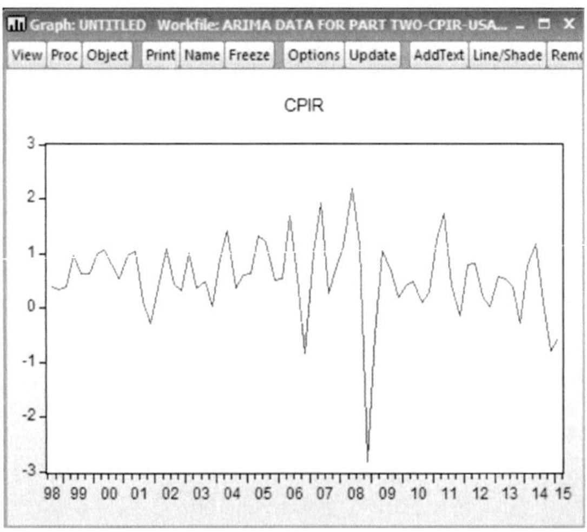

Fig. 7.1 Plot of the CPIR series

The second order autocorrelation coefficient would be computed using the pairs:

$$(51, 54)\ (52, 60)\ (54, 55)\ (60, 61)\ldots$$

We use the notation that r_k is the (auto) correlation between Z_t and Z_{t-k} so r_4 is the (auto) correlation between Z_t and Z_{t-4}. When the AC's are computed for lag 1, lag 2, lag 3 and so and are graphed (r_k against k), the result is called the *sample autocorrelation function* (*ACF*) or *correlogram*. This graph is useful for determining whether a series is stationary and for identifying a tentative ARIMA model.

If a series is non-stationary by virtue of having an upwards trend, then readings a few lags apart will be autocorrelated. If the data are stationary, however, the autocorrelations should all be zero (indicative of random error). This should be a characteristic of the ACF for stationary data. To test whether or not the autocorrelation coefficient is statistically equal to zero, we use, for large samples the t statistic—and meaningful economic time series should involve large samples. When the number of readings is reasonably large and to test the hypothesis that the population autocorrelation coefficient (ρ_k) at lag k is zero, i.e.:

$$H_0 : p_k = 0 \text{ against } H_1 : p_k \neq 0,$$

We adopt *Bartlett's method*. Bartlett derived that if the above null hypothesis is true, then the sample autocorrelations at lag k, r_k, will be closely normally distributed with zero mean and a variance of:

$$\text{Var}(r_k) = n^{-1}\left\{1 + 2\left(r_1^2 + r_2^2 + r_3^2 + \cdots + r_{k-1}^2\right)\right\}.$$

The test statistic is:

$$\frac{r_k}{SD\ of\ r_k}$$

which is distributed as the test statistic with n – 2 degrees of freedom. Given that the t distribution is asymptotically normal, the boundaries of the critical region for the above test are usually taken at ±1.96 (±2).

For example, suppose for a given set of eight readings, that the autocorrelations at lags 1 and 2 were respectively $r_1 = -0.412$ and $r_2 = -0.343$ and that we wished to test if the second order autocorrelation coefficient was significantly different from zero, i.e.:

$$H_0 : p_2 = 0 \text{ against } H_1 : p_2 \neq 0$$

We may compute that:

$$\text{Var } (r_2) = 8^{-1}\left\{1 + 2(-0.412)^2\right\} = 0.1674 \text{ therefore SD of } r_2 = 0.4092$$

The test statistic under H_0 becomes:

$$(-0.343 - 0)/0.4092 = -0.838$$

which is well distant from the critical region boundary for t ($v = 6$). We, therefore, fail to reject H_0 and conclude that ρ_k is zero. It may also be shown that the first order autocorrelation coefficient is also not significant.

The curved lines in an ACF plot represent 95% confidence limits for the autocorrelation coefficients, r_k, based on Bartlett's variance formula given by the Var(r_k) equation. These serve as indications as to whether autocorrelations are statistically significant.

It should be noted that if the researcher is sure that the time series data are stationary, then the (r_k) in Bartlett's variance formula are in (theory) zero. This leads to Quenouille's formula for the variance of the r_k in the instance of stationary data, that:

$$\mathrm{Var}(r_k) = n^{-1}$$

Most computer packages including EViews have both Bartlett's and Quenouille's variance formulae as available options. It can be shown that the autocorrelation for an AR(1) model will in theory be:

$$r_k = \text{Ø}_1^k$$

Whenever the series is stationary, it may be shown that the sum of the AR coefficients:

$$\text{Ø}_1 + \text{Ø}_2 + \text{Ø}_3 \ldots$$

will be less than one. In the case of an AR(1) model, this implies that $\varnothing_1$will be less than one, so the AC's will be decreasing in absolute value as the lag increases, i.e. $\text{Ø}_1 > \text{Ø}_2^2 > \text{Ø}_3^2 > \ldots > \text{Ø}_k^2$, which simply says that the relationship weakens as we go back over time. Further the autocorrelations decline fairly rapidly.

It can be shown that the autocorrelation coefficients for a moving average process of order 1, MA (1), in theory are:

$$r_k = \frac{-\theta_1}{1 + \theta_1^2} \; for\ k = 1$$

$$r_k = 0 \; for\ k = 2$$

7.3.2 Partial Autocorrelation Functions (PACF)

Partial autocorrelation coefficients PAC's are closely related to the AC's. They also take on values between −1 and 1. A diagram of PAC's against the lag k is called the *partial autocorrelation function (PACF).* A partial autocorrelation is the measure of the relationship between two variables when the effect of other variables has been removed or held constant. With temporal data, r_{kk} is the partial autocorrelation between Z_t and Z_{t-k} when the effect of the intervening variables $Z_{t-1}, Z_{t-2}, \ldots, Z_{t-k+1}$ has been removed. This adjustment is to see if the correlation between Z_t and Z_{t-k} is due to the intervening variables or if indeed there is something else causing the relationship. As is discussed in the next section, the behavior of the PAC's along with the AC's for a stationary time series is used to identify a tentative ARIMA model.

The formula for the partial autocorrelation coefficient is quite complex, but numerical values are computed by available statistical packages. It was shown by Quenouille that:

$$\mathrm{Var}(r_{kk}) = n^{-1},$$

So it is possible to examine the hypothesis, such as:

$$H_0 : \rho_{kk} = 0 \text{ versus } H_1 : \rho_{kk} \neq 0$$

For example, suppose that $r_{33} = -0.0318$ based on eight readings. This is the correlation between Z_t and Z_{t-3} when the effects of Z_{t-1} and Z_{t-2} have been removed. The test statistic is again distributed as the t statistic:

$$\frac{r_{kk}}{SD\ of\ r_{kk}}$$

So −0.318/0.354 = −0.898 which is not statistically significant. We, therefore, fail to reject the hypothesis that ρ_{kk} is zero. Again, for large n the boundaries of the critical region are usually taken at ±1.96 (±2). If the data are stationary, then the partial autocorrelations should, in theory, be zero. The two horizontal lines again represent the 95% confidence interval. If the PAC's fail to die out, it indicates that the data are not stationary.

7.3.3 *Patterns of the ACF and PACF*

It is possible to use the ACF and PACF to recognise patterns that characterise moving average (MA), autocorrelation (AR) and mixed (ARMA) models, when the assumption of stationarity has been satisfied. It should be appreciated that we are focusing on theoretical models, but this does facilitate recognition of similar patterns in actual time series data. By comparing actual ACF's and PACF's to the theoretical patterns, we shall be able to identify the specific type of BJ model that will adequately represent the data.

There are general guidelines:

- If the autocorrelations decay and the partial autocorrelations have spikes, the process can be captured by an AR model, where the order equals the number of significant spikes. The ACF should show exponentially declining values.
- If the partial autocorrelations decay and the autocorrelations have spikes, the process is best captured by an MA model, where the order equals the number of significant spikes. The PACF should show exponentially declining values.

- If both the autocorrelation and partial autocorrelations are characterized by irregular patterns on the ACF and PACF, the process is best captured by an ARMA model, where the order equals the number of significant spikes. It may be necessary to invoke several cycles of the identification-estimation-diagnosis process.

AC patterns for moving average models are among the easiest to recognise in practice.

The Box-Jenkins method is just one procedure available for forecasting. There can be no doubt that ARIMA models can be constructed to fit a wide variety of patterns and this can be done with minimum effort as long as a computer is available. Like all other time series models, ARIMA models suffer limitations. They generally fail to capture turning points on time and they provide the decision maker with little explanation. For example, they do not provide information on the potential impact of policies such as pricing actions or advertising programmes. However, multivariate Box-Jenkins models partially overcome these problems. As in the present example, competing ARIMA models may have little to choose between them. Identifying the appropriate order of a mixed model, for example can be difficult.

In order to clarify the choice between different univariate time series models, there have been several 'competitions' to compare the forecasting accuracy of different methods. However, the results of these competitions have not always been consistent. Given the different analysts and data sets used this is perhaps not surprising. The Box-Jenkins approach has not been consistently the best. Regression methods do rather better on average than univariate models, but again, this is not consistently the case.

A final point is that, although there is an advantage in being able to choose from the broad class of ARIMA models, there are also dangers in that considerable experience is need to interpret the ACF and PACF and other indicators. Moreover, when variation in a series is dominated by trend and seasonality, the effectiveness of the fitted ARIMA model is mainly determined by the differencing procedure employed rather than by the identification of the autocorrelation and/or moving average structures of the differenced series. In some situations, a large expenditure of time and effort can be justified and then the Box-Jenkins approach is worth considering. However, for routine sales forecasting, simple methods are more likely to be understood by managers and workers who have to utilize or implement the results.

Whilst noting that the Box-Jenkins approach has been one of the most influential developments in time series analysis, Box-Jenkins models are only worth considering when the following conditions are satisfied:

- The analyst is competent to implement it
- The objectives justify the complexity and
- The variation in the series is not dominated by trend and seasonality

As discussed earlier in this chapter, the BJ methodology follows a four-step procedure:

Step 1: Identification:
We need to find the appropriate values of p, d and q. The main tools in this search are the **correlogram and partial correlogram**.

Step 2: Estimation:
When the right model is identified the next step is to estimate the parameters of the chosen model. In some cases the use of ordinary least-squares (OLS) method is appropriate, but in some cases we have to resort to nonlinear (in parameter) estimation methods.

Step 3: Diagnostic checking:
One simple test of this is to see if the residuals from the fitted model are white noise or not (stationary).

Step 4: Forecasting:
The ultimate test of a successful ARIMA model is forecasting performance, within the sample period as well as outside the sample period.

7.3.3.1 An ARMA Model of Consumer Price Index Rate (CPIR) of the USA (1998Q3-2015Q1)

Step One-The Stationary Process of the CPIR Series:
Quarterly data of consumer price index rate of the USA from 1995q3 to 2015q1 is collected. We check first whether the series is stationary or not. Open the data of CPIR from the workfile. Go to **View** and choose **graph** to plot the CPIR data shown in Fig. 7.1 which appears to be stationary.

Correlogram of CPIR:

(a) H_0: CPIR has no unit root
(b) H_1: CPIR has a unit root
(c) $\alpha = 5\%$ or 0.05

There are no test statistics to calculate. We look at the Prob column in Fig. 7.2.

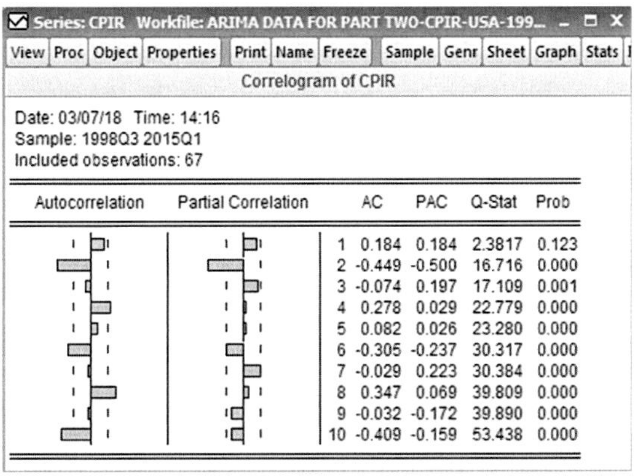
Series: CPIR Workfile: ARIMA DATA FOR PART TWO-CPIR-USA-199...

Correlogram of CPIR

Date: 03/07/18 Time: 14:16
Sample: 1998Q3 2015Q1
Included observations: 67

	AC	PAC	Q-Stat	Prob
1	0.184	0.184	2.3817	0.123
2	-0.449	-0.500	16.716	0.000
3	-0.074	0.197	17.109	0.001
4	0.278	0.029	22.779	0.000
5	0.082	0.026	23.280	0.000
6	-0.305	-0.237	30.317	0.000
7	-0.029	0.223	30.384	0.000
8	0.347	0.069	39.809	0.000
9	-0.032	-0.172	39.890	0.000
10	-0.409	-0.159	53.438	0.000

Fig. 7.2 Correlogram of CPIR

(d) The p-values are mostly 0.000
(e) Prob = 0.000 < 0.05 therefore we reject the H_0
(f) CPIR is not stationary

Formal Approach-The ADF Test:

1. H_0: CPIR has a unit root and is not stationary
2. H_1: CPIR has no unit root and is stationary.
3. $\alpha = 5\%$ or 0.05

According to Fig. 7.3, the ADF test statistic is −8696612

4. The p-values = 0.0000
5. Prob = 0.0000 < 0.05 therefore we reject the H_0.
6. CPIR does not have a unit root and it is stationary.

The BJ methodology is based on stationary time series, we will work with CPIR stationary series. Now we determine which model of above models fits our data of CPIR.

Fig. 7.3 CPIR unit root test

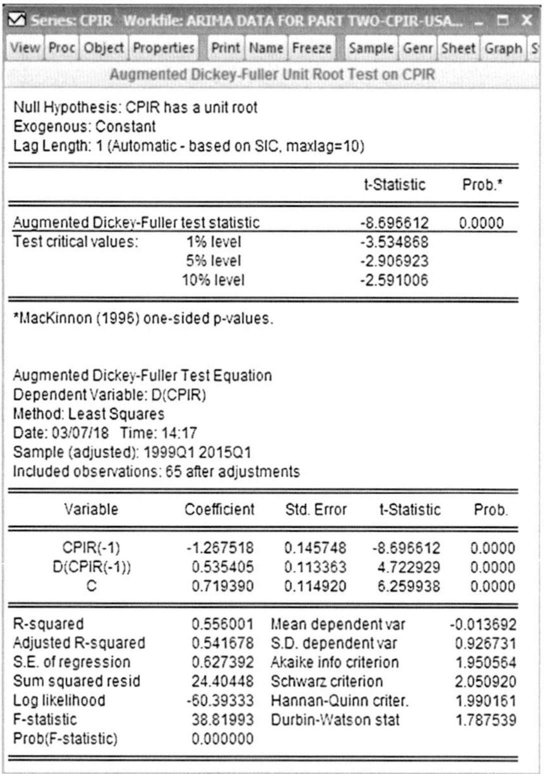

Series: CPIR Workfile: ARIMA DATA FOR PART TWO-CPIR-USA...

View | Proc | Object | Properties | Print | Name | Freeze | Sample | Genr | Sheet | Graph | S

Augmented Dickey-Fuller Unit Root Test on CPIR

Null Hypothesis: CPIR has a unit root
Exogenous: Constant
Lag Length: 1 (Automatic - based on SIC, maxlag=10)

		t-Statistic	Prob.*
Augmented Dickey-Fuller test statistic		-8.696612	0.0000
Test critical values:	1% level	-3.534868	
	5% level	-2.906923	
	10% level	-2.591006	

*MacKinnon (1996) one-sided p-values.

Augmented Dickey-Fuller Test Equation
Dependent Variable: D(CPIR)
Method: Least Squares
Date: 03/07/18 Time: 14:17
Sample (adjusted): 1999Q1 2015Q1
Included observations: 65 after adjustments

Variable	Coefficient	Std. Error	t-Statistic	Prob.
CPIR(-1)	-1.267518	0.145748	-8.696612	0.0000
D(CPIR(-1))	0.535405	0.113363	4.722929	0.0000
C	0.719390	0.114920	6.259938	0.0000

R-squared	0.556001	Mean dependent var	-0.013692
Adjusted R-squared	0.541678	S.D. dependent var	0.926731
S.E. of regression	0.627392	Akaike info criterion	1.950564
Sum squared resid	24.40448	Schwarz criterion	2.050920
Log likelihood	-60.39333	Hannan-Quinn criter.	1.990161
F-statistic	38.81993	Durbin-Watson stat	1.787539
Prob(F-statistic)	0.000000		

Step Two-Identification:

The Correlogram of CPIR up to 15 Lags

Open the CPIR data on EViews and go to **View**. From the list choose: **Correlogram**. Change lags number to 15 and hit **OK** as shown in Fig. 7.4

According to Fig. 7.5, the ACF (autocorrelation function) shows correlation of current CPIR with its values at various lags. The PACF (partial autocorrelation function) shows the correlation between observations that are k periods apart after controlling for the effects of intermediate lags (i.e. lags less than k). This methodology uses these two correlation coefficients in order to identify the type of ARMA model that may be appropriate four our forecasting purpose.

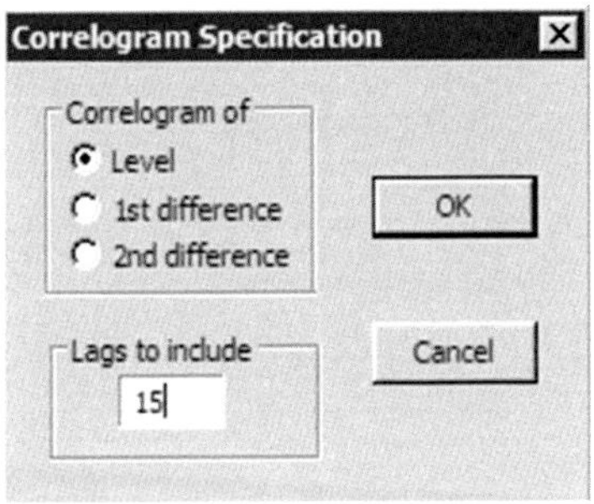

Fig. 7.4 Correlogram specification

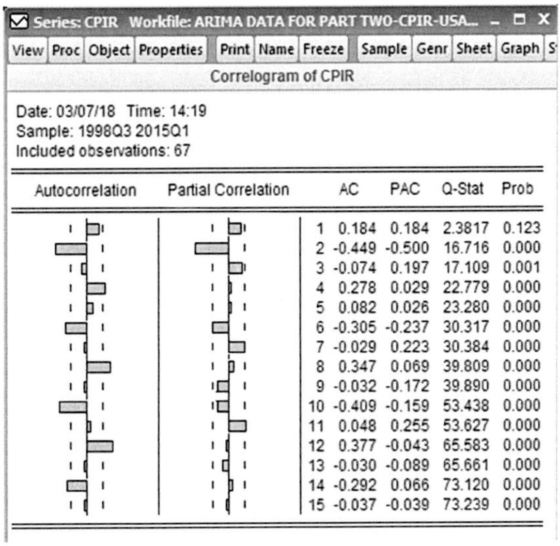

Series: CPIR Workfile: ARIMA DATA FOR PART TWO-CPIR-USA...

View Proc Object Properties Print Name Freeze Sample Genr Sheet Graph

Correlogram of CPIR

Date: 03/07/18 Time: 14:19
Sample: 1998Q3 2015Q1
Included observations: 67

	AC	PAC	Q-Stat	Prob
1	0.184	0.184	2.3817	0.123
2	-0.449	-0.500	16.716	0.000
3	-0.074	0.197	17.109	0.001
4	0.278	0.029	22.779	0.000
5	0.082	0.026	23.280	0.000
6	-0.305	-0.237	30.317	0.000
7	-0.029	0.223	30.384	0.000
8	0.347	0.069	39.809	0.000
9	-0.032	-0.172	39.890	0.000
10	-0.409	-0.159	53.438	0.000
11	0.048	0.255	53.627	0.000
12	0.377	-0.043	65.583	0.000
13	-0.030	-0.089	65.661	0.000
14	-0.292	0.066	73.120	0.000
15	-0.037	-0.039	73.239	0.000

Fig. 7.5 Correlogram of CPIR (15 LAGS)

We can obtain the **autocorrelation function (ACF)** at lag k as follow:

$\rho_k = \frac{\gamma_k}{\gamma_0} = \frac{Covariance\ \ at\ \ lag\ \ k}{Variance}$

In practice, we compute the ACF from a given sample, denoted as $\widehat{\rho_k}$, which is based on the sample covariance at lag k and the sample variance. The length of the lag, k. is a matter of consideration. We can use the Akaike or Schwarz information criterion to determine the lag length. But, a rule of thumb is to compute ACF up to one-quarter to one-third the length of the time series. However, this may not be practically possible if the series sample size is large. For example, if we have 2355 observations. One quarter of this would be about 589 lags. We will not show the ACF at all these lags, but we will consider only the first 50 lags of the ACF.

In our case we have 67 observations and 15 lags may be enough. A plot of $\widehat{\rho_k}$ against k, the lag length, is called the (sample) **correlogram**. We can test the statistical significance of each autocorrelation coefficient by computing its standard error. The statistician Bartlett has shown that if a time series is purely random, the sample autocorrelation, $\widehat{\rho_k}$, is approximately (i.e. in large samples) distributed as follows:

$$\widehat{\rho} \sim N\left(0, \frac{1}{n}\right)$$

Our sample size is 67. Therefore, the variance is 1/67 or about 0.01492 and the standard error is:

$$\text{S.E.} = \sqrt{\frac{1}{n}} = \sqrt{\frac{1}{67}} = \sqrt{0.01492} = 0.1222$$

Then we use the confidence interval formula for a standard normal distribution, $\rho \sim (0, 1)$, in which the population mean is zero and the variance is 1.

The confidence interval formula is:

$$\mu \pm t_{\frac{\alpha}{2}} \text{SE}$$

We use this formula based on 5% level of significance. This leads to:

$$0 \pm 1.96\left(\sqrt{1/\text{n}}\right)$$

Therefore the 95% confidence interval for the true correlation coefficients is about:

$0 \pm 1.96\ (0.1222) = (-0.2395 \text{ to } 0.2395)$

Correlation coefficients lying outside these bounds are statistically significant at the 5% level.

On this basis, both AC and PAC correlations at lags 2 are statistically significant.

We fit first an AR model at lag 2.

Step Three-Estimation:

The AR Model

Based on what we have found we now consider the following AR model:

$$Y_t = \beta_0 + \beta_1 Y_{t-2} + u_t$$

Go to **Quick-Estimate Equation** and write: CPIR C AR(2) and hit OK as shown in Fig. 7.6.

Remember: we use here the whole sample (Figs. 7.7 and 7.8).

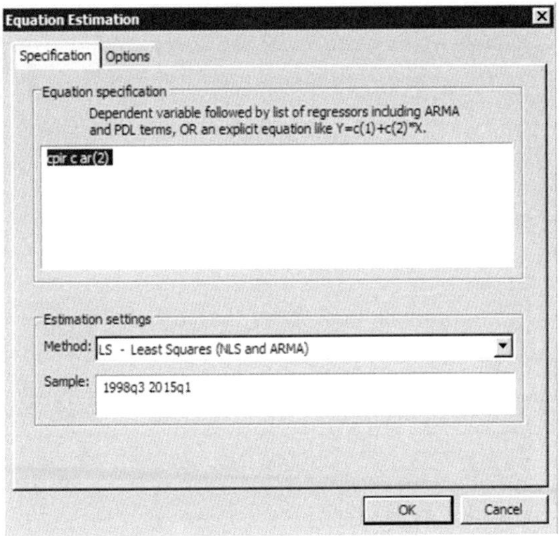

Fig. 7.6 CPIR ARIMA AR(2) Equation estimation

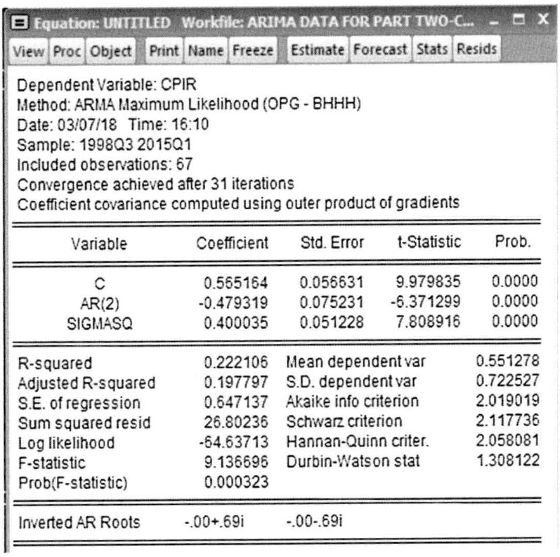

Equation: UNTITLED Workfile: ARIMA DATA FOR PART TWO-C...

View | Proc | Object | Print | Name | Freeze | Estimate | Forecast | Stats | Resids

Dependent Variable: CPIR
Method: ARMA Maximum Likelihood (OPG - BHHH)
Date: 03/07/18 Time: 16:10
Sample: 1998Q3 2015Q1
Included observations: 67
Convergence achieved after 31 iterations
Coefficient covariance computed using outer product of gradients

Variable	Coefficient	Std. Error	t-Statistic	Prob.
C	0.565164	0.056631	9.979835	0.0000
AR(2)	-0.479319	0.075231	-6.371299	0.0000
SIGMASQ	0.400035	0.051228	7.808916	0.0000

R-squared	0.222106	Mean dependent var	0.551278
Adjusted R-squared	0.197797	S.D. dependent var	0.722527
S.E. of regression	0.647137	Akaike info criterion	2.019019
Sum squared resid	26.80236	Schwarz criterion	2.117736
Log likelihood	-64.63713	Hannan-Quinn criter.	2.058081
F-statistic	9.136696	Durbin-Watson stat	1.308122
Prob(F-statistic)	0.000323		

Inverted AR Roots	-.00+.69i	-.00-.69i

Fig. 7.7 CPIR ARIMA AR(2) modelling output

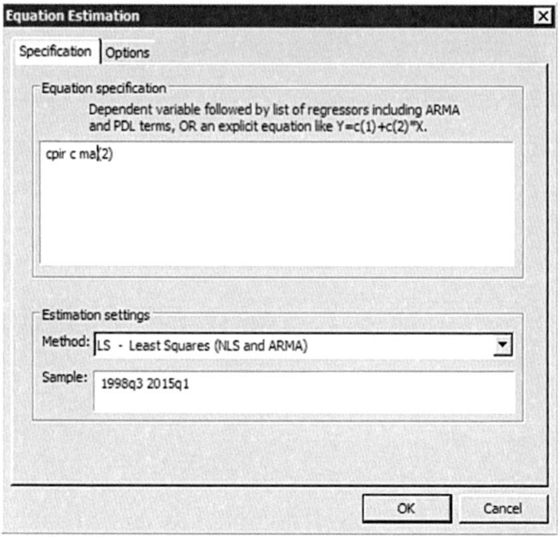

Fig. 7.8 CPIR ARIMA MA(2) Equation estimation

Through this output it can be seen that the coefficient of AR(2) is statistically significant (p-value: 0.0000). So, the model in this Table is a candidate for further consideration. We take notes of two criteria of this output **AR(2):**

Akaike info criterion	2.019019
Schwarz criterion	2.117736

Notes:

1. This process may involve several steps in order to find AR significant coefficients for our final regression. All significant correlations found through the correlogram of the series should be used for running the first AR regression. Then we exclude those ARs which are not statistically significant and re-estimate the regression by using only significant ARs. This process will continue up to reaching a regression result comprises of only significant ARs. This regression will be our final AR regression and we use this for our forecasting proposes if the model is found better than the models below.
2. Assume we have run two models in order to find significant ARs and we have to choose between the two preceding models. In order to decide which model is more appropriate than the other we use the Akaike or Schwarz information criterion to make the choice. **Remember that on the basis of the information criteria, we choose the model with the lowest value of these criteria.**

Fig. 7.9 CPIR ARIMA MA(2) modelling output

Equation: UNTITLED Workfile: ARIMA DATA FOR PART TWO-C...

View | Proc | Object | Print | Name | Freeze | Estimate | Forecast | Stats | Resids

Dependent Variable: CPIR
Method: ARMA Maximum Likelihood (OPG - BHHH)
Date: 03/07/18 Time: 16:16
Sample: 1998Q3 2015Q1
Included observations: 67
Convergence achieved after 39 iterations
Coefficient covariance computed using outer product of gradients

Variable	Coefficient	Std. Error	t-Statistic	Prob.
C	0.569267	0.052815	10.77847	0.0000
MA(2)	-0.433509	0.084323	-5.141059	0.0000
SIGMASQ	0.419422	0.055592	7.544659	0.0000

R-squared	0.184406	Mean dependent var	0.551278
Adjusted R-squared	0.158919	S.D. dependent var	0.722527
S.E. of regression	0.662633	Akaike info criterion	2.064766
Sum squared resid	28.10129	Schwarz criterion	2.163484
Log likelihood	-66.16967	Hannan-Quinn criter.	2.103829
F-statistic	7.235227	Durbin-Watson stat	1.320061
Prob(F-statistic)	0.001469		

Inverted MA Roots	.66	-.66

It needs to be stressed once again that the decision which AR model is appropriate is a process. We use all ARs found statistically significant from the Correlogram Table. We try to find an AR model in which most AR coefficients are statistically significant. The model we found in this way will be a candidate model for further consideration.

The MA Model

We use one lagged MA terms at lag 2 as we did for the AR model above.
We now consider the following MA model:

$$Y_t = C_0 + C_1 u_{t-2}$$

Go to Quick-Estimate Equations-write: CPIR C MA(2) MA(4) and hit OK.

Remember we use the whole sample.

According to Fig. 7.9, the coefficient of MA(2) is statistically significant at 5% level (p-value is 0.0000). This make this model a candidate for further consideration.

Notes:

The notes highlighted above for the AR model need to be considered one by one for the MA model, too.

We take notes of two criteria of this output **MA(2):**

Akaike info criterion	2.064766
Schwarz criterion	2.163484

Which model should we choose?

AR (2) OR MA (2)

From the AR model output we found these two criteria:

Akaike info criterion	2.019019
Schwarz criterion	2.117736

From the MA model output we found these two criteria:

Akaike info criterion	2.064766
Schwarz criterion	2.163484

Since the values of the Akaike and Schwarz information criteria were lowest for the AR model, we choose this over the MA model, although the difference between the two is not great.

The ARMA Model

Now we develop a model using both AR and MA terms. We run first a regression of all significant correlation coefficients for both AR and MA. Our aim is to find an ARMA model with the highest possible significant coefficients. For this purpose we run some experimentation to obtain the best model.

Experiment One:

Go to Quick-Estimate Equation-Write: CPIR C AR(2) MA(2)-hit Ok.

Remember we use the whole sample (Figs. 7.10 and 7.11).

This table shows the coefficients of both AR(2), MA(2) are statistically significant and the model is candidate for further consideration.

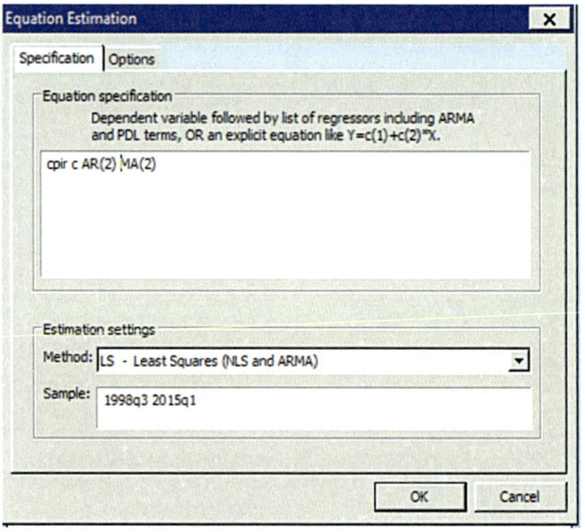

Fig. 7.10 CPIR ARIMA AR(2) MA(2) Equation Estimation

Fig. 7.11 CPIR ARIMA AR(2) MA(2) modelling output

Equation: UNTITLED Workfile: ARIMA DATA FOR PART TWO-C...

View | Proc | Object | Print | Name | Freeze | Estimate | Forecast | Stats | Resids

Dependent Variable: CPIR
Method: ARMA Maximum Likelihood (OPG - BHHH)
Date: 03/07/18 Time: 16:17
Sample: 1998Q3 2015Q1
Included observations: 67
Convergence achieved after 64 iterations
Coefficient covariance computed using outer product of gradients

Variable	Coefficient	Std. Error	t-Statistic	Prob.
C	0.559509	0.087988	6.358954	0.0000
AR(2)	-0.991233	0.036835	-26.91013	0.0000
MA(2)	0.904434	0.128991	7.011597	0.0000
SIGMASQ	0.347670	0.055276	6.289668	0.0000

R-squared	0.323933	Mean dependent var	0.551278
Adjusted R-squared	0.291739	S.D. dependent var	0.722527
S.E. of regression	0.608066	Akaike info criterion	1.937031
Sum squared resid	23.29391	Schwarz criterion	2.068654
Log likelihood	-60.89053	Hannan-Quinn criter.	1.989114
F-statistic	10.06199	Durbin-Watson stat	1.408796
Prob(F-statistic)	0.000017		

Inverted AR Roots	-.00+1.00i	-.00-1.00i
Inverted MA Roots	-.00+.95i	-.00-.95i

We take notes of two criteria of this output ARMA **(2) model:**

Akaike info criterion	1.937031
Schwarz criterion	2.068654

Which model should we choose?
AR (2) OR MA (2) OR ARMA [AR(2), MA(2)]
From the AR model output we found these two criteria:

Akaike info criterion	2.019019
Schwarz criterion	2.117736

From the MA model output we found these two criteria:

Akaike info criterion	2.064766
Schwarz criterion	2.163484

From the ARIMA output we received these two criteria:

Akaike info criterion	1.937031
Schwarz criterion	2.068654

Since the values of the Akaike and Schwarz information criteria were lowest for the ARMA model, we can choose this over the AR and MA models, although the difference between them is not great. The residuals from this model need to be

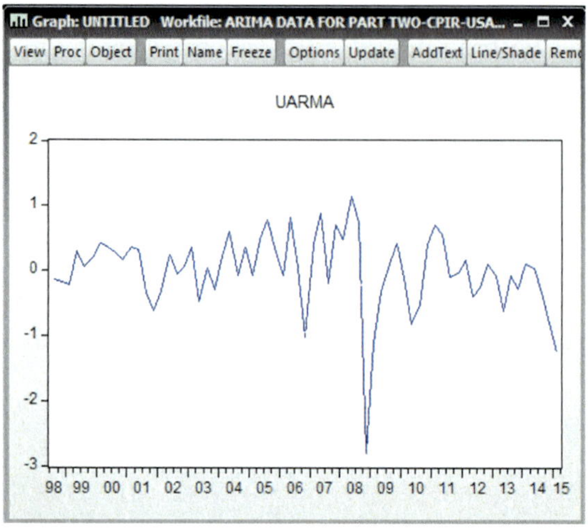

Fig. 7.12 Plot of the ARMA residuals

tested for unit root and if it is found that there is no unit root which means that the residuals from this model are randomly distributed and stationary.

To sum up:

It would seem that ARMA (2) is probably an appropriate model to depict the behaviour of the first differences of the logs of daily closing IBM prices over the sample period.

Step Four: Diagnostic checking:

We run the ARMA model again and we save the residuals of this regression. After having the ARMA model in the screen of the EViews, go to Quick-Generate series-write; UARMA=RESID and hit OK. Open the data of UARMA from the workfile. Go to View and choose graph then hit OK.

According to Fig. 7.12, the residuals of this model are stationary.

Formal Approach-The ADF Test:

1. H_0: series has a unit root and it is not stationary
2. H_1: series has no unit root and it is stationary.
3. $\alpha = 5\%$ or 0.05

According to Fig. 7.13, the ADF test statistic is −5.776453.

1. The p-values = 0.0000
2. Prob = 0.0000 < 0.05 therefore we reject the H_0
3. The series has no unit root and it is stationary.

The residuals from this regression are randomly distributed and stationary.

Step Five-Forecasting with ARMA Model

Once a particular model is fitted, we can use it for forecasting. There are two types of forecast: **static** and **dynamic**.

Series: UARMA Workfile: ARIMA DATA FOR PART TWO-CPIR-U...

View | Proc | Object | Properties | Print | Name | Freeze | Sample | Genr | Sheet | Graph | S

Augmented Dickey-Fuller Unit Root Test on UARMA

Null Hypothesis: UARMA has a unit root
Exogenous: Constant
Lag Length: 0 (Automatic - based on SIC, maxlag=10)

		t-Statistic	Prob.*
Augmented Dickey-Fuller test statistic		-5.776453	0.0000
Test critical values:	1% level	-3.533204	
	5% level	-2.906210	
	10% level	-2.590628	

*MacKinnon (1996) one-sided p-values.

Augmented Dickey-Fuller Test Equation
Dependent Variable: D(UARMA)
Method: Least Squares
Date: 03/07/18 Time: 16:27
Sample (adjusted): 1998Q4 2015Q1
Included observations: 66 after adjustments

Variable	Coefficient	Std. Error	t-Statistic	Prob.
UARMA(-1)	-0.719096	0.124488	-5.776453	0.0000
C	-0.005959	0.071465	-0.083377	0.9338

R-squared	0.342696	Mean dependent var	-0.016664
Adjusted R-squared	0.332426	S.D. dependent var	0.710342
S.E. of regression	0.580386	Akaike info criterion	1.779589
Sum squared resid	21.55829	Schwarz criterion	1.845942
Log likelihood	-56.72643	Hannan-Quinn criter.	1.805808
F-statistic	33.36741	Durbin-Watson stat	1.829169
Prob(F-statistic)	0.000000		

Fig. 7.13 UARMA unit root test

Static Forecasting:

In static forecasts, we use the actual current and lagged values of the forecast variable, blue zeros in the below diagram.

Dynamic Forecasting:

In dynamic forecasts, after the first period forecast, we use the previously forecast values of the forecast variable, red zeros in the diagram below.

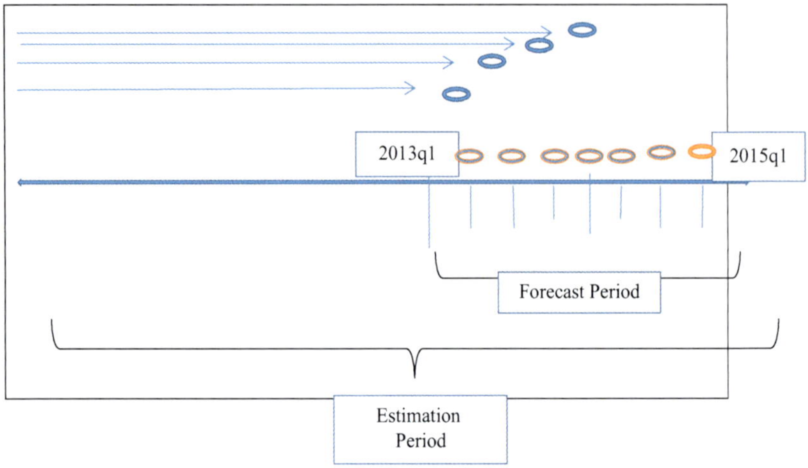

1. Dynamic Forecasting:
   The model found useful for our forecasting purposes was ARMA(2) model. We run this model as indicated below. Go to Quick-Estimate Equation-write: CPIR C AR(2) MA(2), change the sample period up to forecasting period to: 1998q3 2013q1 and as shown in Fig. 7.14. Click OK to generate the Regression output on Fig. 7.15. From the menu list of the regression output click Forecast and choose Dynamic forecast under Method as shown in Fig. 7.16. Change the Forecast sample to the period of forecast indicated above: 2013q1 2015q1 and click OK. From the Workfile it can be seen that the EViews automatically generate: cpirfulldynamic to the list of the variables as shown on Fig. 7.17.

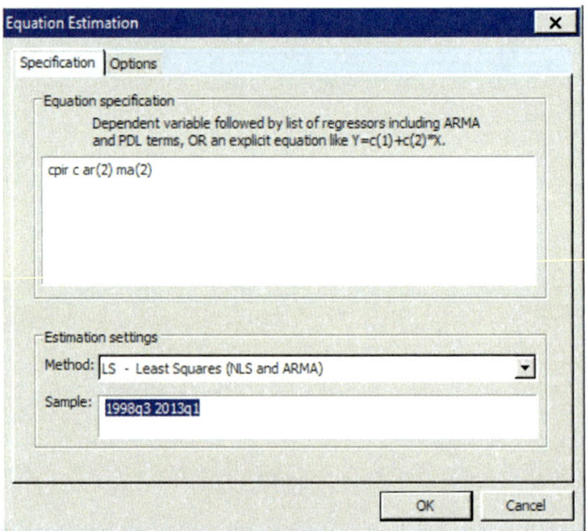

Fig. 7.14 CPIR ARIMA Equation estimation

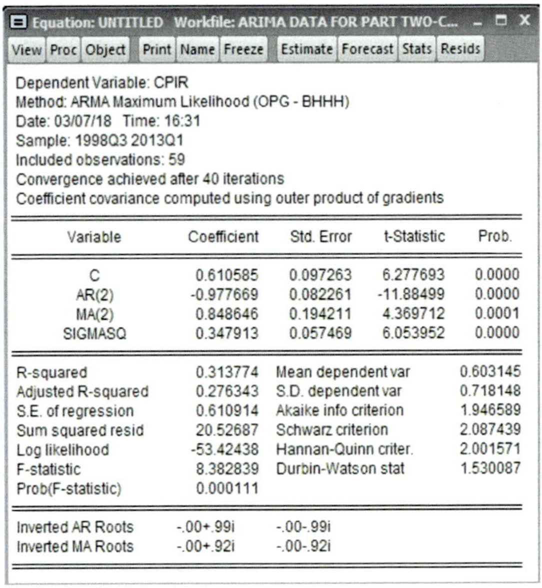

Equation: UNTITLED Workfile: ARIMA DATA FOR PART TWO-C...

View | Proc | Object | Print | Name | Freeze | Estimate | Forecast | Stats | Resids

Dependent Variable: CPIR
Method: ARMA Maximum Likelihood (OPG - BHHH)
Date: 03/07/18 Time: 16:31
Sample: 1998Q3 2013Q1
Included observations: 59
Convergence achieved after 40 iterations
Coefficient covariance computed using outer product of gradients

Variable	Coefficient	Std. Error	t-Statistic	Prob.
C	0.610585	0.097263	6.277693	0.0000
AR(2)	-0.977669	0.082261	-11.88499	0.0000
MA(2)	0.848646	0.194211	4.369712	0.0001
SIGMASQ	0.347913	0.057469	6.053952	0.0000

R-squared	0.313774	Mean dependent var	0.603145
Adjusted R-squared	0.276343	S.D. dependent var	0.718148
S.E. of regression	0.610914	Akaike info criterion	1.946589
Sum squared resid	20.52687	Schwarz criterion	2.087439
Log likelihood	-53.42438	Hannan-Quinn criter.	2.001571
F-statistic	8.382839	Durbin-Watson stat	1.530087
Prob(F-statistic)	0.000111		

Inverted AR Roots	-.00+.99i	-.00-.99i
Inverted MA Roots	-.00+.92i	-.00-.92i

Fig. 7.15 New CPIR ARIMA AR(2) MA(2) modelling output

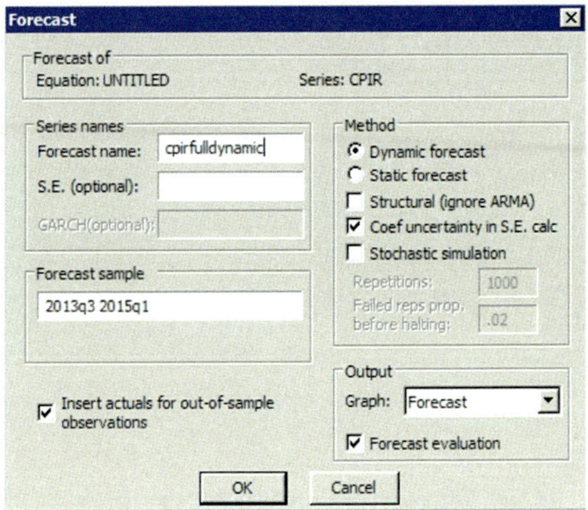

Fig. 7.16 CPIR dynamic forecast

EViews provides a graph of the forecast period and a summary table as per Fig. 7.18:

The blue line is the forecast line and the broken red lines show the confidence interval for this period. This figure gives the actual and forecast values of CPIR, as well as the confidence interval of forecast. The accompanying table gives the same measures of the quality of the forecast that we saw before, namely, the **root mean square**, **mean absolute error**, **mean absolute percentage error** and the

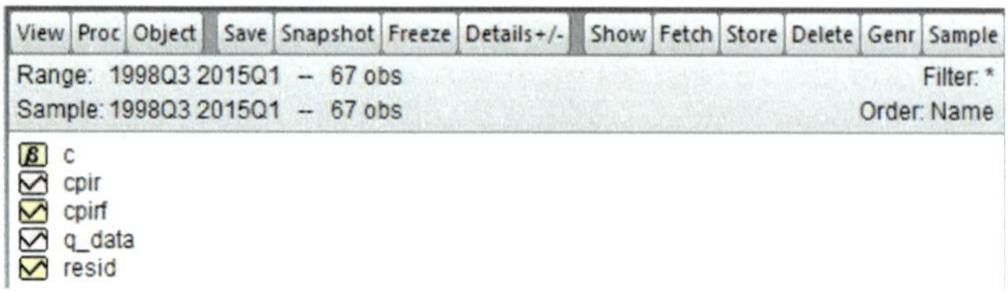

Fig. 7.17 CPIRF Dynamic Forecast

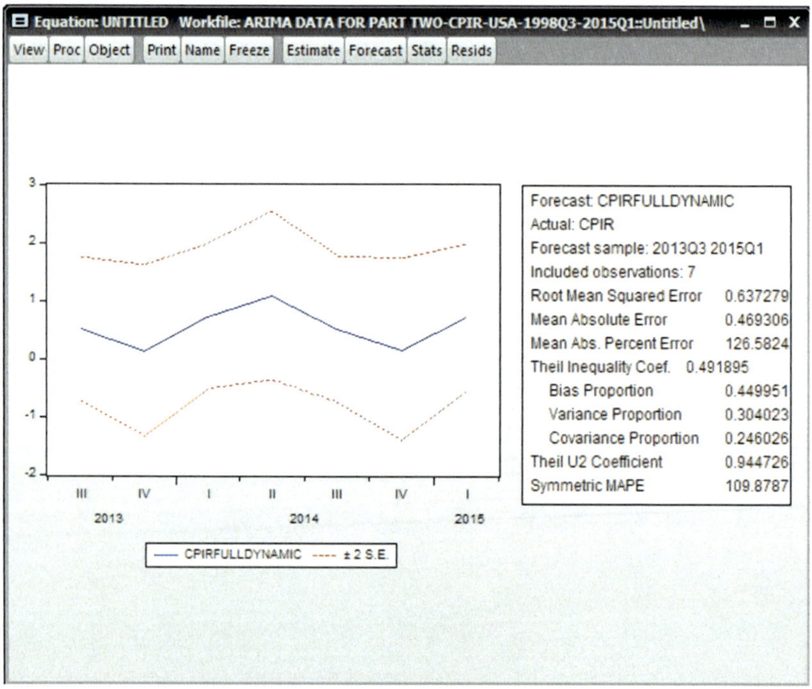

Fig. 7.18 CPIR dynamic forecast plot

Theil Inequality Coefficient. The latter coefficient is practically small, suggesting that the fitted model is quite good. This can also be seen from Fig. 7.17, which shows how closely the actual and forecast values track each other. In order to see the forecast process in a more precise way we take these steps: Highlight the name of cpir and cpirfulldynamic-right click and choose Open-from the Open list choose as Group as shown in Fig. 7.19:

When you have the data series go to View and choose Graph as shown in Fig. 7.20. Figure 7.21 shows clearly the movement of the CPIR actual (red) and the forecast line (blue). The forecast line is a very accurate for two time lags but it

Fig. 7.19 Open as a Group option

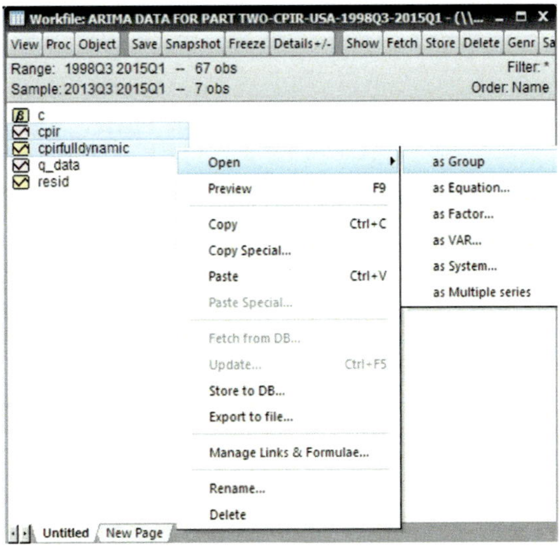

Fig. 7.20 Graph option

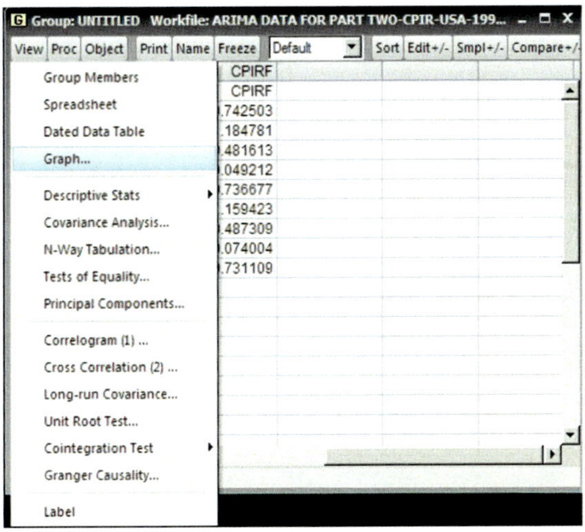

diverges afterward. This reconfirms the statistically significance of coefficients of AR(2) and MA(2) in ARMA(2) model.

In order to see what we have found we can widen the length from 2 to 3 years as follows: Go to the Command part of the EViews and write: smpl 2012q3 2015q1 and hit Enter:

Figure 7.22 reconfirms the above findings.

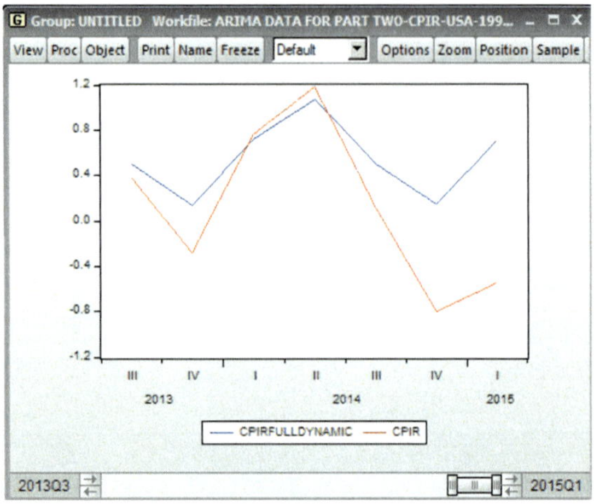

Fig. 7.21 Multiple charts

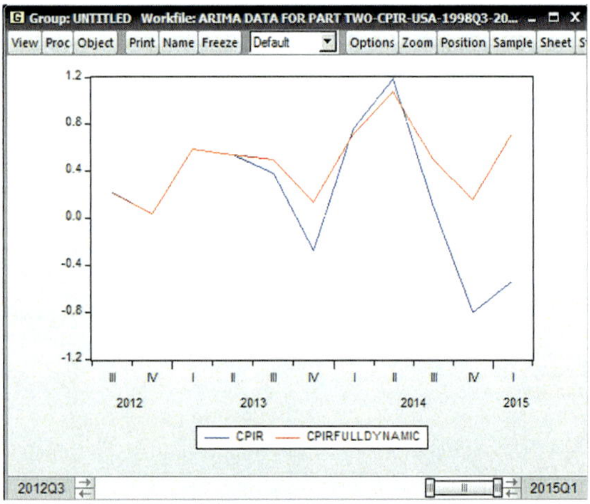

Fig. 7.22 Multiple charts (sample)

2. Static Forecasting:

Return back to the ARIMA output for the forecasting period in Fig. 7.15. From the menu list of the ARIMA output choose Forecast. Under Method, choose Static forecast. Change Forecast name from Series names to: cpirfullStatic. Change the Forecast sample to the period of forecast indicated above: 2013q1 2015q1 and hit OK as shown in Fig. 7.23.

From the Workfile it can be seen that the EViews automatically generate: cpirfullstatic to the list of variables as shown in Fig. 7.24.

EViews provides a graph of the forecast period and a summary table as per Fig. 7.25: The blue line is the forecast line and the broken red lines show the

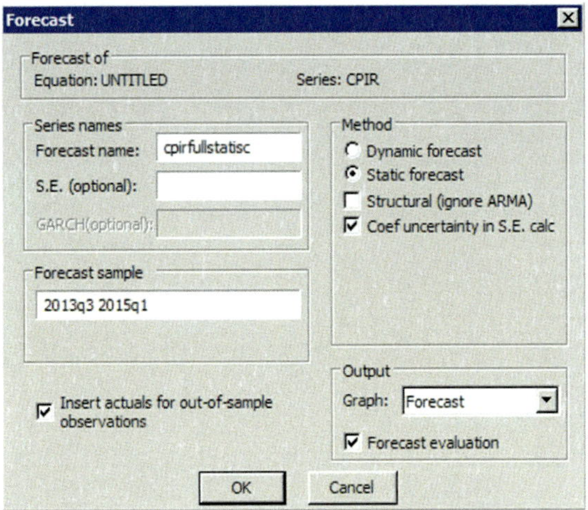

Fig. 7.23 Static Forecast

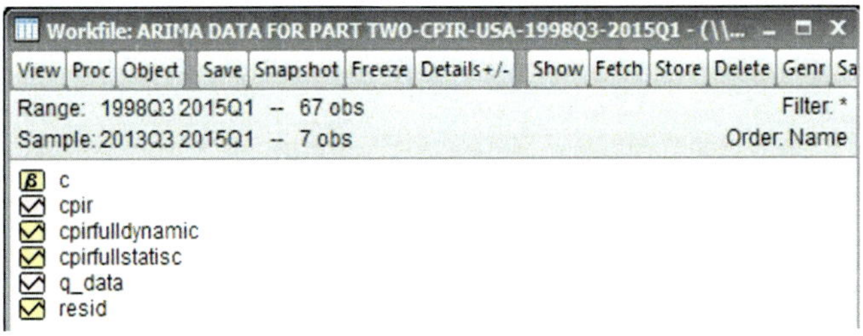

Fig. 7.24 New variable added to the data set

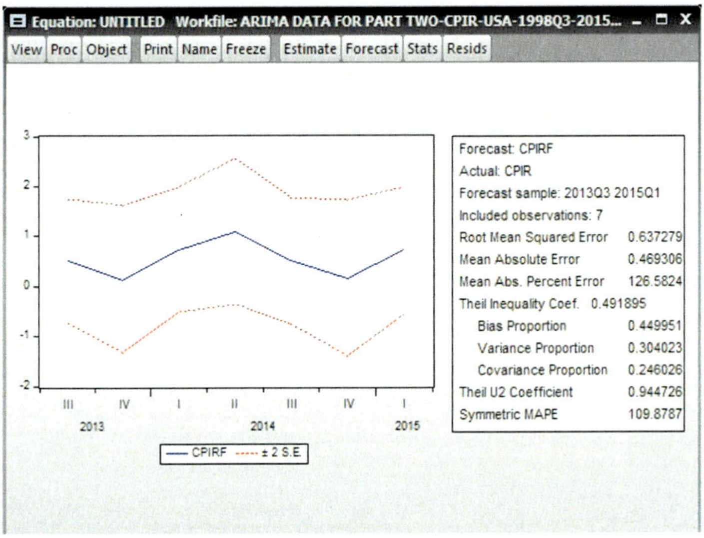

Fig. 7.25 Forecast plots & summary table

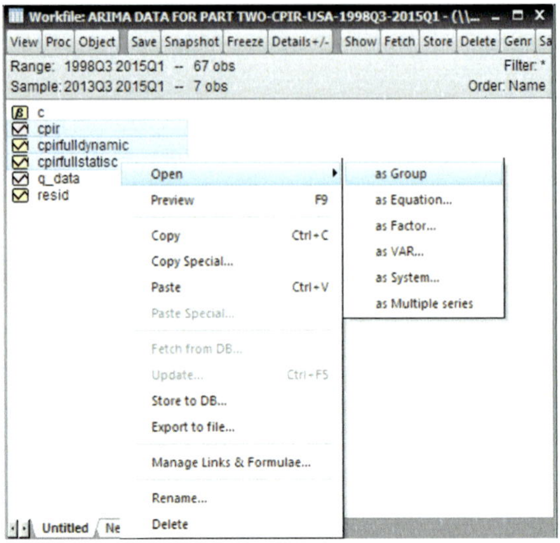

Fig. 7.26 Open variables as a Group

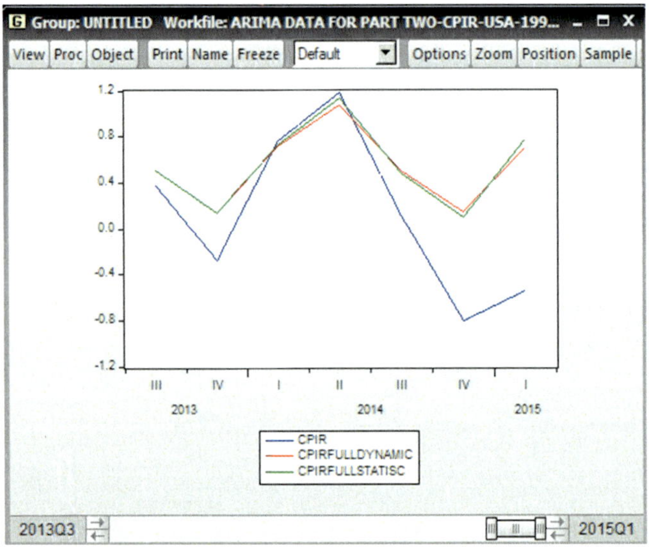

Fig. 7.27 Data plots

confidence interval for this period. In order to see the forecast process in a more precise way we take these steps:

Highlight the name of cpir, cpirdynamic and cpirfulldynamic-right click and choose Open-from the Open list choose as Group (Fig. 7.26):

When you have the data series go to View and choose Graph as shown in Fig. 7.20. Under the Graph menue choose the line graph option and click OK (Fig. 7.27).

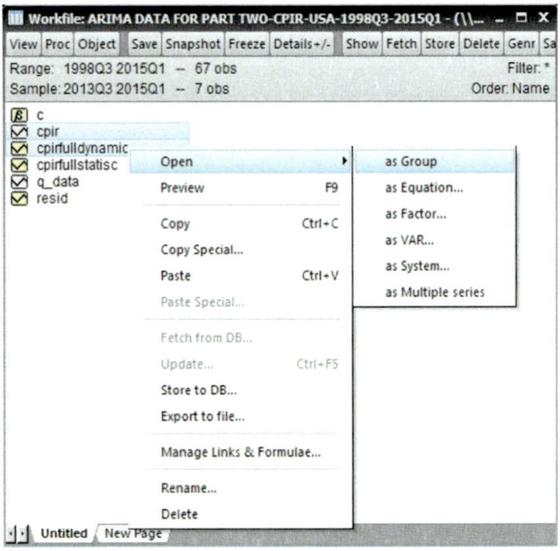

Fig. 7.28 Open as a Group

The graph shows clearly the movement of the CPIR actual (blue) and the static forecast line (green). The forecast line is a very accurate for two time lags but it diverges afterward. This reconfirms the statistically significant of coefficients of AR(2) and MA(2) in ARMA(2) model.

3. The Comparison between the Dynamic Forecasting Period and the Whole Sample Period:

   Go to the Workfile list and highlight the cpir and cpirfulldynamic- right click the highlight-choose Open-from the list choose: as Group (Figs. 7.28 and 7.29):

   From View list choose Graph as shown in Fig. 7.20:

   This graph was presented above.

   Now go to the Command part of the EViews and write: smpl @all and hit Enter:

   This is the graph of the whole period including the forecast period (Fig. 7.30). From this graph it can be seen the later periods is more useful for our forecasting of this case. The huge diverge of 2008–2009 may not be a useful period for our forecast purposes.

4. Regression of the Later Period for Forecasting:

   Now we run a regression by using the later period-2011q3-2013q1. **We use this period as the estimation period.**

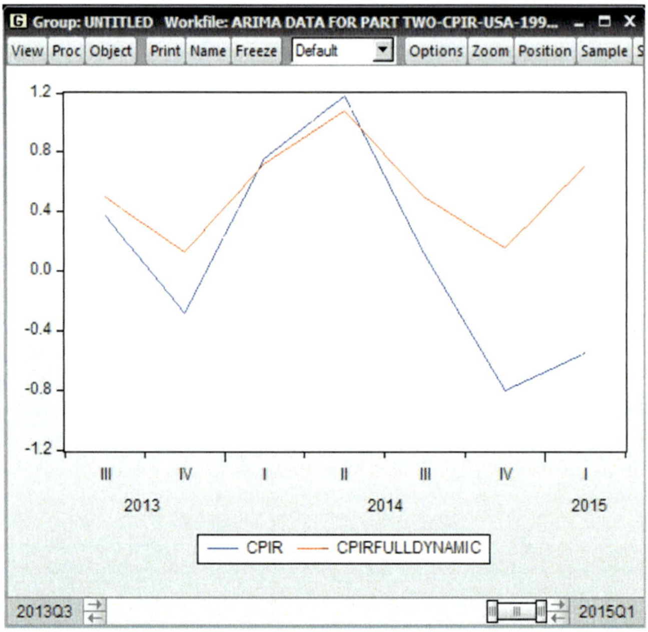

Fig. 7.29 Data plots

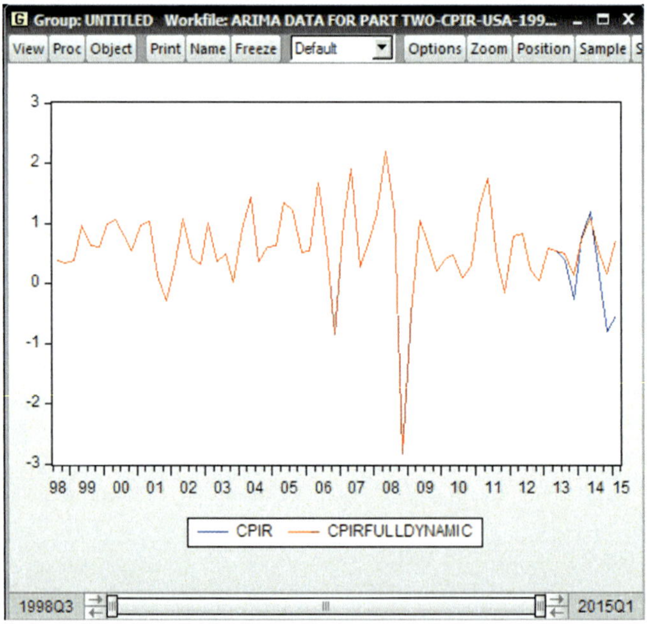

Fig. 7.30 Forecast vs. real

Go to Quick-Estimate Equation-write cpir c ar(2) ma(2) and for the Sample: write: 2011q3 2013q1 and hit OK.

From the menu list of the regression output choose Forecast (Figs. 7.31 and 7.32). Choose Dynamic forecast from the Method. Change Forecast name

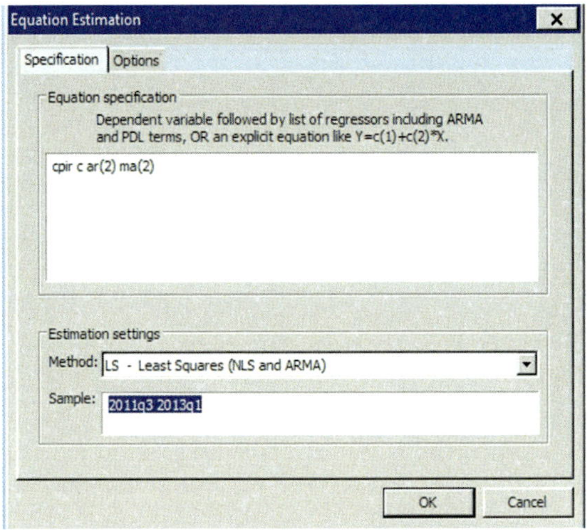

Fig. 7.31 Equation estimation

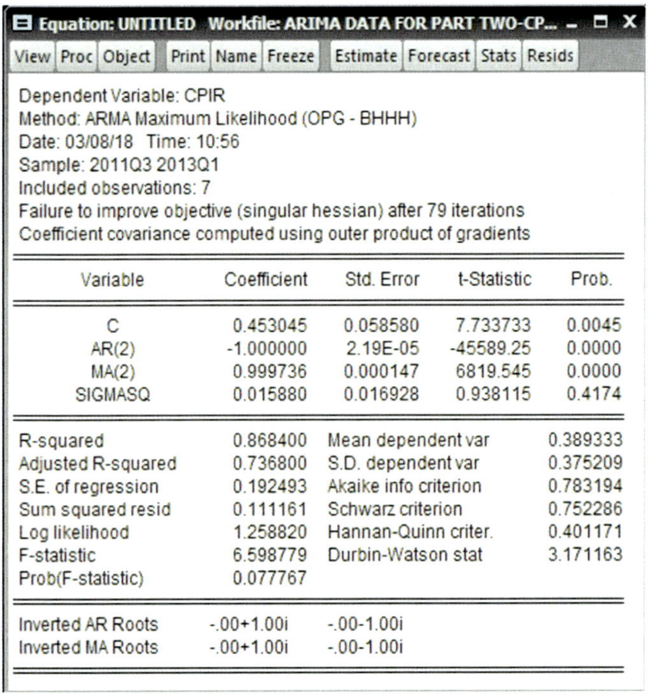

Equation: UNTITLED Workfile: ARIMA DATA FOR PART TWO-CP...

View | Proc | Object | Print | Name | Freeze | Estimate | Forecast | Stats | Resids

Dependent Variable: CPIR
Method: ARMA Maximum Likelihood (OPG - BHHH)
Date: 03/08/18 Time: 10:56
Sample: 2011Q3 2013Q1
Included observations: 7
Failure to improve objective (singular hessian) after 79 iterations
Coefficient covariance computed using outer product of gradients

Variable	Coefficient	Std. Error	t-Statistic	Prob.
C	0.453045	0.058580	7.733733	0.0045
AR(2)	-1.000000	2.19E-05	-45589.25	0.0000
MA(2)	0.999736	0.000147	6819.545	0.0000
SIGMASQ	0.015880	0.016928	0.938115	0.4174

R-squared	0.868400	Mean dependent var	0.389333
Adjusted R-squared	0.736800	S.D. dependent var	0.375209
S.E. of regression	0.192493	Akaike info criterion	0.783194
Sum squared resid	0.111161	Schwarz criterion	0.752286
Log likelihood	1.258820	Hannan-Quinn criter.	0.401171
F-statistic	6.598779	Durbin-Watson stat	3.171163
Prob(F-statistic)	0.077767		

Inverted AR Roots	-.00+1.00i	-.00-1.00i
Inverted MA Roots	-.00+1.00i	-.00-1.00i

Fig. 7.32 ARIMA modelling output

from Series names to: cpirf2011dynamic. Change the Forecast sample to the period of forecast indicated above: 2013q1 2015q1 and hit OK.

From the Workfile it can be seen that EViews automatically generate: cpirf2011dynamic to the list of variables. EViews provides a graph of the forecast period and a summary table as follows (Figs. 7.33, 7.34, and 7.35):

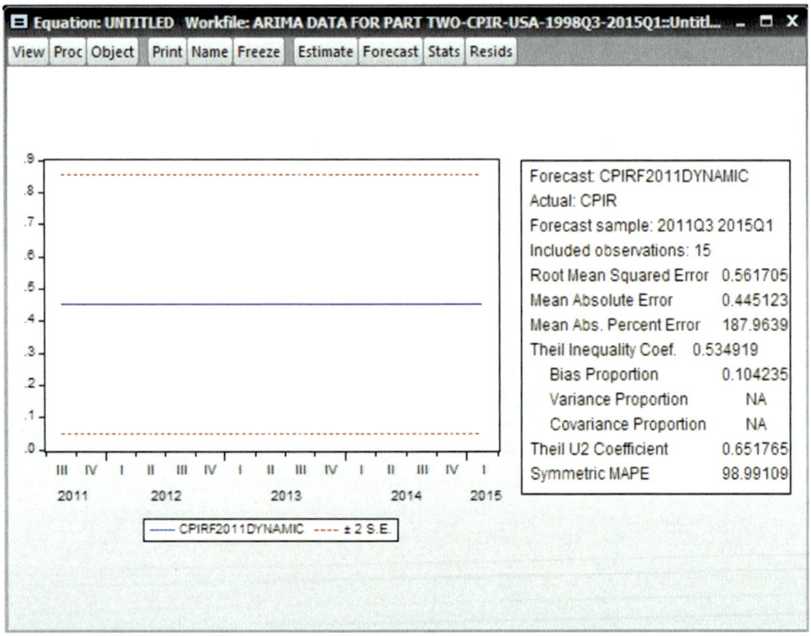

Fig. 7.33 Forecast data

Fig. 7.34 Open as a Group

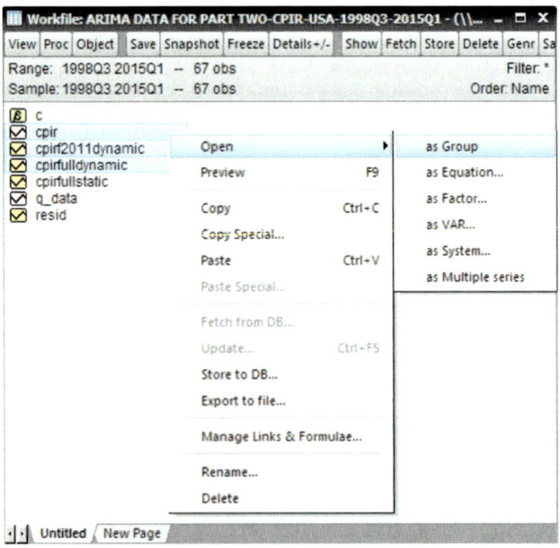

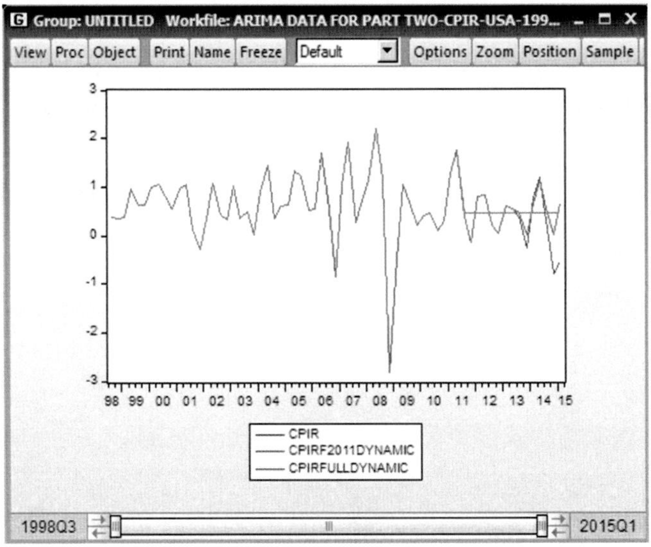

Fig. 7.35 Forecast vs. real

Highlight the name of cpir, cpirf2011dynamic and cpirfulldynamic-right click and choose Open-from the Open list choose as Group:

From the View list choose: Graph.

The graph does not show the later period clearly. In order to have full picture of the later period we do as follows:

Go to the Command part of the EViews and write: smpl 2012q3 2015q1 hit Enter:

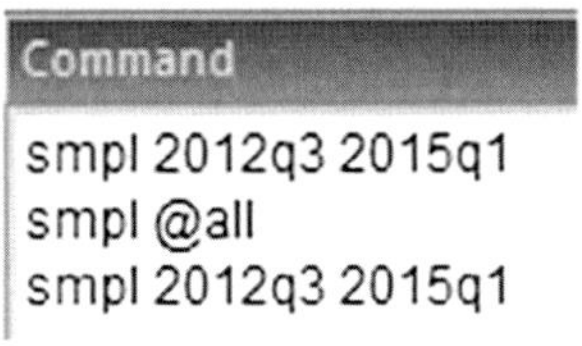

The graph above changes to this graph (Fig. 7.36):

The fulldynamic forecast, the green line, is a good forecast line for two time lags of the ARMA(2) model. The statistical significant of the coefficients of AR(2) and MA(2) reported in Fig. 7.32 confirm what is found from the forecasting processes.

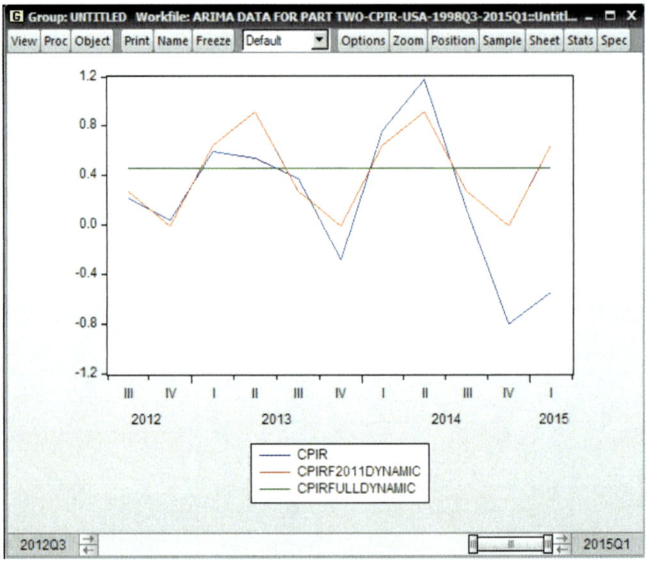

Fig. 7.36 CPIR static and dynamic forecast

Chapter 8
Modelling Volatility in Finance and Economics: ARCH, GARCH and EGARCH Models

In time series analyses, just as in regression, it is assumed that the residuals (or errors) are homoscedastic. In a seminal article, Engle (1982) suggested that heteroscedasticity of residuals might well occur in certain time series contexts. Engle had noticed that in studies of forecasting, especially in speculative markets such as foreign exchange rates and stock market returns, large and small errors tended to occur in clusters. The evidence is that in the context of financial time series, **volatility clustering** is common. Volatility clustering describes the tendency of large changes (of either sign) in, for example, asset prices to follow other large changes; small changes (of either sign) tend to follow small changes. In other words, the current level of volatility tends to be positively (auto) correlated with its level during the immediately preceding time periods.

Figure 8.1 considers a series of daily exchange rates between the US$ and the Deutsche Mark (DM) from 1 January 1980 to 21 May 1987 analysed by Verbeek (2004). Excluding days for which no prices are quoted (New Year's day etc.) this results in a total of 1867 readings (for reasons that are unimportant here, Verbeek considered daily changes in the logarithm of the exchange rate). Figure 8.1 shows the existence of periods of low and high volatility.

8.1 The ARCH Class of Models

The concept of volatility in economic and financial time series led to the development of the **autoregressive conditional heteroscedastic (ARCH)** class of models by Engle (1982). Suppose we fit say an AR(1) model to a time series:

$$Z_t = \mu + \varphi_1 Z_{t-1} + u_t \tag{8.1}$$

in which u_t is the error term with zero mean. Equation (8.1) does not have to be built into it any mechanism for coping with potential volatility. However, volatility

A. Aljandali, M. Tatahi, *Economic and Financial Modelling with EViews*, Statistics and Econometrics for Finance, https://doi.org/10.1007/978-3-319-92985-9_8

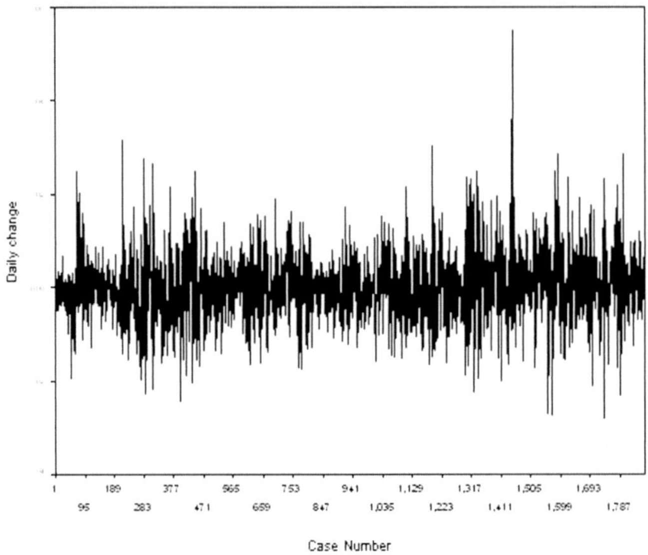

Fig. 8.1 Daily changes in log exchange rate, US$/DM

clustering may well be suggested by an examination of the ACF of u_t. Engel introduced the term **conditional variance** of the error term, denoted by σ_t^2.

Conditional variance means that σ_t^2 is estimated given information about the errors in previous time periods. In particular, if the conditional variance at time t is related to the squared error at time (t − 1), we have what is called an ARCH (1) process and:

$$\sigma_t^2 = \alpha_0 + \alpha_1 u_{t-1}^2$$

The ARCH(1) model says that when a big **shock** happens in period (t − 1) it is more likely that u_{t-1} has a large (absolute) value as well. That is, when u_{t-1}^2 is large, the variance at the next time period (t) will be large. It is crucial that the right hand side of the above equation is positive. In that an AR(1) model was initially fitted, we say that we have an AR(1) process with ARCH(1) errors, written as AR(1)-ARCH (1). The AR(1) part is referred to as **the mean equation**; the ARCH(1) part is referred to as **the variance equation**. Of course, other models may have been initially fitted e.g. ARMA(1,2)-ARCH(1). The originally fitted model need not be of the ARIMA type; it could be a regression model or any other time series model. In an ARCH(q) process the conditional variance at time t is influenced by the squared errors at times (t − 1), (t − 2), . . ., (t − q) and:

$$\sigma_t^2 = \alpha_0 + \alpha_1 u_{t-1}^2 + \alpha_2 u_{t-2}^2 + \ldots + \alpha_q u_{t-q}^2. \tag{8.2}$$

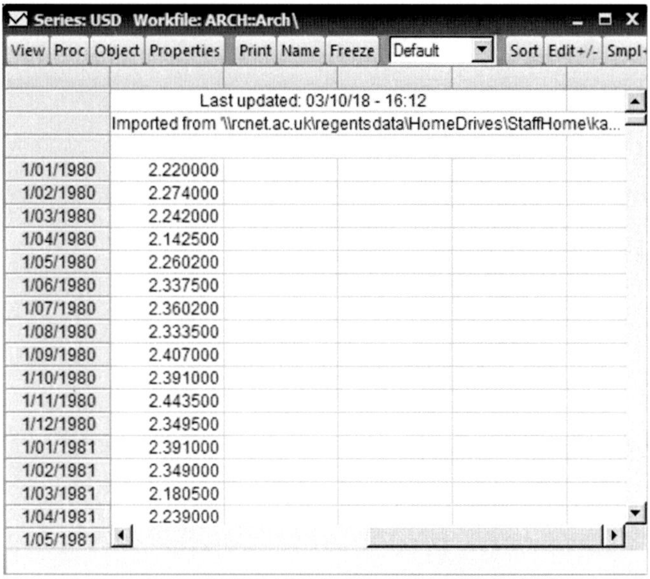

Fig. 8.2 US dollar/£ Sterling exchange rates

In equation (8.2), the conditional variance depends on q **lags** of the squared errors. (Note that in many texts on financial time series, σ_t^2 is often denoted by the notation h_t). The data file used in the current example is denominated ARCH.SAV and made available online for readers. This contains monthly exchange rates of various currencies against the UK pound Sterling. We will focus on the variable **USD**—US dollars to one pound Sterling. The data extend from 1980M1 to 2008M10 inclusive. Open the file in EViews (remember it is a **foreign file**), date it and you will have the information contained in Fig. 8.2.

Generating Rates of Change

Pesaran and Pesaran (1997) used these data to examine **the proportional monthly rate of change in the dollar/sterling rate**, rather than USD itself. The reason for this procedure in financial forecasting is that the proportional rate of change is often stationary, whereas the values of the original variable (here USD) tend not to be. The proportional monthly rate of change is obviously given by $(USD_t - USD_{t-1})/USD_{t-1}$. (For example, if a variable has value 58 at time t and a value of 56 at time $t - 1$, the proportional change in that variable is $2/56 = 0.0357$ or in percentage terms, 3.57%).

This proportional change is often approximated (especially in financial analyses) by ln(value at time t/value at time $t - 1$) or here $\ln(USD_t/USD_{t-1})$. Using the figures

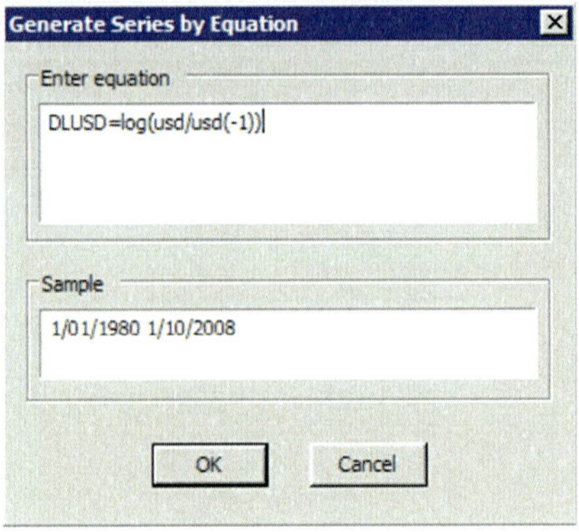

Fig. 8.3 The Generate Series by Equation dialogue box

at the end of the previous paragraph, we estimate the proportional change as ln (58/56) = 0.0351 or 3.51%. Peseran and Peseran used this logarithmic approximation, which we shall too shall employ here.

In EViews, the notation USD(−1) represents the one period lagged variable USD_{t-1}. From the main EViews menu, click:

Quick
  Generate Series

which produces the Generate Series by Equation dialogue box of Fig. 8.3. Enter the formula for the logarithmic approximation as shown and click the OK button to generate the series. Figure 8.4 shows (i) the proportional monthly changes in USD computed via $(USD_t - USD_{t-1})/USD_{t-1}$—variable name **PROPCHA** and (ii) the above logarithmic approximation of these proportional monthly changes computed via $\ln(USD_t/USD_{t-1})$—variable name **DLUSD**. We shall use the latter.

8.2 Testing for ARCH Effects

The first step in testing for potential ARCH effects is to fit the **mean equation** (here an ARIMA model) for **DLUSD** = $\ln(USD_t/USD_{t-1})$. The ADF test when applied to **DLUSD** revealed the variable to be stationary (ADF test statistic = −17.249, p = 0.000), so no differencing is required. We will apply ARIMA models to the data from 1980M1 to 2007M12 and reserve the data for the first 8 months of 2008 to examine the adequacy of the models fitted. The model selected by Peseran and Peseran for the mean equation was an AR(1) process without a constant, despite the fact that the AR(1) term was marginally not significant (AR(1) = 0.065, p = 0.1032, see Fig. 8.5).

We now have to consider whether our AR(1) model contains ARCH errors as per equation (2). In this equation, we test the null hypothesis:

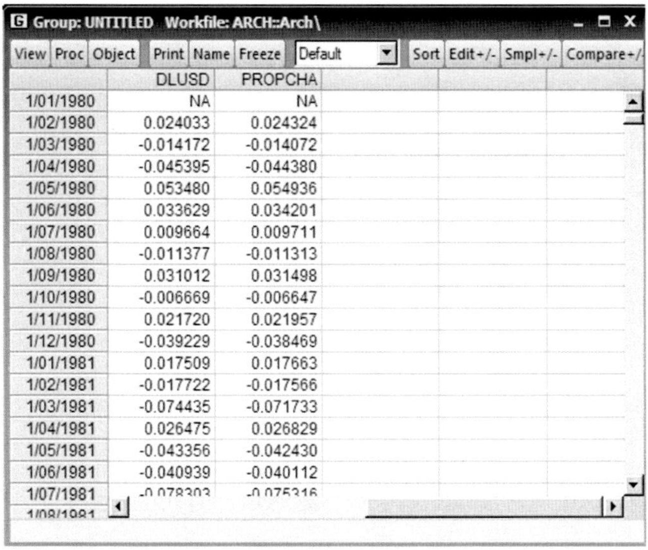

Group: UNTITLED Workfile: ARCH::Arch\

View | Proc | Object | Print | Name | Freeze | Default | Sort | Edit+/- | Smpl+/- | Compare+/-

	DLUSD	PROPCHA
1/01/1980	NA	NA
1/02/1980	0.024033	0.024324
1/03/1980	-0.014172	-0.014072
1/04/1980	-0.045395	-0.044380
1/05/1980	0.053480	0.054936
1/06/1980	0.033629	0.034201
1/07/1980	0.009664	0.009711
1/08/1980	-0.011377	-0.011313
1/09/1980	0.031012	0.031498
1/10/1980	-0.006669	-0.006647
1/11/1980	0.021720	0.021957
1/12/1980	-0.039229	-0.038469
1/01/1981	0.017509	0.017663
1/02/1981	-0.017722	-0.017566
1/03/1981	-0.074435	-0.071733
1/04/1981	0.026475	0.026829
1/05/1981	-0.043356	-0.042430
1/06/1981	-0.040939	-0.040112
1/07/1981	-0.078303	-0.075316
1/08/1981		

Fig. 8.4 The actual (PROPCHA) and approximated (DLUSD) proportionate monthly rate of change in the dollar/Sterling exchange rates

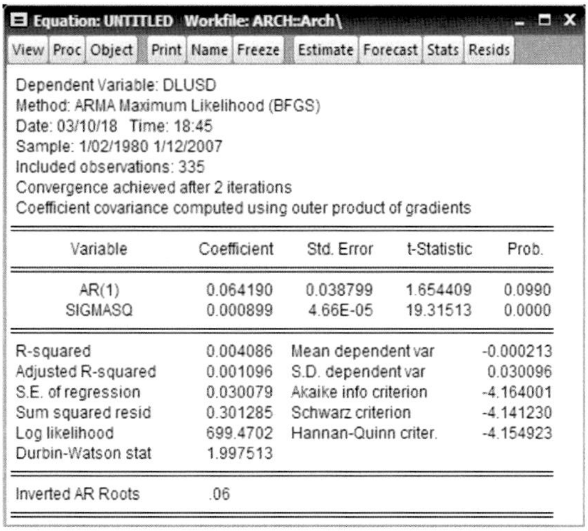

Equation: UNTITLED Workfile: ARCH::Arch\

View | Proc | Object | Print | Name | Freeze | Estimate | Forecast | Stats | Resids

Dependent Variable: DLUSD
Method: ARMA Maximum Likelihood (BFGS)
Date: 03/10/18 Time: 18:45
Sample: 1/02/1980 1/12/2007
Included observations: 335
Convergence achieved after 2 iterations
Coefficient covariance computed using outer product of gradients

Variable	Coefficient	Std. Error	t-Statistic	Prob.
AR(1)	0.064190	0.038799	1.654409	0.0990
SIGMASQ	0.000899	4.66E-05	19.31513	0.0000

R-squared	0.004086	Mean dependent var	-0.000213
Adjusted R-squared	0.001096	S.D. dependent var	0.030096
S.E. of regression	0.030079	Akaike info criterion	-4.164001
Sum squared resid	0.301285	Schwarz criterion	-4.141230
Log likelihood	699.4702	Hannan-Quinn criter.	-4.154923
Durbin-Watson stat	1.997513		

Inverted AR Roots .06

Fig. 8.5 The mean equation for DLUSD

$$H_0 : \alpha_1 = \alpha_2 = \alpha_3 = \ldots = \alpha_q = 0.$$

If all of these coefficients in equation (2) are jointly zero (i.e. we fail to reject the null), then there are no ARCH effects up to q lags.

Let us now test for ARCH effects up to a lag of 12 months. From the menu bar in Fig. 8.5, click:

Equation: UNTITLED Workfile: ARCH::Arch\

View | Proc | Object | Print | Name | Freeze | Estimate | Forecast | Stats | Resids

Heteroskedasticity Test: ARCH

F-statistic	1.692042	Prob. F(12,310)	0.0674
Obs*R-squared	19.85548	Prob. Chi-Square(12)	0.0699

Test Equation:
Dependent Variable: RESID^2
Method: Least Squares
Date: 03/10/18 Time: 18:46
Sample (adjusted): 1/02/1981 1/12/2007
Included observations: 323 after adjustments

Variable	Coefficient	Std. Error	t-Statistic	Prob.
C	0.000613	0.000169	3.626151	0.0003
RESID^2(-1)	0.224679	0.056773	3.957528	0.0001
RESID^2(-2)	-0.024761	0.058153	-0.425789	0.6706
RESID^2(-3)	0.040738	0.058169	0.700331	0.4842
RESID^2(-4)	0.038234	0.058185	0.657103	0.5116
RESID^2(-5)	0.010821	0.058193	0.185950	0.8526
RESID^2(-6)	0.051296	0.058112	0.882699	0.3781
RESID^2(-7)	-0.051660	0.058115	-0.888933	0.3747
RESID^2(-8)	0.036821	0.058189	0.632790	0.5273
RESID^2(-9)	0.027153	0.058075	0.467550	0.6404
RESID^2(-10)	0.005665	0.058045	0.097605	0.9223
RESID^2(-11)	-0.014804	0.058010	-0.255195	0.7987
RESID^2(-12)	-0.026703	0.056636	-0.471488	0.6376

Fig. 8.6 Testing for ARCH effects up to a lag of 12

View
  Residual Diagnostics
    Heteroskedasticity Tests...

Choose ARCH under test Type. You will be asked how many lags you want, so type in 12 in the box provided and click the OK button to generate the results presented in Fig. 8.6.

Under the heading 'ARCH test' in Fig. 8.6, we are testing the null hypothesis that:

$$H_0 : \alpha_1 = \alpha_2 = \alpha_3 = \ldots = \alpha_{12} = 0.$$

This is tested by either of two statistics called F and Obs*R squared (which is a chi-square statistic). The significance levels attached to these are 0.0674 and 0.0699 respectively, so we fail to reject the null since $p > 0.05$. We conclude that there are no ARCH effects up to a lag of 12 months.

Let us now consider whether there are ARCH effects up to a lag of 1 month. In Fig. 8.6, click:

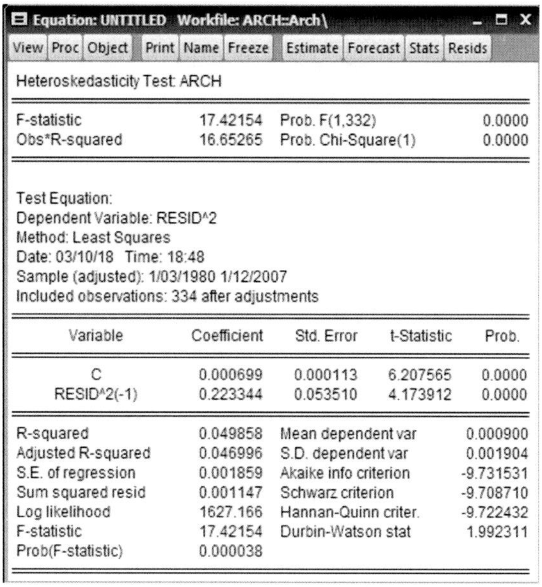
Equation: UNTITLED Workfile: ARCH::Arch\

View | Proc | Object | Print | Name | Freeze | Estimate | Forecast | Stats | Resids

Heteroskedasticity Test: ARCH

F-statistic	17.42154	Prob. F(1,332)	0.0000
Obs*R-squared	16.65265	Prob. Chi-Square(1)	0.0000

Test Equation:
Dependent Variable: RESID^2
Method: Least Squares
Date: 03/10/18 Time: 18:48
Sample (adjusted): 1/03/1980 1/12/2007
Included observations: 334 after adjustments

Variable	Coefficient	Std. Error	t-Statistic	Prob.
C	0.000699	0.000113	6.207565	0.0000
RESID^2(-1)	0.223344	0.053510	4.173912	0.0000

R-squared	0.049858	Mean dependent var	0.000900
Adjusted R-squared	0.046996	S.D. dependent var	0.001904
S.E. of regression	0.001859	Akaike info criterion	-9.731531
Sum squared resid	0.001147	Schwarz criterion	-9.708710
Log likelihood	1627.166	Hannan-Quinn criter.	-9.722432
F-statistic	17.42154	Durbin-Watson stat	1.992311
Prob(F-statistic)	0.000038		

Fig. 8.7 Testing for ARCH effects up to a lag of 1

View
Residual Diagnostics
Heteroskedasticity Tests…

Choose ARCH under Test Type and when requested, enter that 1 lag is required. This generates the results in Fig. 8.7. Both the F statistic (p = 0.000) and the chi-square statistic (p = 0.000) indicate that we reject H_0: $\alpha_1 = 0$ so we conclude that there are significant ARCH effects involving a lag of 1 month. (We are paralleling the approach adopted by Peseran and Peseran. However, if you conduct the ARCH LM test up to four lags, significant results suggest that there may be ARCH effects up to this level).

We will assume that **DLUSD** may be modelled by an AR(1)-ARCH(1) process and now proceed to estimate the model and generate forecasts based on it. From the main EViews menu, click:

Quick
Estimate Equation

and you are presented with the usual dialogue box. Enter the model as:

DLUSD AR1and change the sample period to 1980M1 up to 2007M12. This time we select the Method as:

ARCH - Autoregressive Conditional Heteroscedasticity

and click the OK button to reveal the dialogue box of Fig. 8.8. At the top, you will see the mean equation as already specified. The default model has ARCH 1 and GARCH 1 (see next section for GARCH modelling). Change the GARCH parameter to 0, but leave the ARCH parameter as 1 (i.e. a lag of 1 month). The default is to

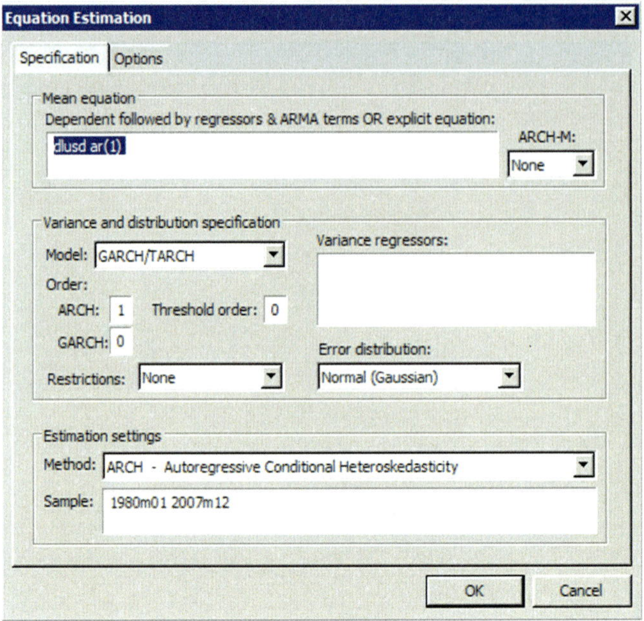

Fig. 8.8 Defining the AR(1)-ARCH(1) model for DLUSD

assume that the errors from this AR(1)-ARCH(1) process are normally distributed. Click the OK button to produce the results in Fig. 8.9.

From Fig. 8.9, **the mean equation** is:

$$\text{DLUSD}_t = 0.0051(\text{DLUSD}_{t-1}).$$

Also from Fig. 8.9 and equation (8.2), **the variance equation** (an ARCH(1) process) is:

$$\sigma_t^2 = \alpha_0 + \alpha_1 u_{t-1}^2$$
$$\sigma_t^2 = 0.0005 + 0.1714 u_{t-1}^2$$

in which EViews adopts the notation RESID$(-1)^2$ for u_{t-1}^2, the squared residual at time $(t - 1)$. Both the intercept term ($p = 0.000$) and the coefficient of the squared residual ($p = 0.0004$) are significantly different from zero i.e. $p < 0.05$ in both instances. Note that the coefficient of DLUSD_{t-1} in Fig. 8.9 is different from that in Fig. 8.5. In Fig. 8.5 the coefficient had the marginally insignificant value of 0.0641 **when we were trying to identify the mean equation alone**. (This was estimated by **ordinary least squares**). In Fig. 8.9, **when ARCH effects are now introduced**, this coefficient is estimated as 0.0051 which is not significant. (In the ARCH case, model estimation is via a different technique called **maximum likelihood**).

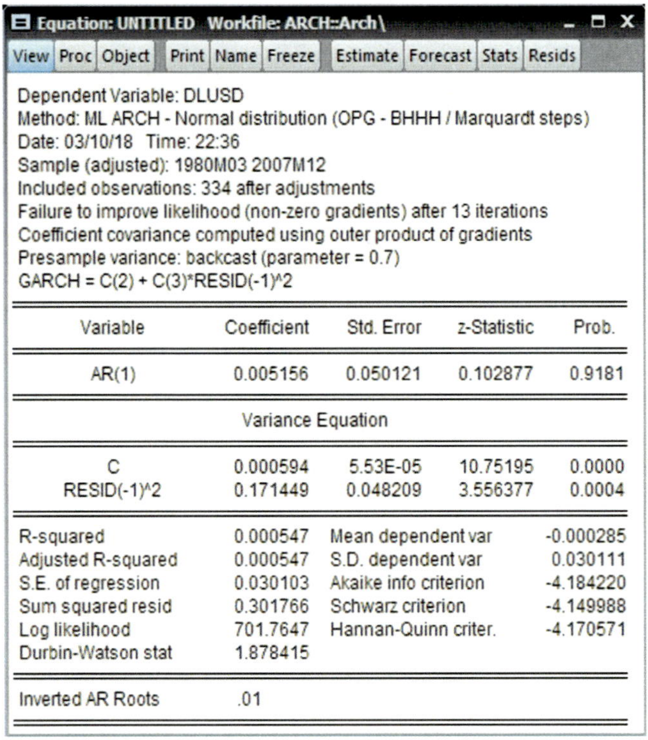
Equation: UNTITLED Workfile: ARCH::Arch\

View Proc Object Print Name Freeze Estimate Forecast Stats Resids

Dependent Variable: DLUSD
Method: ML ARCH - Normal distribution (OPG - BHHH / Marquardt steps)
Date: 03/10/18 Time: 22:36
Sample (adjusted): 1980M03 2007M12
Included observations: 334 after adjustments
Failure to improve likelihood (non-zero gradients) after 13 iterations
Coefficient covariance computed using outer product of gradients
Presample variance: backcast (parameter = 0.7)
GARCH = C(2) + C(3)*RESID(-1)^2

Variable	Coefficient	Std. Error	z-Statistic	Prob.
AR(1)	0.005156	0.050121	0.102877	0.9181
		Variance Equation		
C	0.000594	5.53E-05	10.75195	0.0000
RESID(-1)^2	0.171449	0.048209	3.556377	0.0004

R-squared	0.000547	Mean dependent var	-0.000285
Adjusted R-squared	0.000547	S.D. dependent var	0.030111
S.E. of regression	0.030103	Akaike info criterion	-4.184220
Sum squared resid	0.301766	Schwarz criterion	-4.149988
Log likelihood	701.7647	Hannan-Quinn criter.	-4.170571
Durbin-Watson stat	1.878415		

Inverted AR Roots .01

Fig. 8.9 Results for the AR(1)-ARCH(1) model

Note: We have adhered to the reporting from Peseran and Peseran's original work. However, given the above results, We would suggest returning to the mean equation and searching for an alternative form.

Forecasting from an ARCH Model

From Fig. 8.9, click:

View
  Residual Tests
    Histogram—Normality test

To generate the Jarque-Bera test for normality of residuals. The null of normality is rejected ($p = 0.000$), suggesting model mis-specification and the user may wish to alter the presumed distribution of the residuals in Fig. 8.8. We can test whether there are **further ARCH effects** in our AR(1)-ARCH(1) model by clicking:

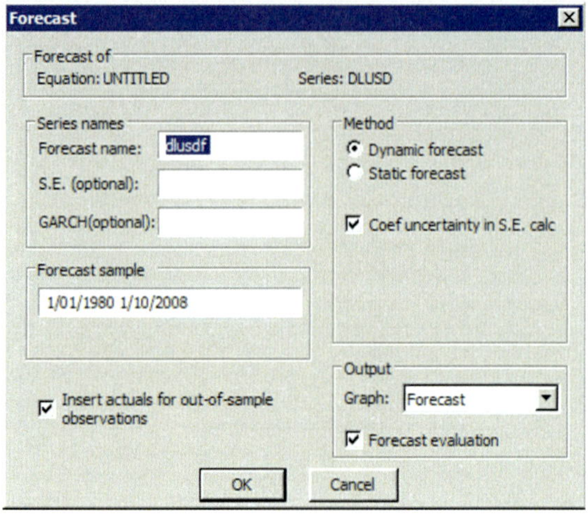

Fig. 8.10 The Forecast dialogue box

View
 Residual Diagnostics
  ARCH LM Test. . .

and after choosing a variety of lags, non-significant F and chi-square statistics indicated no further ARCH effects.

Recall that the first 8 months of 2008 were reserved to test the adequacy of the model in generating forecasts. In other words, we have the actual values of DLUSD for 2008M1 to 2008M8 and we can compare them with the forecasted values for this time period derived from the AR(1)-ARCH(1) model. Click the Forecast button in Fig. 8.9 to produce the Forecast dialogue box of Fig. 8.10.

By default, the forecasted values of DLUSD will be saved under the 'forecast name' of DLUSDF and the user can naturally change this name. At the bottom of Fig. 8.10, EViews has set the 'forecast sample ' (or forecast period) as 1980M1 to 2008M10; change this to 2008M1 to 2008M10. On the right hand side of Fig. 8.10, you will notice that there are two methods for generating forecasts—**dynamic forecasts** (the default) or **static forecasts**. Dynamic forecasts in the forecast period use *previously forecasted* values of the variable at hand. Static forecasts in the forecast period produce results using the *observed values* of the variable at hand. Should there be no observed values, EViews reports NA. Finally, the default output includes a forecast graph and an evaluation of the forecasts.

Click the OK button to generate Fig. 8.11, wherein static forecasts were generated. Repeating this process but for all the months 1980M1 to 2008M10 produced the static forecasts presented in Fig. 8.12. We saved these forecasts under the 'forecast name' ALLDLUSDF.

In Figs. 8.11 and 8.12, there are some of the frequently used measures of forecast adequacy. These are the root mean square error (RMSE), the mean absolute error (MAE), the mean absolute percentage error (MAPE) and Thiel's Inequality

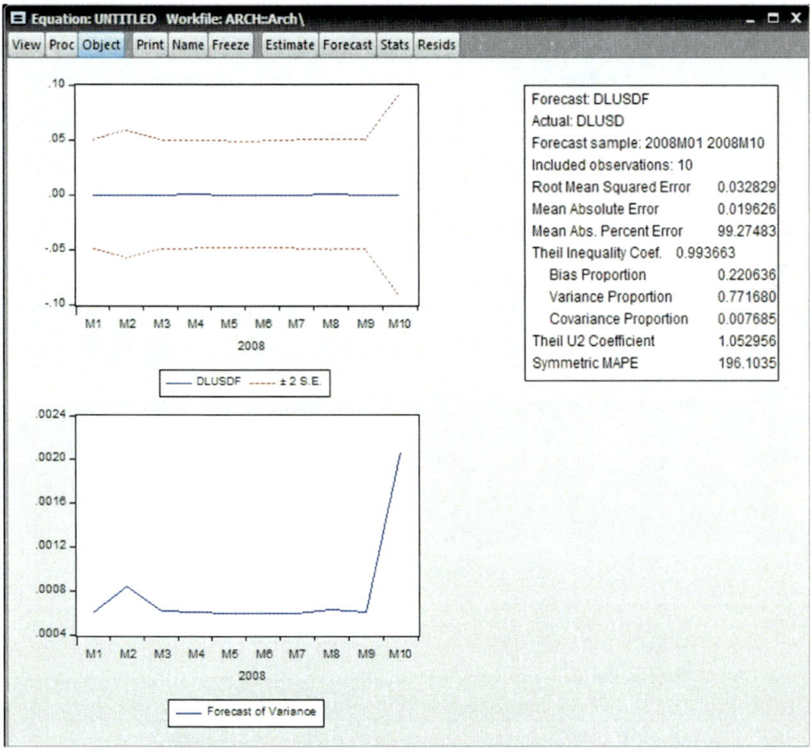

Fig. 8.11 Six month static forecasts from the AR(1)-ARCH(1) model

Coefficient. 'Absolute' is the same as 'modulus'. The RMSE and MAE depend on the scale of the variable being forecast. As such, they are used as relative measures to compare forecasts for the same series across different forecasting models. The smaller the RMSE/MAE, the better is the forecasting ability of that model. The user of a forecasting model might use the MAPE as well, for example, it may be more useful to know that the MAPE of a model is 5% than to know that its MAE is 439.6.

Theil's Inequality Coefficient is also known as Thiel's U statistic. Thiel's Coefficient or U statistic lies between 0 and 1 with 0 indicating a perfect fit i.e. no error in the forecasts. The coefficient may be interpreted as follows:

- If $U = 1$, the forecasting method being used is as good as the Naïve 1 model
- If $U < 1$, the forecasting method being used is better than the Naïve 1 model. The smaller is U the better
- If $U > 1$, there is no point in using whatever forecasting method is being employed, since the Naïve 1 model will produce more accurate results.

In addition, EViews reports the following in respect of Theil's coefficient:

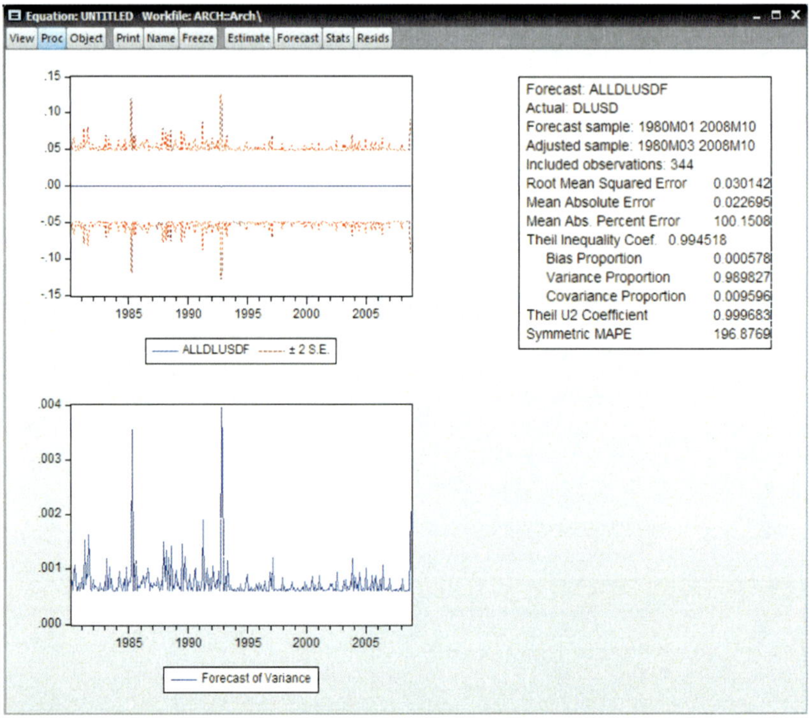

Fig. 8.12 Static forecasts for the entire time period

- The **bias proportion** which tells us how far the mean of the forecast is from the mean of the actual series
- The **variance proportion** which tells us how far the variation of the forecast is from the variation of the actual series
- The **covariance proportion** which measures the remaining unsystematic forecasting errors.

Note that the bias, variance and covariance proportions sum to one.

8.3 Problems with ARCH Models in Practice

ARCH provides a framework for the analysis and development of time series involving volatility. As such, ARCH methods have had wise application over the last 10 years, particularly in the field of financial volatility. However, ARCH models are now used with decreasing frequency, due to a number of difficulties:

- How should the value of q—the number of lags—be determined? There is no clear best approach.
- The value of q required to capture all of the impact on the conditional variance might be very large. This would result in a complex ARCH model that is not parsimonious.
- The larger is the value of q, the greater is the possibility that a negative conditional variance could be the result.

An extension of the ARCH(q) process was developed independently by Bollerslev (1986) and Engle (1982) and is called **the generalised ARCH process or GARCH**. GARCH models have become extremely widely employed in financial analyses in recent years.

8.4 GARCH Models

The GARCH model allows the conditional variance to be dependent upon its own previous lags as well as the squared error terms of the ARCH models. In its simplest case, we have:

$$\sigma_t^2 = \alpha_0 + \alpha_1 u_{t-1}^2 + \beta_1 \sigma_{t-1}^2. \tag{8.3}$$

Equation (8.3) is referred to as a GARCH(1,1) model. The GARCH(1,1) model states that the current conditional variance is a function of an intercept term, information about volatility during the previous time period $\left(\alpha_1 u_{t-1}^2\right)$ and the fitted conditional variance from the model during the previous period $\left(\beta_1 \sigma_{t-1}^2\right)$.

A GARCH(q,p) model has the form:

$$\sigma_t^2 = \alpha_0 + \alpha_1 u_{t-1}^2 + \alpha_2 u_{t-2}^2 + \ldots + \alpha_q u_{t-q}^2 + \beta_1 \sigma_{t-1}^2 + \beta_2 \sigma_{t-2}^2 + \ldots + \beta_p \sigma_{t-p}^2. \tag{8.4}$$

An ARCH(1) model is equivalent to a GARCH(1,0) model. **Equation** (8.4) **parallels the ARIMA model structure that we have seen earlier. The σ_i^2 are the autoregressive (AR) part and the u_j^2 are the moving average (MA) part.**

Practical applications of the GARCH model have indicated that low values for q and p are typically required, often p = q = 1. We shall now see whether the AR (1) model of proportional changes in US$/GBP exchange rates used in the previous sections has a GARCH(1,1) error structure associated with it. This latter model is widely applied to financial series in which volatility is present. The scheme to be analysed is:

$$\begin{gathered} \mathrm{DLUSD_t} = \gamma\left(\mathrm{DLUSD_{t-1}}\right) \text{ with} \\ \sigma_t^2 = \alpha_0 + \alpha_1 u_{t-1}^2 + \beta_1 \sigma_{t-1}^2. \end{gathered} \tag{8.5}$$

which can be denoted by an AR(1)-GARCH(1,1) process and γ, α_0, α_1 and β_1 are parameters to be estimated.

From the main EViews menu, click:

Quick
  Estimate Equation

and specify the mean equation as per equation (8.5) via:

DLUSD AR(1)

Select the sample to be 1980M1 to 2007M12 and, as for ARCH modelling of the previous sections, we select the Method as:

ARCH - Autoregressive Conditional Heteroscedasticity

This generates the dialogue box of Fig. 8.8. This time, we select one ARCH term and one GARCH term, so that the dialogue box is as per Fig. 8.13. The results of applying the AR(1)–GARCH(1,1) are presented in Fig. 8.14.

Note that introducing GARCH terms results in the mean equation being slightly different than it was for just an ARCH process, when AR1 = 0.0051, p = 0.9181. In Fig. 8.14, the first GARCH term denoted by GARCH(−1) is clearly not significant, so a GARCH process is inappropriate for the variable DLUSD. Static forecasts for this model are presented in Fig. 8.15. It may be noted that in terms of the RMSE and MAE measures at least that the GARCH(1,1) model generates worse forecasts than does the ARCH(1) model (compare with Fig. 8.12).

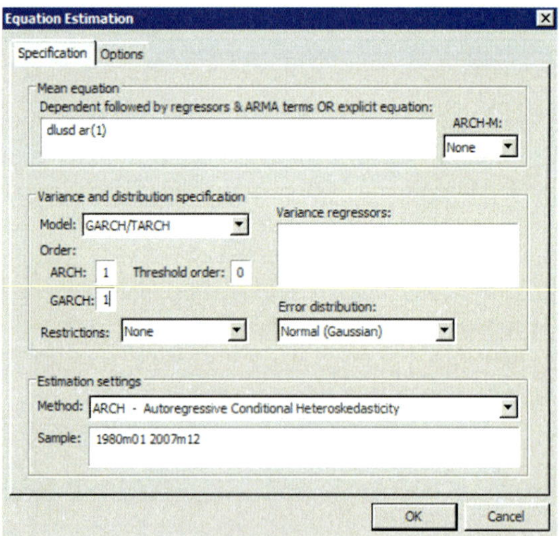

Fig. 8.13 Specifying a GARCH(1,1) model

Equation: UNTITLED Workfile: ARCH::Arch\

View | Proc | Object | Print | Name | Freeze | Estimate | Forecast | Stats | Resids

Dependent Variable: DLUSD
Method: ML ARCH - Normal distribution (OPG - BHHH / Marquardt steps)
Date: 03/10/18 Time: 22:57
Sample (adjusted): 1980M02 2007M12
Included observations: 335 after adjustments
Failure to improve likelihood (non-zero gradients) after 126 iterations
Coefficient covariance computed using outer product of gradients
Presample variance: backcast (parameter = 0.7)
GARCH = C(2) + C(3)*RESID(-1)^2 + C(4)*GARCH(-1)

Variable	Coefficient	Std. Error	z-Statistic	Prob.
AR(1)	0.004571	0.069773	0.065507	0.9478
	Variance Equation			
C	0.000589	0.000490	1.202114	0.2293
RESID(-1)^2	0.149110	0.085355	1.746940	0.0806
GARCH(-1)	0.370810	0.454181	0.816436	0.4143

R-squared	0.000429	Mean dependent var	-0.000388
Adjusted R-squared	0.000429	S.D. dependent var	0.030132
S.E. of regression	0.030126	Akaike info criterion	-4.167527
Sum squared resid	0.303127	Schwarz criterion	-4.121986
Log likelihood	702.0608	Hannan-Quinn criter.	-4.149371
Durbin-Watson stat	1.871255		
Inverted AR Roots	.00		

Fig. 8.14 Results of applying the GARCH(1,1) model

8.5 Application: Modelling Volatility & Estimating a GARCH (1, 1) Model

The aim is to find influential factors behind the volatility of a financial asset or a financial instrument. More explicitly, we want to identify common internal and external factors, which have led the volatility to be present for a period of time. We carry out a GARCH (1, 1) model. We estimate that one ARCH stands for the first '1' in the bracket and one GARCH stands for the second '1' in the bracket. The main attention will be given to the residual square one time lag and to the residual variances one time lag, as shown above. Two equations will be considered, one is the main equation and the other one is the variance equation. The aim is to run the main equation to get the residual, and use the residual variance to estimate the variance equation. The main equation we use in this example is a CAPM model, in which the return of the Royal Bank of Scotland (RBS) is assumed to be a function of the return of the UK market, FTSE 100, as follow:

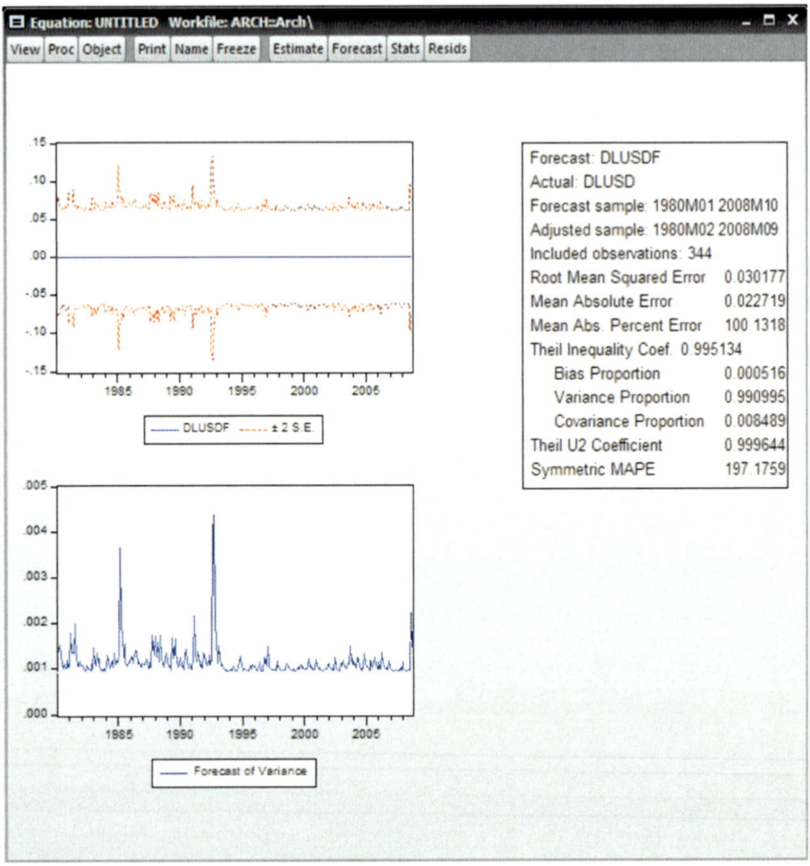

Fig. 8.15 Forecasts from the GARCH(1,1) model

$$\widehat{RRBS}_t = \beta_0 + \beta_1 RFTSE_t + \varepsilon_t$$

The variance equation will include these items:

- Item one is the intercept.
- Item two is the residual square one time lag, which represents a 'shock' or a 'news' or an 'error'. We call it the ARCH term. If this term is statistically significant, it means a shock has influenced the variance of the residual, GARCH
- Item three is the variance of the residual one time lag. If this item is statistically significant, it means variance has not remained constant and homoscedasticity is present. We treat item two and three together.
- Note: Factor one, two and three are regarded as internal factors.
- Two more items, which we call external factors, are used. One is the return of Dow Jones Industrial average, DJ. The second is the return of Frankfurt Stock

Exchange, Dax. These two external factors will indicate whether the volatility is caused by the internal factors or by external ones, or by both.

To sum up:

1. We will consider whether or not in the variance equation, both the lagged squared error term and the lagged conditional variance term, are individually significant. If lagged conditional variance affects current conditional variance, there is clear evidence that there is a pronounced ARCH effect.
2. External factors will be treated individually, depending on whether they are statistically significant or not. If any of these two external factors is statistically significant, we can conclude that there is evidence that the variance (GARCH) has been affected by that factor and the volatility has partially been caused by that factor.

The variance equation is:

$$h_t \ \widehat{or} \ \sigma_t^2 = \beta_0 + \beta_1 \varepsilon_{t-1}^2 + \beta_2 \sigma_{t-1}^2 + \beta_3 RDJ + \beta_4 RDAX$$

Daily end prices of the Royal Bank of Scotland (RBS) shares, the FTSE 100 (FTSE) Index shares, the Dow Jones Industrial Average Index shares (DJ) and the Frankfurt Stock Exchange share composite (Dax) are collected. A sample of 1000 observations for each of these variables for a period from 19-Mar-2008 to 05-Mar-2012 is deployed. Data of the prices of the above variables are transferred to EViews and the returns are found as follows:

Step One: All the data (prices) are converted to the natural logs as follows:

LRBS=LOG (RBS)
LFTSE=LOG (FTSE)
LDJ=LOG (DJ)
LDAX=LOG (DAX)

Step Two: The returns of the above variables are found as follows:

RRBS=LRBS-LRBS (−1)
RFTSE=LFTSE-LFTSE (−1)
RDJ=LDJ-LDJ (−1)
RDAX=LDAX-LDAX (−1)

Note:
The data for running the above models must be stationary. This is a necessary condition in order to avoid the resulting regression from being spurious (nonsense) regression.

Time Series One: RRBS Informal Methods
According to Fig. 8.16, the series appears to be stationary.

(a) Correlogram of the series, RRBS:
Double click the RRBS form the Workfile to get the data of the series RRBS.

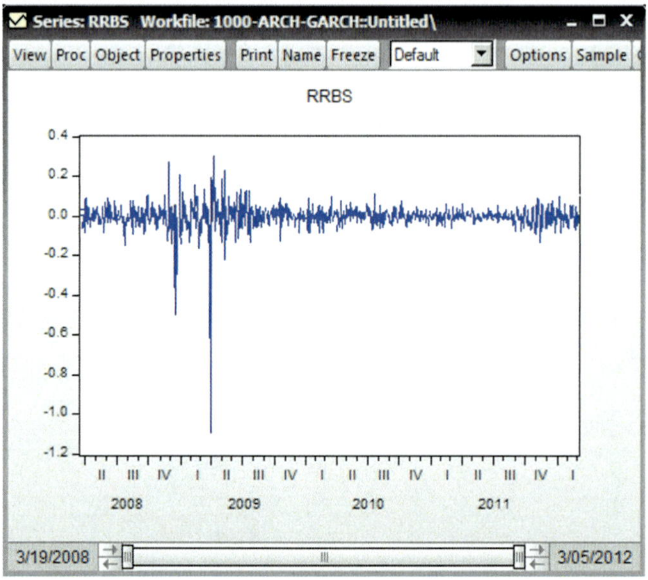

Fig. 8.16 Plot of RBS series

Go to View—Correlogram—Select a correlogram of the Level and then OK

1. H_0: The series has no unit root
2. H_1: The series has a unit root
3. $\alpha = 5\%$ or 0.05

There are no test statistics to calculate. We look at the Prob column in Fig. 8.17.

4. The p-values = 0.000.
5. 0.000 < 0.05. We reject the H_0; we accept the H_1.
6. The series is not stationary.

Formal Method:
Augmented Dickey-Fuller (ADF) Test for RRBS

Follow these steps in EViews:
Double click on the variable, RRBS, from the Workfile.
From the Window containing the data of RRBS, choose View-Unit Root test-
From Test type, choose Augmented Dickey-Fuller
From Test for unit root in, choose level
From Include in test equation, choose Trend and Intercept if the plot of the series indicates it is trended. Otherwise choose only the Intercept
And then click OK.

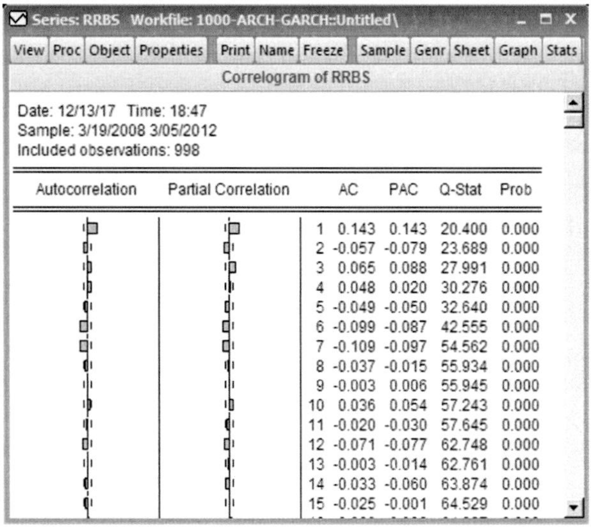

Series: RRBS Workfile: 1000-ARCH-GARCH::Untitled\

View | Proc | Object | Properties | Print | Name | Freeze | Sample | Genr | Sheet | Graph | Stats

Correlogram of RRBS

Date: 12/13/17 Time: 18:47
Sample: 3/19/2008 3/05/2012
Included observations: 998

Autocorrelation	Partial Correlation		AC	PAC	Q-Stat	Prob
		1	0.143	0.143	20.400	0.000
		2	-0.057	-0.079	23.689	0.000
		3	0.065	0.088	27.991	0.000
		4	0.048	0.020	30.276	0.000
		5	-0.049	-0.050	32.640	0.000
		6	-0.099	-0.087	42.555	0.000
		7	-0.109	-0.097	54.562	0.000
		8	-0.037	-0.015	55.934	0.000
		9	-0.003	0.006	55.945	0.000
		10	0.036	0.054	57.243	0.000
		11	-0.020	-0.030	57.645	0.000
		12	-0.071	-0.077	62.748	0.000
		13	-0.003	-0.014	62.761	0.000
		14	-0.033	-0.060	63.874	0.000
		15	-0.025	-0.001	64.529	0.000

Fig. 8.17 Correlogram of RRBS

1. H_0: series has a unit root and it is not stationary
2. H_1: series has no unit root and it is stationary.
3. $\alpha = 5\%$ or 0.05

According to Fig. 8.18, ADF test statistic is -16.66555.

4. The p-values = 0.0000.
5. Prob = 0.0000 < 0.05 therefore we reject the H_0
6. RRBS has no unit root and is stationary.

Time Series Two: RFTSE Informal Methods

According to Fig. 8.19, the RFTSE series is stationary.

(a) **Correlogram of the series RFTSE**

   1. H_0: RFTSE has no unit root
   2. H_1: RFTSE has a unit root
   3. $\alpha = 5\%$ or 0.05

      There are no test statistics to calculate. We look at the Prob column in Fig. 8.20.

   4. The p-values = 0.000.
   5. Prob = 0.000 < 0.05 therfore we reject the H_0
   6. RFTSE is not stationary.

Formal Method:
Augmented Dickey-Fuller (ADF) Test for RFTSE

1. H_0: RFTSE has a unit root and is not stationary
2. H_1: RFTSE has no unit root and is stationary.
3. $\alpha = 5\%$ or 0.05

Series: RRBS Workfile: 1000-ARCH-GARCH::Untitled\

View | Proc | Object | Properties | Print | Name | Freeze | Sample | Genr | Sheet | Graph | Stats

Augmented Dickey-Fuller Unit Root Test on RRBS

Null Hypothesis: RRBS has a unit root
Exogenous: Constant
Lag Length: 2 (Automatic - based on SIC, maxlag=21)

		t-Statistic	Prob.*
Augmented Dickey-Fuller test statistic		-16.66555	0.0000
Test critical values:	1% level	-3.436703	
	5% level	-2.864233	
	10% level	-2.568256	

*MacKinnon (1996) one-sided p-values.

Augmented Dickey-Fuller Test Equation
Dependent Variable: D(RRBS)
Method: Least Squares
Date: 12/13/17 Time: 18:48
Sample (adjusted): 3/26/2008 3/05/2012
Included observations: 995 after adjustments

Variable	Coefficient	Std. Error	t-Statistic	Prob.
RRBS(-1)	-0.843316	0.050602	-16.66555	0.0000
D(RRBS(-1))	0.004959	0.041348	0.119927	0.9046
D(RRBS(-2))	-0.088121	0.031632	-2.785794	0.0054
C	-0.002532	0.001915	-1.322239	0.1864

R-squared	0.436388	Mean dependent var	1.58E-05
Adjusted R-squared	0.434681	S.D. dependent var	0.080077
S.E. of regression	0.060208	Akaike info criterion	-2.778020
Sum squared resid	3.592345	Schwarz criterion	-2.758311
Log likelihood	1386.065	Hannan-Quinn criter.	-2.770528
F-statistic	255.7668	Durbin-Watson stat	2.003156
Prob(F-statistic)	0.000000		

Fig. 8.18 RRBS unit root test

According to Fig. 8.21, ADF test statistic is −24.57451.

4. The p-values = 0.0000.
5. Prob = 0.0000 < 0.05 therfore we reject the H_0
6. RFTSE has no unit root and is stationary.

Time Series Three: RDJ Informal Methods

According to Fig. 8.22, RDJ appears to be stationary, with very high volatility for some time.

(a) Correlogram of RDJ

1. H_0: RDJ has no unit root
2. H_1: RDJ has a unit root
3. $\alpha = 5\%$ or 0.05

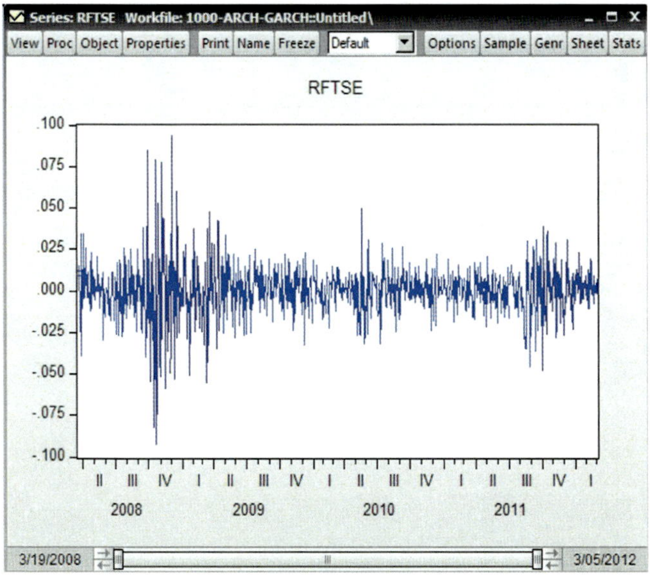

Fig. 8.19 Plot of RFTSE series

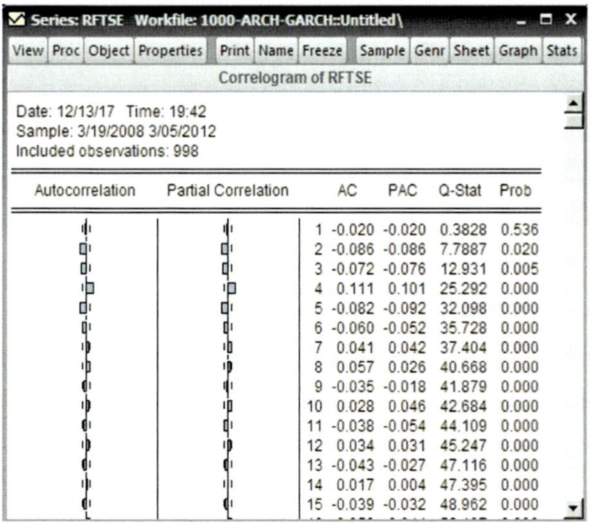

Fig. 8.20 Correlogram of RFTSE

Series: RFTSE Workfile: 1000-ARCH-GARCH::Untitled\

View | Proc | Object | Properties | Print | Name | Freeze | Sample | Genr | Sheet | Graph

Augmented Dickey-Fuller Unit Root Test on RFTSE

Null Hypothesis: RFTSE has a unit root
Exogenous: Constant, Linear Trend
Lag Length: 1 (Automatic - based on SIC, maxlag=2)

		t-Statistic	Prob.*
Augmented Dickey-Fuller test statistic		-24.57451	0.0000
Test critical values:	1% level	-3.967289	
	5% level	-3.414332	
	10% level	-3.129289	

*MacKinnon (1996) one-sided p-values.

Augmented Dickey-Fuller Test Equation
Dependent Variable: D(RFTSE)
Method: Least Squares
Date: 12/13/17 Time: 19:45
Sample (adjusted): 3/25/2008 3/05/2012
Included observations: 996 after adjustments

Variable	Coefficient	Std. Error	t-Statistic	Prob.
RFTSE(-1)	-1.109564	0.045151	-24.57451	0.0000
D(RFTSE(-1))	0.087028	0.031609	2.753262	0.0060
C	-0.000738	0.001022	-0.722492	0.4702
@TREND("3/19/2008")	1.54E-06	1.77E-06	0.868302	0.3854

R-squared	0.514334	Mean dependent var	-1.80E-05
Adjusted R-squared	0.512865	S.D. dependent var	0.023002
S.E. of regression	0.016054	Akaike info criterion	-5.421669
Sum squared resid	0.255679	Schwarz criterion	-5.401975
Log likelihood	2703.991	Hannan-Quinn criter.	-5.414183
F-statistic	350.1847	Durbin-Watson stat	2.011086
Prob(F-statistic)	0.000000		

Fig. 8.21 RFTSE unit root test

There are no test statistics to calculate. We look at the Prob column in Fig. 8.23.

4. The p-values = 0.000.
5. Prob = 0.000 < 0.05 therfore reject the H_0
6. RDJ is not stationary.

Formal method:
Augmented Dickey-Fuller (ADF) Test for RDJ

1. H_0: RDJ has a unit root and is not stationary
2. H_1: RDJ has no unit root and is stationary.
3. $\alpha = 5\%$ or 0.05

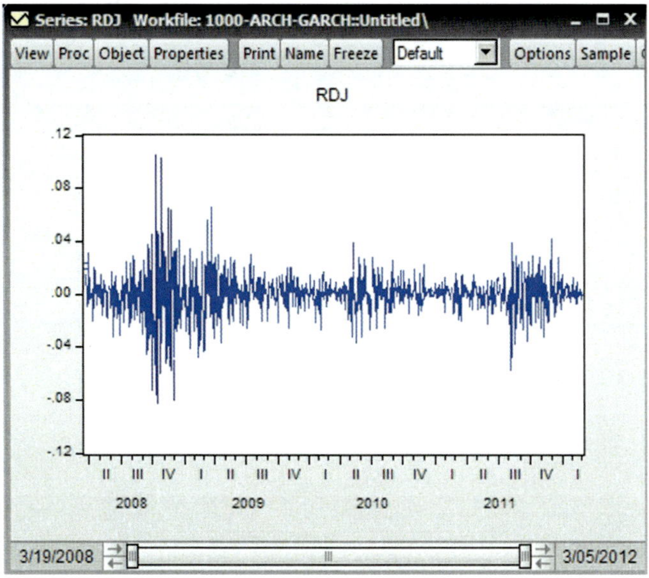

Fig. 8.22 Plot of RDJ series

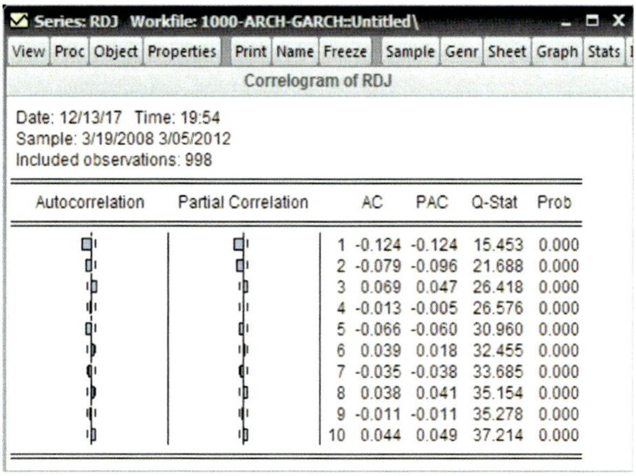

Series: RDJ Workfile: 1000-ARCH-GARCH::Untitled\

View Proc Object Properties Print Name Freeze Sample Genr Sheet Graph Stats

Correlogram of RDJ

Date: 12/13/17 Time: 19:54
Sample: 3/19/2008 3/05/2012
Included observations: 998

Autocorrelation	Partial Correlation		AC	PAC	Q-Stat	Prob
		1	-0.124	-0.124	15.453	0.000
		2	-0.079	-0.096	21.688	0.000
		3	0.069	0.047	26.418	0.000
		4	-0.013	-0.005	26.576	0.000
		5	-0.066	-0.060	30.960	0.000
		6	0.039	0.018	32.455	0.000
		7	-0.035	-0.038	33.685	0.000
		8	0.038	0.041	35.154	0.000
		9	-0.011	-0.011	35.278	0.000
		10	0.044	0.049	37.214	0.000

Fig. 8.23 Correlogram of RDJ

According to Fig. 8.24, ADF test statistic is −26.07087.

4. The p-values = 0.0000.
5. Prob = 0.0000 < 0.05 therefore reject the H_0
6. RDJ has no unit root and is stationary.

Series: RDJ Workfile: 1000-ARCH-GARCH::Untitled\

View | Proc | Object | Properties | Print | Name | Freeze | Sample | Genr | Sheet | Graph | Stats

Augmented Dickey-Fuller Unit Root Test on RDJ

Null Hypothesis: RDJ has a unit root
Exogenous: Constant
Lag Length: 1 (Automatic - based on SIC, maxlag=2)

		t-Statistic	Prob.*
Augmented Dickey-Fuller test statistic		-26.07087	0.0000
Test critical values:	1% level	-3.436696	
	5% level	-2.864230	
	10% level	-2.568255	

*MacKinnon (1996) one-sided p-values.

Augmented Dickey-Fuller Test Equation
Dependent Variable: D(RDJ)
Method: Least Squares
Date: 12/13/17 Time: 19:56
Sample (adjusted): 3/25/2008 3/05/2012
Included observations: 996 after adjustments

Variable	Coefficient	Std. Error	t-Statistic	Prob.
RDJ(-1)	-1.233619	0.047318	-26.07087	0.0000
D(RDJ(-1))	0.095987	0.031541	3.043275	0.0024
C	4.61E-05	0.000508	0.090797	0.9277

R-squared	0.567094	Mean dependent var	-1.62E-05
Adjusted R-squared	0.566222	S.D. dependent var	0.024327
S.E. of regression	0.016022	Akaike info criterion	-5.426709
Sum squared resid	0.254905	Schwarz criterion	-5.411939
Log likelihood	2705.501	Hannan-Quinn criter.	-5.421094
F-statistic	650.4010	Durbin-Watson stat	1.991301
Prob(F-statistic)	0.000000		

Fig. 8.24 RDJ unit root test

Time Series Three: RDAX Informal Methods

The series appears to be stationary, with very high volatility for some time (Fig. 8.25).

(b) **Correlogram of the series RDAX**

1. H_0: RDAX has no unit root
2. H_1: RDAX has a unit root
3. $\alpha = 5\%$ or 0.05

   There are no test statistics to calculate. We look at the Prob column in Fig. 8.26.
4. Some p-values are bigger than 0.05 while others are smaller than 0.05.
5. Majority of p-values are <0.05 therefore we reject the H_0
6. We conclude that RDAX is not stationary.

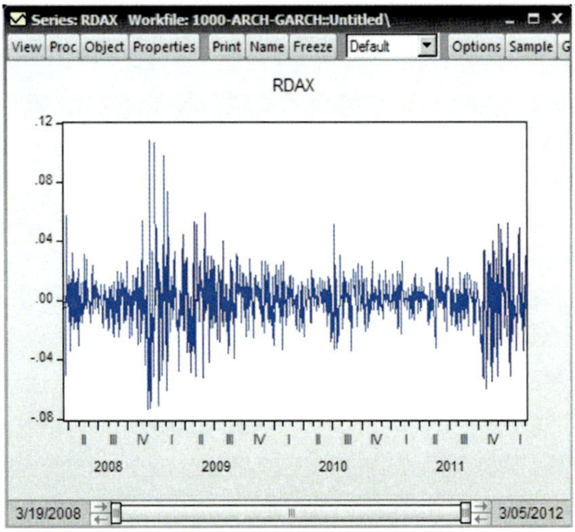

Fig. 8.25 Plot of RDAX series

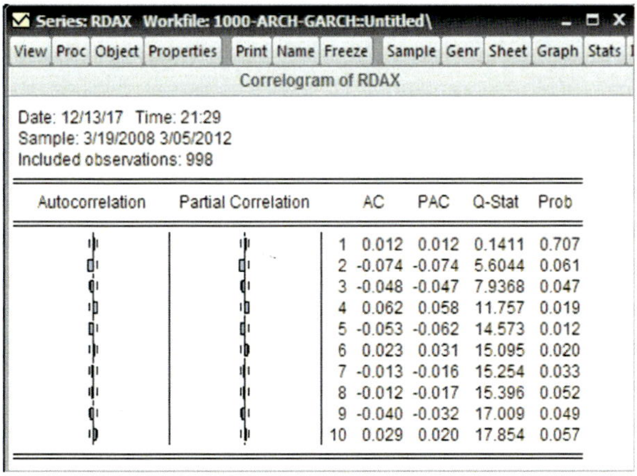

Series: RDAX Workfile: 1000-ARCH-GARCH::Untitled\

Correlogram of RDAX

Date: 12/13/17 Time: 21:29
Sample: 3/19/2008 3/05/2012
Included observations: 998

	AC	PAC	Q-Stat	Prob
1	0.012	0.012	0.1411	0.707
2	-0.074	-0.074	5.6044	0.061
3	-0.048	-0.047	7.9368	0.047
4	0.062	0.058	11.757	0.019
5	-0.053	-0.062	14.573	0.012
6	0.023	0.031	15.095	0.020
7	-0.013	-0.016	15.254	0.033
8	-0.012	-0.017	15.396	0.052
9	-0.040	-0.032	17.009	0.049
10	0.029	0.020	17.854	0.057

Fig. 8.26 Correlogram of RDAX

Formal method:
Augmented Dickey-Fuller (ADF) Test for RDJ

1. H_0: RDJ has a unit root and is not stationary
2. H_1: RDJ has no unit root and is stationary.
3. $\alpha = 5\%$ or 0.05

   According to Fig. 8.27, the ADF test statistic is −31.11467.

4. The p-values = 0.0000.
5. Prob = 0.0000 < 0.05 therefore we reject the H_0
6. We conclude that RDJ has no unit root and is stationary.

To sum up: All series are found to be stationary.

Series: RDAX Workfile: 1000-ARCH-GARCH::Untitled\

View | Proc | Object | Properties | Print | Name | Freeze | Sample | Genr | Sheet | Graph | St

Augmented Dickey-Fuller Unit Root Test on RDAX

Null Hypothesis: RDAX has a unit root
Exogenous: Constant, Linear Trend
Lag Length: 0 (Automatic - based on SIC, maxlag=2)

		t-Statistic	Prob.*
Augmented Dickey-Fuller test statistic		-31.11467	0.0000
Test critical values:	1% level	-3.967280	
	5% level	-3.414327	
	10% level	-3.129286	

*MacKinnon (1996) one-sided p-values.

Augmented Dickey-Fuller Test Equation
Dependent Variable: D(RDAX)
Method: Least Squares
Date: 12/13/17 Time: 21:30
Sample (adjusted): 3/24/2008 3/05/2012
Included observations: 997 after adjustments

Variable	Coefficient	Std. Error	t-Statistic	Prob.
RDAX(-1)	-0.988201	0.031760	-31.11467	0.0000
C	-0.000518	0.001160	-0.446439	0.6554
@TREND("3/19/2008")	7.46E-07	2.01E-06	0.371098	0.7106

R-squared	0.493408	Mean dependent var	3.38E-05
Adjusted R-squared	0.492388	S.D. dependent var	0.025643
S.E. of regression	0.018270	Akaike info criterion	-5.164098
Sum squared resid	0.331793	Schwarz criterion	-5.149340
Log likelihood	2577.303	Hannan-Quinn criter.	-5.158488
F-statistic	484.0648	Durbin-Watson stat	1.988837
Prob(F-statistic)	0.000000		

Fig. 8.27 RDAX unit root test

8.6 Cointegration

Cointegration Test Steps:

1. Run regression using the original series as per Fig. 8.28
2. Save the residuals

> Go to Quick-Generate series (after running the regression)-write in the process window: U=resid-then OK. Now we have U in the Workfile.

Fig. 8.28 RRBS Regression output

Equation: UNTITLED Workfile: 1000-ARCH-GARCH::Untitled\

View | Proc | Object | Print | Name | Freeze | Estimate | Forecast | Stats | Resids

Dependent Variable: RRBS
Method: Least Squares
Date: 12/13/17 Time: 21:32
Sample (adjusted): 3/20/2008 3/05/2012
Included observations: 998 after adjustments

Variable	Coefficient	Std. Error	t-Statistic	Prob.
C	-0.003043	0.001935	-1.572473	0.1162
RFTSE	0.028016	0.120315	0.232858	0.8159

R-squared	0.000054	Mean dependent var	-0.003042
Adjusted R-squared	-0.000950	S.D. dependent var	0.061112
S.E. of regression	0.061141	Akaike info criterion	-2.749250
Sum squared resid	3.723328	Schwarz criterion	-2.739419
Log likelihood	1373.876	Hannan-Quinn criter.	-2.745513
F-statistic	0.054223	Durbin-Watson stat	1.713377
Prob(F-statistic)	0.815919		

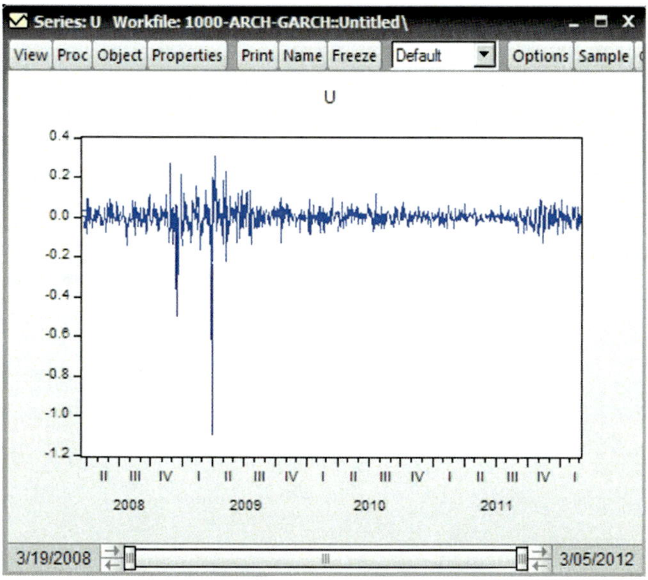

Fig. 8.29 RRBS Regression residuals

3. Plot the residual series

   The plot of the residual series shows that the series is stationary (Fig. 8.29).

4. Run correlogram for the residuals

Fig. 8.30 Correlogram of U series

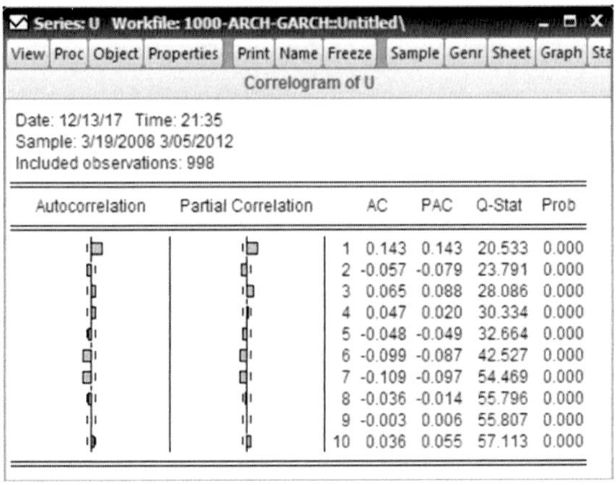
Series: U Workfile: 1000-ARCH-GARCH::Untitled\
View Proc Object Properties Print Name Freeze Sample Genr Sheet Graph Sta
Correlogram of U

Date: 12/13/17 Time: 21:35
Sample: 3/19/2008 3/05/2012
Included observations: 998

	AC	PAC	Q-Stat	Prob
1	0.143	0.143	20.533	0.000
2	-0.057	-0.079	23.791	0.000
3	0.065	0.088	28.086	0.000
4	0.047	0.020	30.334	0.000
5	-0.048	-0.049	32.664	0.000
6	-0.099	-0.087	42.527	0.000
7	-0.109	-0.097	54.469	0.000
8	-0.036	-0.014	55.796	0.000
9	-0.003	0.006	55.807	0.000
10	0.036	0.055	57.113	0.000

1. H_0: The series has no unit root
2. H_1: The series has a unit root
3. $\alpha = 5\%$ or 0.05

   There are no test statistics to calculate. We look at the Prob column in Fig. 8.30.
4. Prob $= 0.000 < 0.05$ therefore we reject the H_0
5. The series has a unit root and is not stationary

Formal method:

Augmented Dickey-Fuller (ADF) Test for U

Before running the test in EViews, we need to decide whether to include the trend or not. If the plot of the series is trended, include the trend. If the plot of the series is not trended, do not include the trend. Include only the intercept.

1. H_0: U has a unit root and is not stationary
2. H_1: U has no unit root and is stationary.
3. $\alpha = 5\%$ or 0.05

According to Fig. 8.31, ADF test statistic is -16.66196.

1. The p-values $= 0.0000$.
2. Prob $= 0.0000 < 0.05$ therefore we reject the H_0
3. U has no unit root and is stationary.

The residual of the main equation is stationary; it means these two series are cointegrated and there is a long-run relationship between them.

Using EViews for ARCH, GARCH and EGARCH

1. **Using Normal (Gaussian) as Error Distribution**

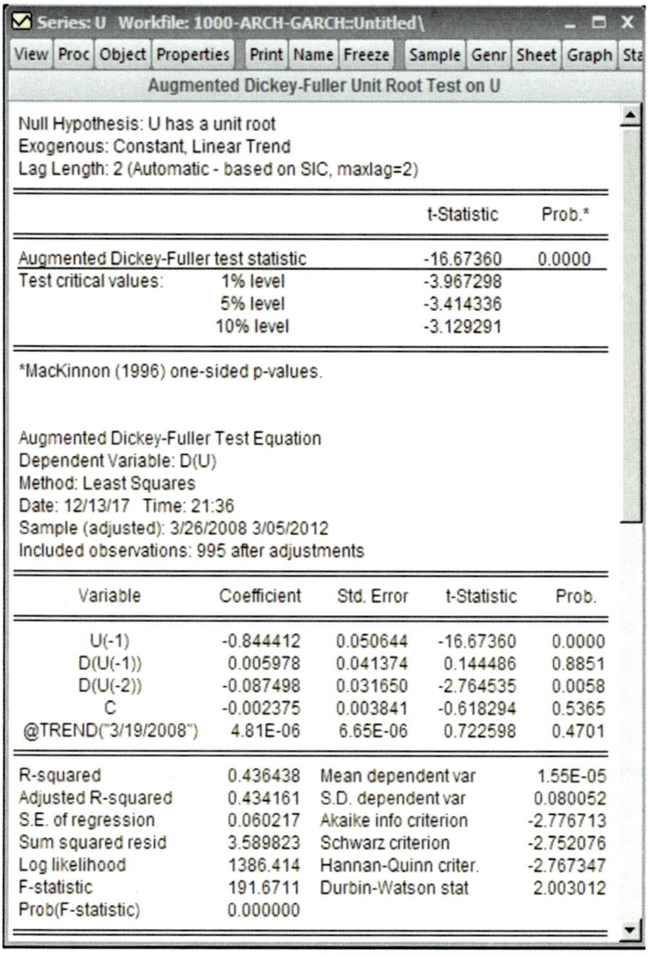
Series: U Workfile: 1000-ARCH-GARCH::Untitled\

View Proc Object Properties Print Name Freeze Sample Genr Sheet Graph Sta

Augmented Dickey-Fuller Unit Root Test on U

Null Hypothesis: U has a unit root
Exogenous: Constant, Linear Trend
Lag Length: 2 (Automatic - based on SIC, maxlag=2)

		t-Statistic	Prob.*
Augmented Dickey-Fuller test statistic		-16.67360	0.0000
Test critical values:	1% level	-3.967298	
	5% level	-3.414336	
	10% level	-3.129291	

*MacKinnon (1996) one-sided p-values.

Augmented Dickey-Fuller Test Equation
Dependent Variable: D(U)
Method: Least Squares
Date: 12/13/17 Time: 21:36
Sample (adjusted): 3/26/2008 3/05/2012
Included observations: 995 after adjustments

Variable	Coefficient	Std. Error	t-Statistic	Prob.
U(-1)	-0.844412	0.050644	-16.67360	0.0000
D(U(-1))	0.005978	0.041374	0.144486	0.8851
D(U(-2))	-0.087498	0.031650	-2.764535	0.0058
C	-0.002375	0.003841	-0.618294	0.5365
@TREND("3/19/2008")	4.81E-06	6.65E-06	0.722598	0.4701

R-squared	0.436438	Mean dependent var	1.55E-05
Adjusted R-squared	0.434161	S.D. dependent var	0.080052
S.E. of regression	0.060217	Akaike info criterion	-2.776713
Sum squared resid	3.589823	Schwarz criterion	-2.752076
Log likelihood	1386.414	Hannan-Quinn criter.	-2.767347
F-statistic	191.6711	Durbin-Watson stat	2.003012
Prob(F-statistic)	0.000000		

Fig. 8.31 U unit root test

Go to- **Quick/Estimate Equation**-write the name of variables of the main model only, as follows:

RRBS C RFTSE

Do not add the name of external factors for running this regression here.

From the method box, choose ARCH and then click **OK** as shown in Fig. 8.32.

How to deal with this Equation Estimation window?

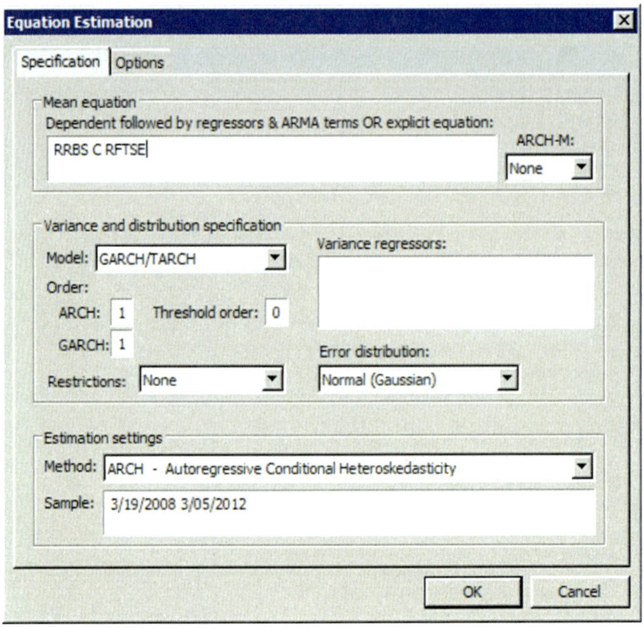

Fig. 8.32 Generating an ARCH model

Window of ARCH has two parts: specification and options.
We deal with the **Specification Part:**

1. The Mean Equation: variables of the main equation, including the intercept, will automatically appear in this box. So, you do not need to make any changes here.
2. Leave the ARCH-M box as it is: None.
3. The variance and distribution specification
   i. From the list of the model, choose GARCH/TARTCH
   ii. Choose 1 for ARCH and 1 for GARCH terms
   iii. Leave the Threshold order 0
   iv. For Restrictions, leave it as it is: None
   v. In the edit box labeled Variance regressors, write the name of external factors: RDJ and RDAX
   vi. From the Error Distribution box, choose Normal (Gaussian).
4. Leave Estimation Settings as it is.
5. Then click OK

Figure 8.33 shows the ARCH and GARCH output: the distribution for this approach is **Normal distribution**.
From The View button, choose Representation option; you will get this:

Equation: UNTITLED Workfile: 1000-ARCH-GARCH::Untitled\

View | Proc | Object | Print | Name | Freeze | Estimate | Forecast | Stats | Resids

Dependent Variable: RRBS
Method: ML ARCH - Normal distribution (BFGS / Marquardt steps)
Date: 12/13/17 Time: 21:45
Sample (adjusted): 3/20/2008 3/05/2012
Included observations: 998 after adjustments
Convergence achieved after 32 iterations
Coefficient covariance computed using outer product of gradients
Presample variance: backcast (parameter = 0.7)
GARCH = C(3) + C(4)*RESID(-1)^2 + C(5)*GARCH(-1) + C(6)*RDJ + C(7)
*RDAX

Variable	Coefficient	Std. Error	z-Statistic	Prob.
C	0.000443	0.000994	0.445519	0.6559
RFTSE	-0.107005	0.064834	-1.650434	0.0989
		Variance Equation		
C	6.60E-05	1.51E-05	4.364780	0.0000
RESID(-1)^2	0.307801	0.014601	21.08090	0.0000
GARCH(-1)	0.728979	0.017488	41.68408	0.0000
RDJ	0.004417	0.001751	2.522244	0.0117
RDAX	-0.000264	0.002042	-0.129155	0.8972

R-squared	-0.004454	Mean dependent var	-0.003042
Adjusted R-squared	-0.005463	S.D. dependent var	0.061112
S.E. of regression	0.061279	Akaike info criterion	-3.479669
Sum squared resid	3.740116	Schwarz criterion	-3.445260
Log likelihood	1743.355	Hannan-Quinn criter.	-3.466590
Durbin-Watson stat	1.712596		

Fig. 8.33 ARCH and GARCH output

Estimation Equation:

=============

RRBS = C(1) + C(2)*RFTSE
GARCH = C(3) + C(4)*RESID(−1)^2 + C(5)*GARCH(−1) + C(6)*RDJ + C(7)*RDAX
Substituted Coefficients:

==============

RRBS = 0.00044059057028 − 0.107028069526*RFTSE
GARCH = 6.59263919847e-05 + 0.307546277944*RESID(−1)^2 + 0.729125918193*GARCH(−1) + 0.00441565197904*RDJ − 0.00026570 $32#637195*RDAX

Analyzing the Result from the ARCH/GARCH output Using Normal distribution

The main equation

The coefficient of the RFTSE, which is C (2), has a p-value of 0.0988. This p-value is not smaller than 0.05. This coefficient is not statistically significant. This proves that this variable has no effect on the share price of RBS.

The variance equation GARCH

The variance of the residual (GARCH) is a function of two internal as well as two external factors. The two internal, or own family factors are: Resid(−1) squared, representing ARCH, past shocks or news, as well as GARCH (−1) the residual variance one time lag, or heteroscedasticity. The two external or exogenous factors are RDJ and RDAX.

1. The coefficient of the Resid(−1) squared, C(4) has a p-value of 0.0000, which is smaller than 0.05. Because of this, we can reject the H_0 that this coefficient is zero. We can conclude that the ARCH or past shock or news has affected the variance of the residual, GARCH.
2. The coefficient of the GARCH (−1), C (5) has a p-value of 0.0000, which is smaller than 0.05. Because of this, we can reject the H_0 that this coefficient is zero. The coefficient of the lagged variance, GARCH (−1), is statistically significant. We conclude that the heroskedasticity is present.
3. Both the lagged squared error term and the lagged conditional variance term are individually significant. As lagged conditional variance affects current conditional variance, there is clear evidence that there is a pronounced ARCH effect.
4. The second proof for this is that the sum of the coefficients of these two items are bigger than 1, 0.307546 + 0.729126 = 1.036672.
5. The coefficient of the RDJ, C (6) has a p-value of 0.0117, which is smaller than 0.05. Because of this, we can reject the H_0 that this coefficient is zero. We can conclude that this external factor, the Dow Jones Industrial Average, has affected the variance of the residual (GARCH) and the volatility can partially be explained by this factor.
6. The coefficient of the RDAX, C (7) has a p-value of 0.8964, which is not smaller than 0.05. Because of this, we cannot reject the H_0 that this coefficient is zero. We can conclude that this external factor, the Frankfurt Stock Exchange price Index has no effect on the variance of the residual (GARCH) and the volatility cannot be partially explained by this factor.

Residual Analysis of this Model

Check for Serial Correlation

From the ARCH output, go to View and then choose residual diagnostics. From the list, choose correlogram residuals squared.

1. H_0: There is no serial correlation
2. H_1: There is serial correlation
3. $\alpha = 5\%$ or 0.05

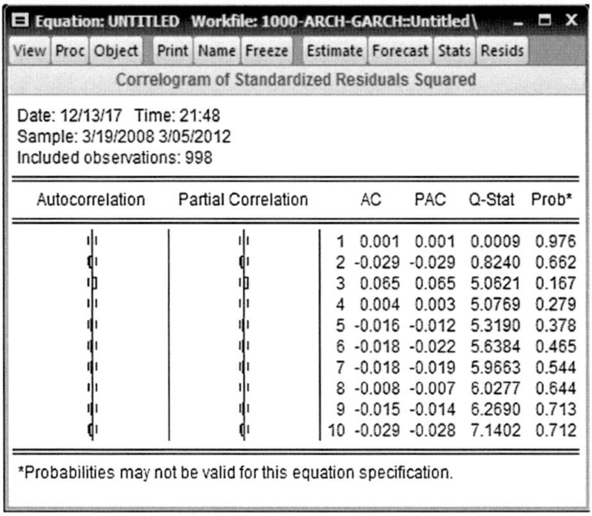

	AC	PAC	Q-Stat	Prob*
1	0.001	0.001	0.0009	0.976
2	-0.029	-0.029	0.8240	0.662
3	0.065	0.065	5.0621	0.167
4	0.004	0.003	5.0769	0.279
5	-0.016	-0.012	5.3190	0.378
6	-0.018	-0.022	5.6384	0.465
7	-0.018	-0.019	5.9663	0.544
8	-0.008	-0.007	6.0277	0.644
9	-0.015	-0.014	6.2690	0.713
10	-0.029	-0.028	7.1402	0.712

Fig. 8.34 Correlogram of the Standardized residuals squared

We look at the Prob column in Fig. 8.34:

4. The p-values are bigger than 0.05
5. Because p-values are bigger than 0.05, we cannot reject the H_0; we accept the H_0.
6. There is no problem of serial correlation.

Check for Hetroskedasticity

From the ARCH output, go to View and then choose residual diagnostics. From the list, choose ARCH LM.

1. H_0: There is no problem of heteroscedasticity
2. H_1: There is problem of heteroscedasticity
3. $\alpha = 5\%$ or 0.05

According to Fig. 8.35, Obs*R-squared = 5.041692

4. Prob. Chi-Square = 0.1688
5. 0.1688 > 0.05 we cannot reject the H_0
6. There is no problem of heteroscedasticity.

Check for Normality

From the ARCH output, go to View and then choose residual diagnostics. From the list, choose Residual Normality.

1. H_0: Residuals are normally distributed
2. H_1: Residuals are not normally distributed
3. $\alpha = 5\%$ or 0.05

Equation: UNTITLED Workfile: 1000-ARCH-GARCH::Untitled\

View | Proc | Object | Print | Name | Freeze | Estimate | Forecast | Stats | Resids

Heteroskedasticity Test: ARCH

F-statistic	1.682332	Prob. F(3,991)	0.1691
Obs*R-squared	5.041692	Prob. Chi-Square(3)	0.1688

Test Equation:
Dependent Variable: WGT_RESID^2
Method: Least Squares
Date: 12/14/17 Time: 08:37
Sample (adjusted): 3/26/2008 3/05/2012
Included observations: 995 after adjustments

Variable	Coefficient	Std. Error	t-Statistic	Prob.
C	0.961465	0.112518	8.544969	0.0000
WGT_RESID^2(-1)	0.003080	0.031698	0.097173	0.9226
WGT_RESID^2(-2)	-0.028815	0.031683	-0.909477	0.3633
WGT_RESID^2(-3)	0.065127	0.031696	2.054726	0.0402

R-squared	0.005067	Mean dependent var	1.000944
Adjusted R-squared	0.002055	S.D. dependent var	3.090904
S.E. of regression	3.087726	Akaike info criterion	5.096759
Sum squared resid	9448.247	Schwarz criterion	5.116469
Log likelihood	-2531.638	Hannan-Quinn criter.	5.104252
F-statistic	1.682332	Durbin-Watson stat	2.000391
Prob(F-statistic)	0.169127		

Fig. 8.35 Heteroskedasticity Test: ARCH

According to Fig. 8.36, Jarque-Bera test statistic = 2362.519

4. P-value = 0.0000
5. 0.0000 < 0.05 we can reject the H_0
6. The residuals are not normally distributed.

2. **Using Student's t with Fixed df as Error Distribution** (Fig. 8.37)

> Go to- **Quick/Estimate Equation**-write the name of variables of the main model only, as follows:
>
> RRBS C RFTSE
>
> Do not add the name of external factors for running this regression here.
>
> From the method box, choose ARCH and then click OK.

How to deal with this Equation Estimation window?
Window of ARCH has two parts: specification and options.
We deal with the **Specification Part:**

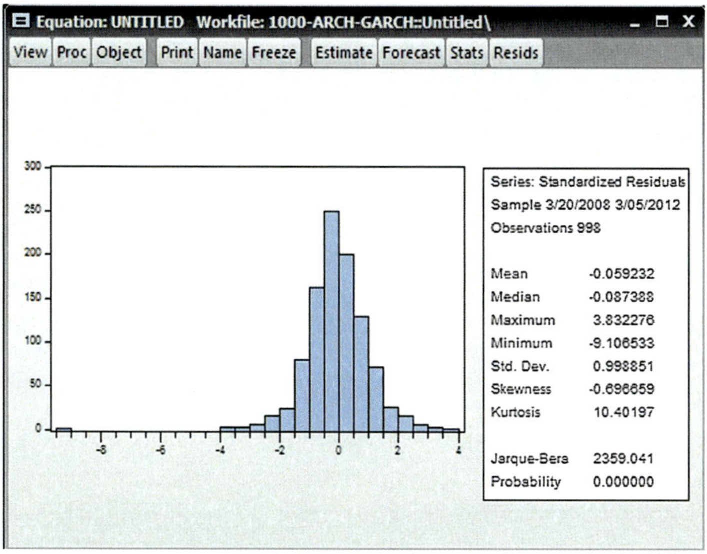

Fig. 8.36 Normality test: ARCH

Fig. 8.37 ARCH model estimation

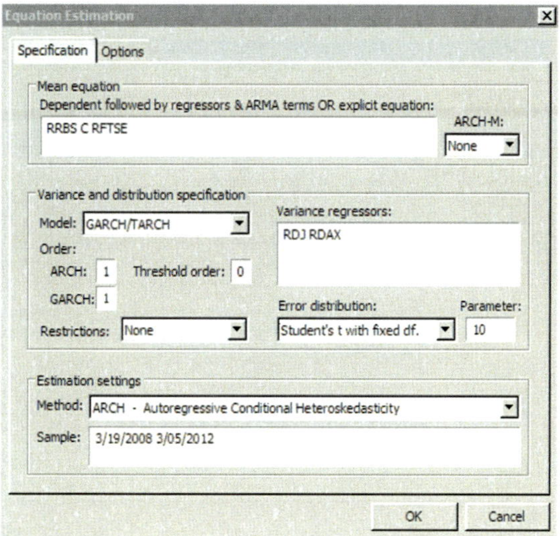

1. The Mean Equation: variables of the main equation, including the intercept, will automatically appear in this box. So, you do not need to make any changes here.
2. Leave the ARCH-M box as it is: None.
3. The variance and distribution specification:

i. From the list of the model, choose GARCH/TARTCH
ii. Choose 1 for ARCH and 1 for GARCH terms
iii. Leave the Threshold order 0
iv. For Restrictions, leave it as it is: None
v. In the edit box labeled Variance regressors, write the name of external factors: RDJ and RDAX
vi. From the Error Distribution box, choose Student's with fixed df.

4. Leave Estimation Settings as it is.
5. Then click OK.

Figure 8.38 shows the ARCH and GARCH output: the distribution for this approach is **Student's t distribution**.

Under View, choose Representation to generate the output below.

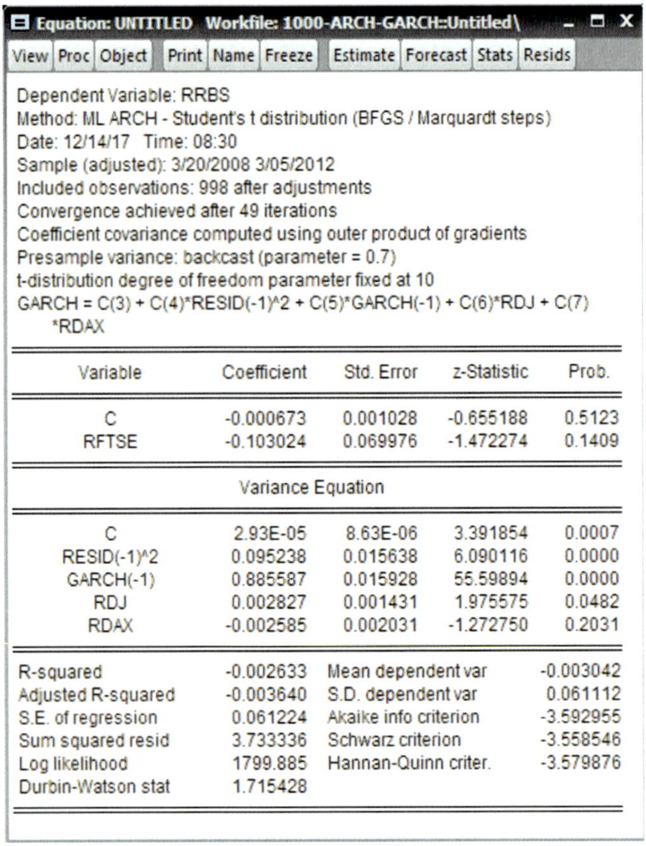
Equation: UNTITLED Workfile: 1000-ARCH-GARCH::Untitled\

View Proc Object Print Name Freeze Estimate Forecast Stats Resids

Dependent Variable: RRBS
Method: ML ARCH - Student's t distribution (BFGS / Marquardt steps)
Date: 12/14/17 Time: 08:30
Sample (adjusted): 3/20/2008 3/05/2012
Included observations: 998 after adjustments
Convergence achieved after 49 iterations
Coefficient covariance computed using outer product of gradients
Presample variance: backcast (parameter = 0.7)
t-distribution degree of freedom parameter fixed at 10
GARCH = C(3) + C(4)*RESID(-1)^2 + C(5)*GARCH(-1) + C(6)*RDJ + C(7)*RDAX

Variable	Coefficient	Std. Error	z-Statistic	Prob.
C	-0.000673	0.001028	-0.655188	0.5123
RFTSE	-0.103024	0.069976	-1.472274	0.1409
Variance Equation				
C	2.93E-05	8.63E-06	3.391854	0.0007
RESID(-1)^2	0.095238	0.015638	6.090116	0.0000
GARCH(-1)	0.885587	0.015928	55.59894	0.0000
RDJ	0.002827	0.001431	1.975575	0.0482
RDAX	-0.002585	0.002031	-1.272750	0.2031

R-squared	-0.002633	Mean dependent var	-0.003042
Adjusted R-squared	-0.003640	S.D. dependent var	0.061112
S.E. of regression	0.061224	Akaike info criterion	-3.592955
Sum squared resid	3.733336	Schwarz criterion	-3.558546
Log likelihood	1799.885	Hannan-Quinn criter.	-3.579876
Durbin-Watson stat	1.715428		

Fig. 8.38 ARCH and GARCH output

Estimation Equation:

==============

RRBS = C(1) + C(2)*RFTSE
GARCH = C(3) + C(4)*RESID(−1)^2 + C(5)*GARCH(−1) + C(6)*RDJ + C(7)*RDAX

Substituted Coefficients:

=============

RRBS = −0.000672926747248 − 0.103030746203*RFTSE
GARCH = 2.92664285293e-05 + 0.0952620258503*RESID(−1)^2 + 0.885561667313*GARCH(−1) + 0.00282713540958*RDJ − 0.0025837 $32#9938799*RDAX

Analysing the Result from the ARCH/GARCH output using Student's t distribution

The main equation

The coefficient of the RFTSE, which is C (2), has a p-value of 0.1409. This p-value is not smaller than 0.05. This coefficient is not statistically significant. This proves that this variable has no effect on the share price of RBS.

The variance equation GARCH

The variance of the residuals (GARCH) is a function of two internal as well as two external factors. The two internal or own family factors are: Resid(−1) squared, representing ARCH, past shocks or news, as well as GARCH (−1), the residuals variance one time lag, or hetroskedasticity. The two external or exogenous factors are RDJ and RDAX.

1. The coefficient of the Resid(−1) squared, C(4) has a p-value of 0.0000, which is smaller than 0.05. Because of this, we can reject the H_0 that this coefficient is zero. We can conclude that the ARCH, past shock or news has affected the variance of the residual, GARCH.
2. The coefficient of the GARCH (−1), C (5) has a p-value of 0.0000, which is smaller than 0.05. Because of this, we can reject the H_0 that this coefficient is zero. The coefficient of the lagged variance, GARCH (−1), is statistically significant. We conclude that the heroskedasticity is present.
3. Both the lagged squared error term and the lagged conditional variance term are individually significant. As lagged conditional variance affects current conditional variance, there is clear evidence that there is a pronounced ARCH effect.
4. The second proof for this is that the sum of the coefficients of these two items is very close to 1, 0.095262 + 0.885562 = 0.980824
5. The coefficient of the RDJ, C (6) has a p-value of 0.0482, which is smaller than 0.05. Because of this, we can reject the H_0 that this coefficient is zero. We can conclude that this external factor, the Dow Jones Industrial Average, has affected the variance of the residual (GARCH) and the volatility can partially be explained by this factor.

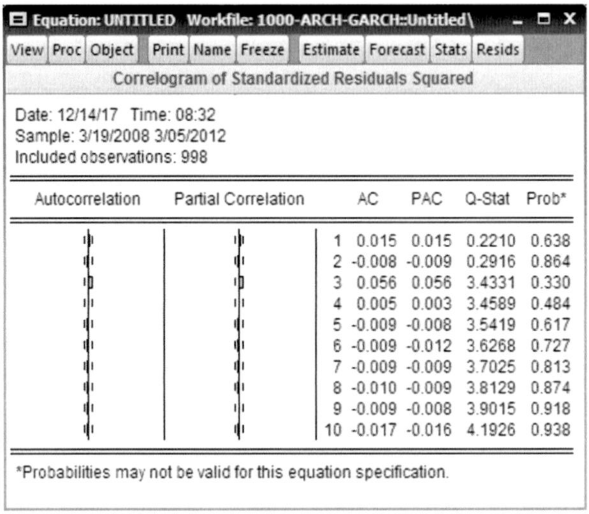

Equation: UNTITLED Workfile: 1000-ARCH-GARCH::Untitled\

View | Proc | Object | Print | Name | Freeze | Estimate | Forecast | Stats | Resids

Correlogram of Standardized Residuals Squared

Date: 12/14/17 Time: 08:32
Sample: 3/19/2008 3/05/2012
Included observations: 998

Autocorrelation	Partial Correlation		AC	PAC	Q-Stat	Prob*
		1	0.015	0.015	0.2210	0.638
		2	-0.008	-0.009	0.2916	0.864
		3	0.056	0.056	3.4331	0.330
		4	0.005	0.003	3.4589	0.484
		5	-0.009	-0.008	3.5419	0.617
		6	-0.009	-0.012	3.6268	0.727
		7	-0.009	-0.009	3.7025	0.813
		8	-0.010	-0.009	3.8129	0.874
		9	-0.009	-0.008	3.9015	0.918
		10	-0.017	-0.016	4.1926	0.938

*Probabilities may not be valid for this equation specification.

Fig. 8.39 Correlogram of Standardized residuals sqaured

6. The coefficient of the RDAX, C (7) has a p-value of 0.2034, which is not smaller than 0.05. Because of this, we cannot reject the H_0 that this coefficient is zero. We can conclude that this external factor, the Frankfurt Stock Exchange price Index has no effect on the variance of the residuals (GARCH) and the volatility cannot partially be explained by this factor.

Residual Analysis of this Model

Check for Serial Correlation

From the ARCH output, go to View and then choose residual diagnostics. From the list, choose correlogram residuals squared (Fig. 8.39).

1. H_0: No serial correlation
2. H_1: There is serial correlation
3. $\alpha = 5\%$ or 0.05

We look at the Prob column in Fig. 8.40

4. The p-values are bigger than 0.05
5. Because p-values are bigger than 0.05 therefore we reject the H_0
6. There is no problem of serial correlation.

Check for Hetroskedasticity

From the ARCH output, go to View and then choose residual diagnostics. From the list, choose ARCH LM.

1. H_0: There is no problem of heteroscedasticity
2. H_1: There is problem of heteroscedasticity
3. $\alpha = 5\%$ or 0.05

According to Fig. 8.40, Obs*R-squared = 3.442766

Equation: UNTITLED Workfile: 1000-ARCH-GARCH::Untitled\

View | Proc | Object | Print | Name | Freeze | Estimate | Forecast | Stats | Resids

Heteroskedasticity Test: ARCH

F-statistic	1.146671	Prob. F(3,991)	0.3292
Obs*R-squared	3.441950	Prob. Chi-Square(3)	0.3284

Test Equation:
Dependent Variable: WGT_RESID^2
Method: Least Squares
Date: 12/14/17 Time: 08:35
Sample (adjusted): 3/26/2008 3/05/2012
Included observations: 995 after adjustments

Variable	Coefficient	Std. Error	t-Statistic	Prob.
C	1.125360	0.191137	5.887718	0.0000
WGT_RESID^2(-1)	0.015522	0.031716	0.489408	0.6247
WGT_RESID^2(-2)	-0.009498	0.031718	-0.299438	0.7647
WGT_RESID^2(-3)	0.056241	0.031716	1.773281	0.0765

R-squared	0.003459	Mean dependent var	1.200140
Adjusted R-squared	0.000442	S.D. dependent var	5.665329
S.E. of regression	5.664075	Akaike info criterion	6.310176
Sum squared resid	31793.01	Schwarz criterion	6.329886
Log likelihood	-3135.313	Hannan-Quinn criter.	6.317669
F-statistic	1.146671	Durbin-Watson stat	2.000385
Prob(F-statistic)	0.329186		

Fig. 8.40 Heteroscedasticity Test: ARCH

4. Prob. Chi-Square = 0.3283
5. Prob = 0.3283 > 0.05 therefore we reject the H_0
6. There is no problem of heteroscedasticity

Check for Normality

From the ARCH output, go to View and then choose residual diagnostics. From the list, choose Residual Normality.

1. Residuals are normally distributed
2. Residuals are not normally distributed
3. $\alpha = 5\%$ or 0.05

According to Fig. 8.41, Jarque-Bera test statistic = 17221.68

4. P-value = 0.0000
5. 0.0000 < 0.05; we can reject the H_0
6. The residuals is not distributed normally.

3. **Using Generalised Error (GED) as Error Distribution**

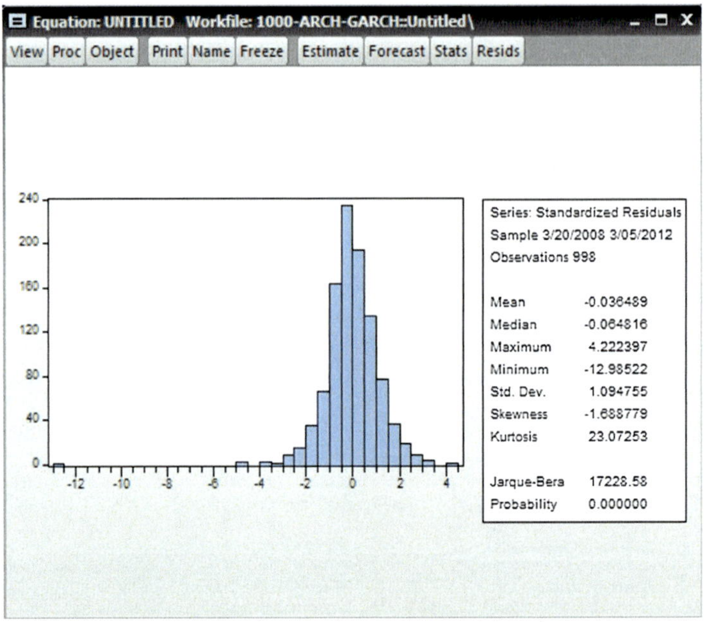

Fig. 8.41 Jarque-Bera test statistic

> Go to- **Quick/Estimate Equation**-write the name of variables of the main model only, as follows:
> RRBS C RFTSE
> Do not add the name of external factors for running this regression here.
> From the method box, choose ARCH and then click **OK** as shown in Fig. 8.42.

How to deal with this Equation Estimation window?

Window of ARCH has two parts: specification and options.

We deal with the **Specification Part:**

1. The Mean Equation: variables of the main equation, including the intercept, will automatically appear in this box. So, you do not need to make any changes here.
2. Leave the ARCH-M box as it is: None.
3. The variance and distribution specification:

    i. From the list of the model, choose GARCH/TARCH
    ii. Choose 1 for ARCH and 1 for GARCH terms
    iii. Leave the Threshold order 0
    iv. For Restrictions, leave it as it is: None

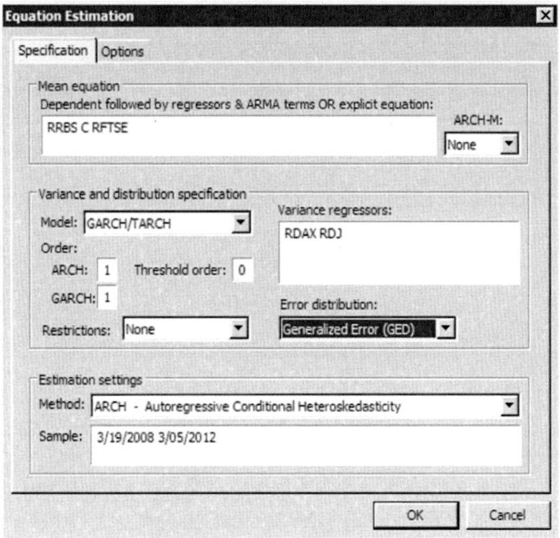

Fig. 8.42 ARCH equation modelling: Generalized Error (GED)

v. In the edit box labeled Variance regressors, write the name of external factors: RDJ and RDAX
vi. From the Error Distribution box, choose Generalize Error (GED).

4. Leave Estimation Settings as it is.
5. Then click OK.

This is the ARCH and GARCH output: the distribution for this approach is **Generalise Error (GED)** (Fig. 8.43).

From The View, choose Representation; you will get this:

Estimation Equation:

=========================

RRBS = C(1) + C(2)*RFTSE

GARCH = C(3) + C(4)*RESID(−1)^2 + C(5)*GARCH(−1) + C(6)*RDJ + C(7)*RDAX

Substituted Coefficients:

=========================

RRBS = −0.00180553895589 − 0.071258180958*RFTSE

GARCH = 3.93856073084e-05 + 0.160036844195*RESID(−1)^2 + 0.840161327051*GARCH(−1) + 0.00336141819334*RDJ − 0.00197366$32#147581*RDAX

Analysing the Result from the ARCH/GARCH output using Generalise Error (GED)

Equation: UNTITLED Workfile: 1000-ARCH-GARCH::Untitled\

View | Proc | Object | Print | Name | Freeze | Estimate | Forecast | Stats | Resids

Dependent Variable: RRBS
Method: ML ARCH - Generalized error distribution (GED) (BFGS / Marquardt steps)
Date: 12/14/17 Time: 08:45
Sample (adjusted): 3/20/2008 3/05/2012
Included observations: 998 after adjustments
Convergence achieved after 47 iterations
Coefficient covariance computed using outer product of gradients
Presample variance: backcast (parameter = 0.7)
GARCH = C(3) + C(4)*RESID(-1)^2 + C(5)*GARCH(-1) + C(6)*RDAX + C(7)*RDJ

Variable	Coefficient	Std. Error	z-Statistic	Prob.
C	-0.001806	0.000924	-1.954190	0.0507
RFTSE	-0.071216	0.062811	-1.133814	0.2569
Variance Equation				
C	3.94E-05	1.57E-05	2.509403	0.0121
RESID(-1)^2	0.160037	0.028660	5.584074	0.0000
GARCH(-1)	0.840161	0.026190	32.07891	0.0000
RDAX	-0.001974	0.003035	-0.650267	0.5155
RDJ	0.003361	0.002220	1.514223	0.1300
GED PARAMETER	1.198182	0.051335	23.34050	0.0000

R-squared	-0.001036	Mean dependent var	-0.003042
Adjusted R-squared	-0.002041	S.D. dependent var	0.061112
S.E. of regression	0.061175	Akaike info criterion	-3.577904
Sum squared resid	3.727388	Schwarz criterion	-3.538579
Log likelihood	1793.374	Hannan-Quinn criter.	-3.562956
Durbin-Watson stat	1.716105		

Fig. 8.43 ARCH and GARCH output

The mean equation

The coefficient of the RFTSE, which is C (2), has a p-value of 0.2566. This p-value is not smaller than 0.05. This coefficient is not statistically significant. This proves that this variable has no effect on the share price of RBS.

The variance equation GARCH

The variance of the residuals (GARCH) is a function of two internal as well as two external factors. The two internal or own family factors are: Resid(−1) squared, representing ARCH, past shocks or news, as well as GARCH (−1), the residual variance one time lag, or hetroskedasticity. The two external or exogenous factors are RDJ and RDAX.

1. The coefficient of the Resid(−1) squared, C(4) has a p-value of 0.0000, which is smaller than 0.05. Because of this, we can reject the H_0 that this coefficient is zero. We can conclude that the ARCH or past shock or news has affected the variance of the residual, GARCH.

2. The coefficient of the GARCH (−1), C (5) has a p-value of 0.0000, which is smaller than 0.05. Because of this, we can reject the H_0 that this coefficient is zero. The coefficient of the lagged variance, GARCH (−1), is statistically significant. We conclude that the heroskedasticity is present.
3. Both the lagged squared error term and the lagged conditional variance term are individually significant. As lagged conditional variance affects current conditional variance, there is clear evidence that there is a pronounced ARCH effect.
4. The second proof for this is that the sum of the coefficients of these two items is bigger than 1, 0.159862 + 0.840300 = 1.000162.
5. The coefficient of the RDJ, C (6) has a p-value of 0.1299, which is bigger than 0.05. Because of this, we cannot reject the H_0 that this coefficient is zero. We can conclude that this external factor, the Dow Jones Industrial Average, has not affected the variance of the residual (GARCH) and the volatility cannot be explained by this factor.
6. The coefficient of the RDAX, C (7) has a p-value of 0.5140, which is not smaller than 0.05. Because of this, we cannot reject the H_0 that this coefficient is zero. We can conclude that this external factor, the Frankfurt Stock Exchange price Index, has no effect on the variance of the residual (GARCH) and the volatility cannot partially be explained by this factor.

Residual Analysis of this Model

Check for Serial Correlation

From the ARCH output, go to View and then choose residual diagnostics. From the list, choose correlogram residuals squared.

1. H_0: No serial correlation
2. H_1: There is serial correlation
3. $\alpha = 5\%$ or 0.05

We look at the Prob column in Fig. 8.44

4. The p-values are bigger than 0.05
5. Because p-values are bigger than 0.05, we cannot reject the H_0; we accept the H_0.
6. There is no problem of serial correlation.

Check for Hetroskedasticity

From the ARCH output, go to View and then choose residual diagnostics. From the list, choose ARCH LM.

1. H_0: There is no problem of heteroscedasticity
2. H_1: There is problem of heteroscedasticity
3. $\alpha = 5\%$ or 0.05

According to Fig. 8.45, Obs*R-squared = 4.460577

4. Prob. Chi-Square = 0.2158
5. Prob = 0.2161 > 0.05 therefore we cannot reject the H_0
6. There is no problem of heteroscedasticity.

Equation: UNTITLED Workfile: 1000-ARCH-GARCH::Untitled\

View | Proc | Object | Print | Name | Freeze | Estimate | Forecast | Stats | Resids

Correlogram of Standardized Residuals Squared

Date: 12/14/17 Time: 08:52
Sample: 3/19/2008 3/05/2012
Included observations: 998

	AC	PAC	Q-Stat	Prob*
1	0.011	0.011	0.1256	0.723
2	-0.014	-0.014	0.3283	0.849
3	0.064	0.064	4.4448	0.217
4	0.003	0.002	4.4566	0.348
5	-0.012	-0.010	4.5999	0.467
6	-0.013	-0.017	4.7617	0.575
7	-0.013	-0.013	4.9306	0.668
8	-0.012	-0.010	5.0689	0.750
9	-0.012	-0.010	5.2206	0.815
10	-0.022	-0.020	5.7092	0.839

*Probabilities may not be valid for this equation specification.

Fig. 8.44 Correlogram of Standardized residuals

Equation: UNTITLED Workfile: 1000-ARCH-GARCH::Untitled\

View | Proc | Object | Print | Name | Freeze | Estimate | Forecast | Stats | Resids

Heteroskedasticity Test: ARCH

F-statistic	1.487550	Prob. F(3,991)	0.2164
Obs*R-squared	4.460577	Prob. Chi-Square(3)	0.2158

Test Equation:
Dependent Variable: WGT_RESID^2
Method: Least Squares
Date: 12/14/17 Time: 09:18
Sample (adjusted): 3/26/2008 3/05/2012
Included observations: 995 after adjustments

Variable	Coefficient	Std. Error	t-Statistic	Prob.
C	0.990384	0.146157	6.776165	0.0000
WGT_RESID^2(-1)	0.012377	0.031700	0.390440	0.6963
WGT_RESID^2(-2)	-0.015131	0.031698	-0.477352	0.6332
WGT_RESID^2(-3)	0.064416	0.031699	2.032072	0.0424

R-squared	0.004483	Mean dependent var	1.055529
Adjusted R-squared	0.001469	S.D. dependent var	4.237121
S.E. of regression	4.234007	Akaike info criterion	5.728187
Sum squared resid	17765.47	Schwarz criterion	5.747896
Log likelihood	-2845.773	Hannan-Quinn criter.	5.735680
F-statistic	1.487550	Durbin-Watson stat	2.000244
Prob(F-statistic)	0.216351		

Fig. 8.45 Heteroscedasticity Test: ARCH

Check for Normality

From the ARCH output, go to View and then choose residual diagnostics. From the list, choose Residual Normality.

1. Residuals are normally distributed
2. Residuals are not normally distributed
3. $\alpha = 5\%$ or 0.05

According to Fig. 8.46, Jarque-Bera test statistic = 8500.638

4. P-value = 0.0000
5. $0.0000 < 0.05$; we can reject the H_0
6. The residuals are not normally distributed.

The Leverage Effects

Suggested by Nelson (1991). The variance equation is given by:

$$\log(\sigma_t^2) = \omega + \beta \log(\sigma_{t-1}^2) + \gamma \frac{u_{t-1}}{\sqrt{\sigma_{t-1}^2}} + \alpha \left[\frac{|u_{t-1}|}{\sqrt{\sigma_{t-1}^2}} - \sqrt{\frac{2}{\pi}} \right]$$

Advantages of the model:

Since we model the log (σ_t^2), even if the parameters are negative, σ_t^2 will be positive.

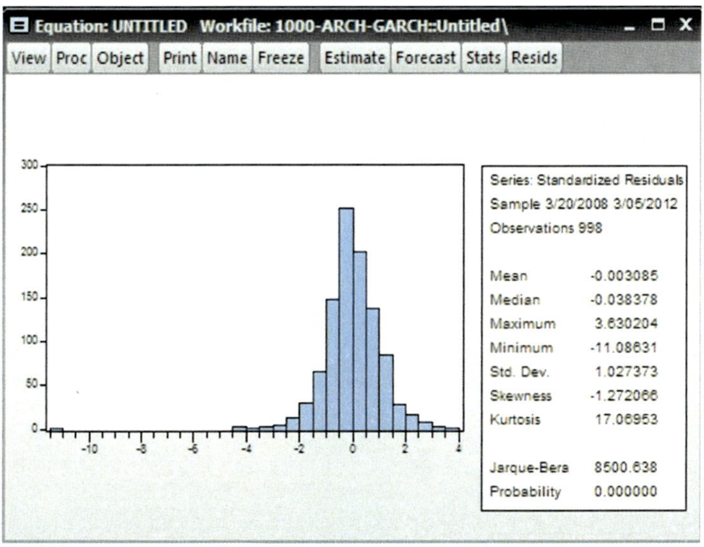

Fig. 8.46 Jarque-Bera test statistic

We can account for the leverage effect: if the relationship between volatility and returns is negative, γ, will be negative.

Using EViews

Go to- Quick/Estimate Equation-write the name of variables of the main model only, as follows:

RRBS C RFTSE

Do not add the name of external factors for running this regression here.

From the method box, choose ARCH and then OK as shown in Fig. 8.47

How to deal with this Equation Estimation window?

Window of ARCH has two parts: specification and options.

We deal with the **Specification Part:**

1. The Mean Equation: variables of the main equation, including the intercept, will automatically appear in this box. So, you do not need to make any changes here.
2. Leave the ARCH-M box as it is: None.
3. The variance and distribution specification:

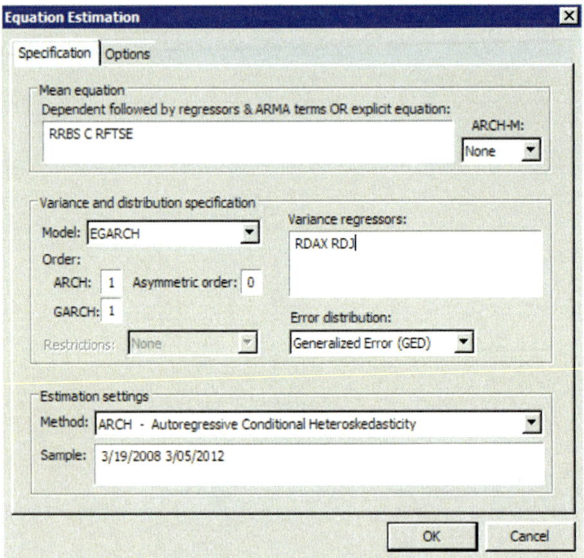

Fig. 8.47 EGARCH equation modelling

Equation: UNTITLED Workfile: 1000-ARCH-GARCH::Untitled\

View | Proc | Object | Print | Name | Freeze | Estimate | Forecast | Stats | Resids

Dependent Variable: RRBS
Method: ML ARCH - Generalized error distribution (GED) (BFGS / Marquardt steps)
Date: 03/11/18 Time: 11:55
Sample (adjusted): 3/20/2008 3/05/2012
Included observations: 998 after adjustments
Failure to improve likelihood (non-zero gradients) after 84 iterations
Coefficient covariance computed using outer product of gradients
Presample variance: backcast (parameter = 0.7)
LOG(GARCH) = C(3) + C(4)*ABS(RESID(-1)/@SQRT(GARCH(-1))) + C(5)*LOG(GARCH(-1)) + C(6)*RDAX + C(7)*RDJ

Variable	Coefficient	Std. Error	z-Statistic	Prob.
C	-0.002177	0.000942	-2.310268	0.0209
RFTSE	-0.078620	0.073720	-1.066469	0.2862
		Variance Equation		
C(3)	-0.012350	0.007440	-1.659941	0.0969
C(4)	-0.019110	0.009996	-1.911821	0.0559
C(5)	0.995959	1.7E-104	5.7E+103	0.0000
C(6)	-5.271494	0.696198	-7.571828	0.0000
C(7)	-1.260094	0.652933	-1.929898	0.0536
GED PARAMETER	1.267589	0.059360	21.35410	0.0000

R-squared	-0.000933	Mean dependent var	-0.003042
Adjusted R-squared	-0.001938	S.D. dependent var	0.061112
S.E. of regression	0.061172	Akaike info criterion	-3.603371
Sum squared resid	3.727004	Schwarz criterion	-3.564046
Log likelihood	1806.082	Hannan-Quinn criter.	-3.588423
Durbin-Watson stat	1.716736		

Fig. 8.48 EGARCH equation output

i. From the list of the model, choose EGARCH
ii. Choose 1 for ARCH and 1 for GARCH terms
iii. Leave the Threshold order 0
iv. For Restrictions, leave it as it is: None
v. In the edit box labeled Variance regressors, write the name of external factors: RDJ and RDAX
vi. From the Error Distribution box, choose Generalize Error (GED).

4. Leave Estimation Settings as it is.
5. Then click OK (Fig. 8.48).

Under View, choose Representation to get the output below:

Estimation Equation:

=========================

RRBS = C(1) + C(2)*RFTSE

LOG(GARCH) = C(3) + C(4)*ABS(RESID(−1)/@SQRT(GARCH(−1))) + C(5)*RESID(−1)/@SQRT(GARCH(−1)) + C(6)*LOG(GARCH(−1)) + C(7)*RDJ

Substituted Coefficients:

=========================

RRBS = −0.002177 − 0.07862*RFTSE

LOG(GARCH) = −0.01235 − 0.019110*ABS(RESID(−1)/@SQRT(GARCH(−1))) − 0.99595*RESID(−1)/@SQRT(GARCH(−1)) −5.271494*LOG(GARCH(−1)) − 1.260094*RDJ

The main focus is on the coefficient of *RESID (−1)/@SQRT (GARCH (−1))*, which is C (5).

This coefficient is positive and it is equal 0.995959. It is statistically significant, because the p-value of this coefficient is 0.0000 and it is smaller than 0.05. We can conclude that the leverage effect has been presented.

Residual Analysis of this Model

Check for Serial Correlation

From the ARCH output, go to View and then choose residual diagnostics. From the list, choose correlogram residuals squared.

1. H_0: No serial correlation
2. H_1: There is serial correlation
3. $\alpha = 5\%$ or 0.05

We look at the Prob column in Fig. 8.49

4. The p-values are bigger than 0.05
5. Because p-values are bigger than 0.05, we cannot reject the H_0; we accept the H_0.
6. There is no problem of serial correlation.

Check for Hetroskedasticity

From the ARCH output, go to View and then choose residual diagnostics. From the list, choose ARCH LM.

1. H_0: There is no problem of heteroscedasticity
2. H_1: There is problem of heteroscedasticity
3. $\alpha = 5\%$ or 0.05

According to Fig. 8.50, Obs*R-squared = 1.811635

4. Prob. Chi-Square = 0.6124
5. Prob = 0.6124 > 0.05; we cannot reject the H_0
6. There is no problem of heteroscedasticity.

Equation: UNTITLED Workfile: 1000-ARCH-GARCH::Untitled\

View | Proc | Object | Print | Name | Freeze | Estimate | Forecast | Stats | Resids

Correlogram of Standardized Residuals Squared

Date: 03/11/18 Time: 11:52
Sample: 3/19/2008 3/05/2012
Included observations: 998

Autocorrelation	Partial Correlation		AC	PAC	Q-Stat	Prob*
		1	0.036	0.036	1.3110	0.252
		2	0.026	0.024	1.9767	0.372
		3	0.021	0.020	2.4386	0.486
		4	0.004	0.001	2.4510	0.653
		5	0.013	0.011	2.6089	0.760
		6	-0.011	-0.012	2.7216	0.843
		7	0.062	0.063	6.6539	0.466
		8	-0.026	-0.031	7.3481	0.500
		9	-0.017	-0.017	7.6320	0.572
		10	-0.028	-0.028	8.3999	0.590

*Probabilities may not be valid for this equation specification.

Fig. 8.49 Correlogram of standardized residuals

Equation: UNTITLED Workfile: 1000-ARCH-GARCH::Untitled\

View | Proc | Object | Print | Name | Freeze | Estimate | Forecast | Stats | Resids

Heteroskedasticity Test: ARCH

F-statistic	0.761671	Prob. F(3,991)	0.5157
Obs*R-squared	2.288958	Prob. Chi-Square(3)	0.5146

Test Equation:
Dependent Variable: WGT_RESID^2
Method: Least Squares
Date: 03/11/18 Time: 11:53
Sample (adjusted): 3/26/2008 3/05/2012
Included observations: 995 after adjustments

Variable	Coefficient	Std. Error	t-Statistic	Prob.
C	0.948845	0.108178	8.771145	0.0000
WGT_RESID^2(-1)	0.034938	0.031759	1.100104	0.2716
WGT_RESID^2(-2)	0.023709	0.031770	0.746284	0.4557
WGT_RESID^2(-3)	0.019711	0.031759	0.620638	0.5350

R-squared	0.002300	Mean dependent var	1.029557
Adjusted R-squared	-0.000720	S.D. dependent var	2.940088
S.E. of regression	2.941146	Akaike info criterion	4.999487
Sum squared resid	8572.484	Schwarz criterion	5.019197
Log likelihood	-2483.245	Hannan-Quinn criter.	5.006980
F-statistic	0.761671	Durbin-Watson stat	2.000101
Prob(F-statistic)	0.515683		

Fig. 8.50 Heteroscedasticity Test: ARCH

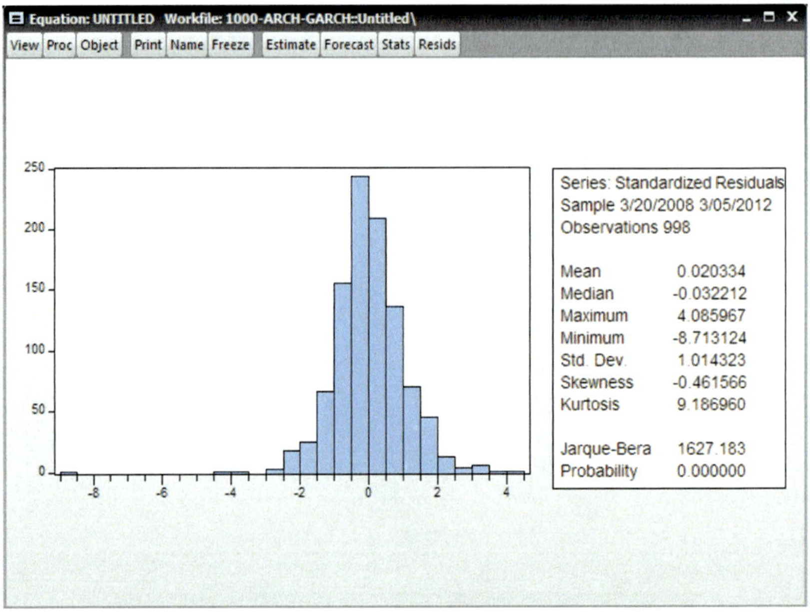

Fig. 8.51 Jarque-Bera test statistic

Check for Normality

From the ARCH output, go to View and then choose residual diagnostics. From the list, choose Residual Normality.

1. The residual is distributed normally
2. The residual is not distributed normally
3. $\alpha = 5\%$ or 0.05

According to Fig. 8.51, Jarque-Bera test statistic = 1627.183

4. P-value = 0.0000
5. Prob = 0.0000 < 0.05; we can reject the H_0
6. Residuals are not normally distributed

8.7 Concluding Remarks

There have been recent developments in ARCH and GARCH modelling. One drawback with the models presented here is that they are **symmetric models**. In symmetric models, only the size of a shock is considered, not its sign. That is to say a big negative shock has the same impact on future volatility as does a big positive shock of the same magnitude. An interesting development on modelling volatility in financial time series permits "good news" and "bad news" to have different impacts on future volatility. Models have been developed that allow for the possibility that an unexpected drop in price of a share, for example, ("bad news") has a larger impact on future volatility than an unexpected increase in price ("good news"). Such developments have taken place in the last 15 years and their analysis is possible in EViews.

The use of the ARCH and generalised ARCH (GARCH) to describe the time varying properties of univariate economic and financial time series was extended to multivariate scenarios by Bollerslev et al. (1988) in their study of returns on US Treasury Bills, gilts and stocks. The authors established a framework for the multivariate generalised autoregressive conditional heteroscedastic (MGARCH) class of models, which extended the concepts about the measurement of time-varying univariate conditional volatility (or conditional variance) into multivariate settings. The underlying reasoning for the development of MGARCH models was that if volatility changed in one market following a shock in another, then a univariate approach to analysis would be mis-specified. In essence, MGARCH models were designed to assist market participants in understanding volatility transmission (or spillover effects) over time and across various sectors of activity (Ewing 2002; Righia and Cerella 2012).

There have been numerous applications of MGARCH models to financial data. Several of the earlier applied studies involved analysis of dynamic hedging strategies (see for example, Baillie and Myers (1991); Kroner and Claessens (1991); Lien and Luo (1994)). More recently, Brooks et al. (2002) compared the effectiveness of hedging based on hedge ratios derived from different types of MGARCH models. The authors concluded that the latter models produced superior performances by way of lower portfolio volatilities than did competing methods such as rolling ordinary least squares hedges. Besides hedging, more recent applications of MGARCH models have involved asset pricing models, portfolio selection, value at risk estimates and volatility spillovers amongst a variety of different assets classes and markets. Reviews of such applications may be found in Bauwens et al. (2006) and Silvennoinen and Teräsvirta (2009). Software for the application of MGARCH models is available in Microfit 5.0 (Pesaran and Pesaran 2009). The authors illustrate the method via a study of returns on six daily currency futures. Application of the MGARCH method showed that the conditional variances of the futures returns moved closely together over time and was consequently evidence of close financial integration, especially in respect of those countries in the Euro area.

Several different formulations have been proposed over time for MGARCH models and a review of them may be found in Chang et al. (2012). Generally speaking, multivariate approaches to volatility analysis extend their univariate equivalents by means of a focus on *conditional (or time-varying) correlation* and the *conditional covariance* (in addition to the *conditional variance* of the univariate models). Indeed, conditional correlations are critical inputs for many of the common tasks of financial management, such as risk assessment, asset pricing models, hedging and portfolio selection (Ledoit et al. 2003).

Recall the formula for the unconditional (i.e. non-time varying) Pearsonian correlation coefficient, r, between two time series $x_1$and x_2:

$$r = \frac{n\sum x_1x_2 - \sum x_1 \sum x_2}{\sqrt{\left[n\sum x_1^2 - \left(\sum x_1\right)^2\right]}.\sqrt{\left[n\sum x_2^2 - \left(\sum x_2\right)^2\right]}} \tag{8.6}$$

The unconditional correlation is simply the correlation between the two series, using data from the first available reading to the last. Multiply the numerator and denominator of (6) each by $1/n^2$ and using the expectation operator, E, the unconditional correlation between the two series becomes:

$$r = \frac{E(x_1, x_2) - E(x_1)E(x_2)}{(SD\ of\ x_1).(SD\ of\ x_2)}, \tag{8.7}$$

where $E(x_1, x_2) = \frac{\sum x_1x_2}{n}$, $E(x_1) = \frac{\sum x_1}{n}$ and SD represents the unconditional standard deviation. The numerator of (8.7) is referred to as the *unconditional covariance* and can be written as $cov(x_1, x_2) = E(x_1, x_2) - E(x_1)E(x_2)$. When the unconditional covariance is zero, the unconditional correlation too is zero. The difference between the unconditional correlation/covariance/variance and the conditional correlation/covariance/variance is that the latter are estimated from information known up to the previous time period $(t - 1)$ and so are time dependent and time varying. The relevance of the time factor in conditional MGARCH models necessitates the use of a time (t) subscript. Therefore, the notation $r_t \mid \Omega_{t-1}$ can be used to represent an estimate of the conditional correlation between two data sets at time t and Ω_{t-1} represents the information set available up to time $(t - 1)$. Similarly, the notation $cov(x_{1t}, x_{2t}) \mid \Omega_{t-1}$ can represent the covariance between the two data sets at time t, conditional on the information set available up to $(t - 1)$.

Many methods have been suggested for estimating conditional correlations. Many procedures estimate the conditional covariances between each pair of time series as well as estimating the individual series' conditional variances and then apply equation (8.7) to derive estimates of the conditional correlations. Over the course of time, a popular approach to this estimation problem involved VECH model derived by Bollerslev et al. (1988). (VECH stands for the "vector half operator" in the language of matrix algebra). However, a problem with VECH models is the large number of parameters to be estimated (De Goeij and Marquering 2004). For example, given just three time series, 78 parameters have to be estimated for the conditional covariance and variance equations (Brooks et al. 2003) and this figure increases dramatically with the number of series and can become infeasible. Estimation of the VECH model is a "formidable task" even in the case of two time series (Brooks 2004: 508). Bollerslev et al. (1988) were able to reduce the number of parameters to be estimated via what they called the diagonal VECH model. This reduces the number of parameters in the above trivariate case to 18. There are two potential statistical problems associated with VECH models. Firstly, the matrix of covariances and variances is required to be what is referred to in matrix algebra as *positive definite*. If this condition is not met, then negative values for variances can result. Secondly, in the absence of this condition, the required equality $\text{cov}(x_{1t}, x_{2t}) = \text{cov}(x_{2t}, x_{1t})$ can break down.

The BEKK model was proposed by Engle and Kroner (1995) and ensures that the variance-covariance matrix is always positive definite. (BEKK models take their name from an earlier paper by Baba et al. (1989)). BEKK along with the DCC-GARCH (Multivariate GARCH Dynamic Conditional Correlation) model of Engle (2002)—which also has a positive definite variance-covariance matrix—are the two most widely applied models of conditional covariances and correlations (Caporin and McAleer 2012). The DCC-GARCH model has computational advantages over BEKK in that the number of parameters to be estimated is independent of the number of time series to be correlated and it is consequently applied to the CEE stock price data in this thesis. Furthermore, Engle (2002) claims (in the two time series scenario at least) that DCC-GARCH models are frequently more accurate than competing estimation methods and he concludes that empirical applications of them involving typical financial applications are "quite encouraging" in terms of their ability to reveal important time-varying features inherent in the data set.

Chapter 9
Limited Dependent Variable Models

The standard regression model assumes that the dependent variable, Y, is measured quantitatively. The independent (or regressor) variables, X_i, may be measured quantitatively or qualitatively. A dummy regressor is a variable that is measured qualitatively. Logit models apply to situations where the dependent variable is dichotomous in nature, taking a 0 or 1 value. For example, the dependent variable, Y, could be whether or not a person is unemployed ("employed" = 1, "unemployed" = 0). The regressors could include X_1 the average national wage rate, X_2 the individual's education, X_3 the national unemployment rate, X_4 family income etc. The question arises as to how we handle models involving dichotomous dependent variables.

The main question of this topic is: how we can model non numeric dependent variables? This chapter enables us to answer questions such as:

- Why firms choose to list their shares on the NASDAQ rather than the NYSE?
- Why some stocks pay dividends while others do not?
- What factors affect whether countries default on their sovereign debt?
- Why some firms choose to issue new stock to finance an expansion while others issue bonds?
- Why some firms choose to engage in stock splits while others do not?

In all these cases the correct form for the dependent variable would be a 0–1 dummy variable as far as there are only two possible outcomes. There are, of course, also situations where it would be more useful to allow the dependent variable to take on other values. When a variable takes only two values we can also call this variable a binary variable.

A. Aljandali, M. Tatahi, *Economic and Financial Modelling with EViews*, Statistics and Econometrics for Finance, https://doi.org/10.1007/978-3-319-92985-9_9

9.1 The Linear Probability Model

A simple and obvious method for dealing with binary dependent variables, is known as the *linear probability model*. This model is based on an assumption that the probability of an event occurring, P_i, is linearly related to a set of explanatory variables.

It is clear that the actual probabilities cannot be observed. As a result of this we would estimate a model where the outcomes, y_i (the series of zeros and ones), would be the dependent variable. This is a linear regression model and would be estimated by OLS. The set of explanatory (dependent) variables could include either quantitative variables or dummies or both. The fitted values from this regression are the estimated probabilities for $y_i = 1$ for each observation i.

The slope estimates for the linear probability model can be interpreted as the change in the probability that the dependent variable will equal 1 for a one-unit change in a given explanatory variable, holding the effect of all other explanatory variables fixed.

To fix ideas, consider the following simple model:

$$\widehat{Y} = \beta_1 + \beta_2 X$$

where X is family income (£ 000's) and Y is dichotomous, such that Y = 1 if the family owns a house and Y = 0 if the family does not own a house. Models such as the above which express the dichotomous Y as a linear function of the regressor variable(s) X are called *linear probability models*. However, there are problems with the assumptions that underpin regression when applying ordinary least squares to linear probability models.

(A) The residuals are not normally distributed. To see this:

$$\text{Residual} = \text{Y} - \widehat{Y} = \text{Y} - \beta_1 - \beta_2 X$$
$$\text{When Y} = 1, \quad \text{Residual} = 1 - \beta_1 - \beta_2 X$$
$$\text{When Y} = 0, \quad \text{Residual} = -\beta_1 - \beta_2 X.$$

Consequently, the residuals cannot follow the normal distribution. (In fact, they are binomially distributed).

(B) It can no longer be maintained that the residuals are homoscedastic. It can be shown that the variance of the residuals depends on the value taken by X and is thus not homoscedastic.

(C) Consider the data in Fig. 9.1. The variable Y is defined as above. If regression is applied to the linear probability model, we would obtain the result that:

$$\widehat{Y} = -0.9457 + 0.1021(INCOME).$$

If a family had an income of £8000 (i.e. X = 8), then there would be a negative probability of home ownership. Indeed, it is possible to have an income that coincides with a probability of home ownership in excess of Consequently,

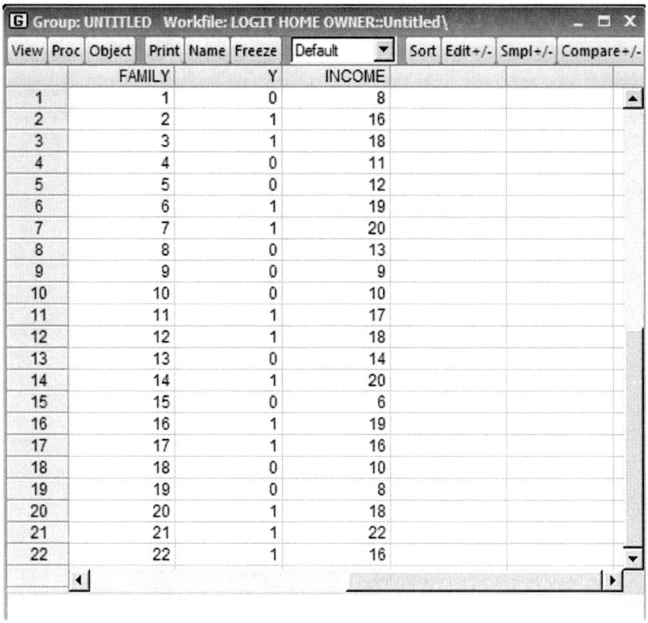

	FAMILY	Y	INCOME
1	1	0	8
2	2	1	16
3	3	1	18
4	4	0	11
5	5	0	12
6	6	1	19
7	7	1	20
8	8	0	13
9	9	0	9
10	10	0	10
11	11	1	17
12	12	1	18
13	13	0	14
14	14	1	20
15	15	0	6
16	16	1	19
17	17	1	16
18	18	0	10
19	19	0	8
20	20	1	18
21	21	1	22
22	22	1	16

Fig. 9.1 Home ownership and income (£000's)

the linear probability model is not recommended when the dependent variable is dichotomous.

(D) The value of the coefficient of determination as a measure of goodness of fit becomes questionable. Corresponding to a given value of X, Y is either 0 or 1. Therefore, all values of Y will either lie along the X-axis or along the line corresponding to Y = 1 (see Fig. 9.2). Consequently, no linear probability model is expected to fit such a scatter well. The coefficient of determination is likely to be much lower than 100% for such models (even if the model is constrained to lie between Y = 0 and Y = 1).

There are ways to overcome some of the problems associated with the linear probability model. However, there remains a fundamental problem that is not very attractive because the model assumes that Y (or probability) increases linearly with X. This implies that the impact of X remains constant throughout. Thus, in the home ownership example, we find that as X increases by a unit (£1000), the probability of home ownership increases by 0.1021. This is the case whether income is £8000, £10,000 or £80000. This seems patently unrealistic. At a very low income, a family will not own a house. At a sufficiently high income say X^*, people will be most likely to own a house. Beyond X^*, income will have little effect on the probability of owning a home. Thus at both ends of the income distribution, the probability of owning a home will be virtually unaffected by a small increase in X. The probability of owning a home is nonlinearly related to income.

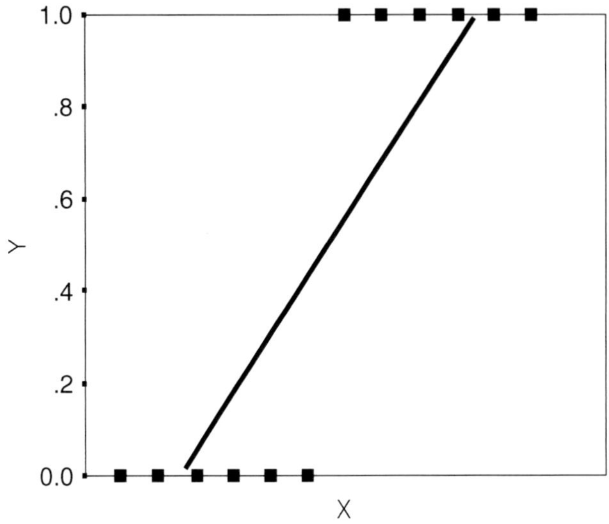

Fig. 9.2 Regression line when Y is dichotomous

Disadvantages of the Linear Probability Model

The linear probability model is simple to estimate and it is simple to interpret. The process will result in too many observations for which the estimated probabilities are exactly zero or one. More importantly, it is simply not plausible to suggest that the worker's probability of being a member of the union is either exactly zero or exactly one. Are we really certain that a worker will definitely never be a member of the union or he or she is a member of the union always? Probably it is not true.

Another reason is that the dependent variable only takes one or two values, for given (fixed in repeated samples) values of the explanatory variables, the disturbance term will also only take on one of two values. As a result of this the error term cannot credibly be assumed to be normally distributed. Moreover, the disturbance term changes systematically with the explanatory variables. This leads the disturbance to be heteroscedastic. It is therefore essential that heteroscedasticity-robust standard errors are always used in the context of limited dependent variable models.

So a different kind of model is usually used for binary dependent variables either a *logit* or a *probit* specification.

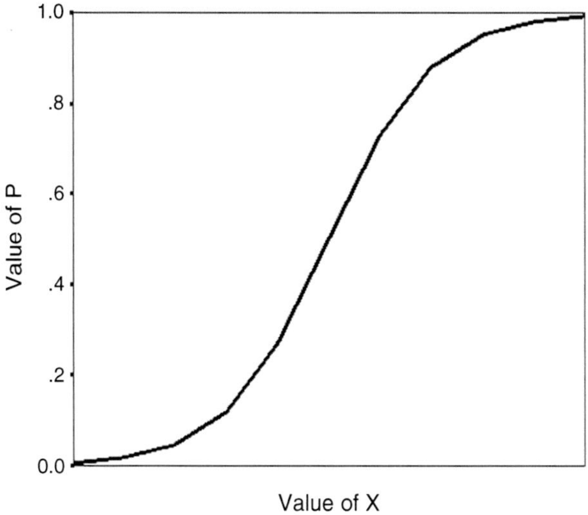

Fig. 9.3 A plot of the logistic distribution function

9.2 The Logit Model

The logit and probit model approaches can overcome the limitation of the LPM that it can produce estimated probabilities that are negative or greater than one. These two model approaches use a function that effectively transforms the regression model so that the fitted values are bounded within the (0,1) interval. The fitted regression model will appear as an S-shape rather than a straight line, as was the case for the LPM.

Now consider the following representation for home ownership, in which P represents the probability that a family owns a home i.e. P(Y = 1):

$$P = \frac{1}{1 + \exp - (\beta_1 + \beta_2 X)} \tag{9.1}$$

in which exp-(X) = e^X. Equation (9.1) is called *the logistic distribution function*. As shown in Fig. 9.3, equation (9.1) permits P to range only between 0 and 1, thus solving one of the problems associated with the linear probability model. If P is the probability of owning a home, then (1 − P) is the probability of not owning a home and:

$$\begin{aligned}1 - P &= 1 - \frac{1}{1 + \exp - (\beta_1 + \beta_2 X)} = \frac{1 + \exp - (\beta_1 + \beta_2 X) - 1}{1 + \exp - (\beta_1 + \beta_2 X)} \\ &= \frac{\exp - (\beta_1 + \beta_2 X)}{1 + \exp - (\beta_1 + \beta_2 X)} = \frac{1/_{\exp(\beta_1 + \beta_2 X)}}{1 + 1/_{\exp(\beta_1 + \beta_2 X)}} = \frac{1/_{\exp(\beta_1 + \beta_2 X)}}{\exp(\beta_1 + \beta_2 X) + 1/_{\exp(\beta_1 + \beta_2 X)}} \\ &= \frac{1}{1 + \exp(\beta_1 + \beta_2 X)}.\end{aligned} \tag{9.2}$$

Therefore, using equations (9.1) and (9.2), we can write:

$$\frac{P}{1 - P} = \frac{1}{1 + \exp - (\beta_1 + \beta_2 X)} \cdot [1 + \exp(\beta_1 + \beta_2 X)]$$

$$\frac{P}{1 - P} = \frac{1}{[\exp(\beta_1 + \beta_2 X) + 1]/_{\exp(\beta_1 + \beta_2 X)}} \cdot [1 + \exp(\beta_1 + \beta_2 X)]$$

$$\frac{P}{1 - P} = \exp(\beta_1 + \beta_2 X)$$

and taking natural logarithms (i.e. base e):

$$\begin{aligned}\ln\left(\frac{P}{1 - P}\right) &= \ln\left[\exp(\beta_1 + \beta_2 X)\right] \\ \ln\left(\frac{P}{1 - p}\right) &= \beta_1 + \beta_2 X,\end{aligned} \tag{9.3}$$

because $\ln(e^X) = X\ln e = X$.

The left hand side of equation (9.3) is called the *logit* and the whole equation is called the *logit model*. The left hand side is the logarithm of the probability that a family owns a home against the probability that it does not. This is called the *logarithm of the odds ratio*. Naturally the logit model of equation (9.3) may be extended to the multivariate case:

$$\ln\left(\frac{P}{1 - p}\right) = \beta_0 + \beta_1 X_1 + \beta_2 X_2 + \beta_3 X_3 + \ldots\ldots$$

9.3 Applying the Logit Model

The logit model of equation (9.3), where X is income (£000's), was applied to the data in Fig. 9.1. (Computer packages use a method called "maximum likelihood" to generate the logit coefficients). The resultant model was:

$$\ln\left(\frac{\widehat{P}}{1-\widehat{P}}\right) = -1.6587 + 0.0792(\text{INCOME}) \tag{9.4}$$

The first family in Fig. 9.1 had an income of £8000 (X = 8). Inserting this value of X into equation (9.4):

$$\ln\left(\frac{\widehat{P}}{1-\widehat{P}}\right) = -1.0251, \text{ whereby} \frac{\widehat{P}}{1-\widehat{P}} = \mathrm{e}^{-1.0251} = 0.3588.$$

Hence, $\widehat{P} = 0.3588 - 0.3588\widehat{P}$
$1.3588\widehat{P} = 0.3588$
$\widehat{P} = 0.2641.$

The logit model estimates that there is a probability of 0.2641 that this family owns its home.

It is possible to compute the **change in probability** of owning a home associated with a one unit (£1000) increase in income for this family who currently earn £8000. The change in probability is given by:

$$\widehat{\beta}_2 \cdot \widehat{P}\left(1 - \widehat{P}\right) = (0.0792)(0.2641)(0.7359) = 0.0139.$$

If this family's income increases by £1000, there is an extra 1.39% chance that they will become a house owner. This extra probability is not constant, but varies with income level. The former was a disadvantage of the linear probability model.

9.4 The Logit Model in EViews

An early, classic application of the logit model was in examining the choice of fertiliser use by Philippine farmers. The dependent variable to be explained is **FERUSE**—a binary variable equal to one if fertiliser is ued and equal to zero otherwise. The explanatory variables are:

- **CREDIT**—the amount of credit (per hectare) held by the farmer,
- **DMARKET**—the distance of the farm to the nearest market,
- **HOURMEET**—no. of hours the farmer spent with an agricultural expert,
- **IRSTAT**—a dummy variable = 1 if irrigation is used, = 0 otherwise and
- **OWNER**—a dummy variable = 1 if the farmer owns the land, = 0 otherwise.

491 farms using the data file titled LOGIT FERTILISER. (There is an extra variable in this file called **QFER**, which records the amount of fertiliser used if the farmer indeed uses it). The data are imported into EViews via:

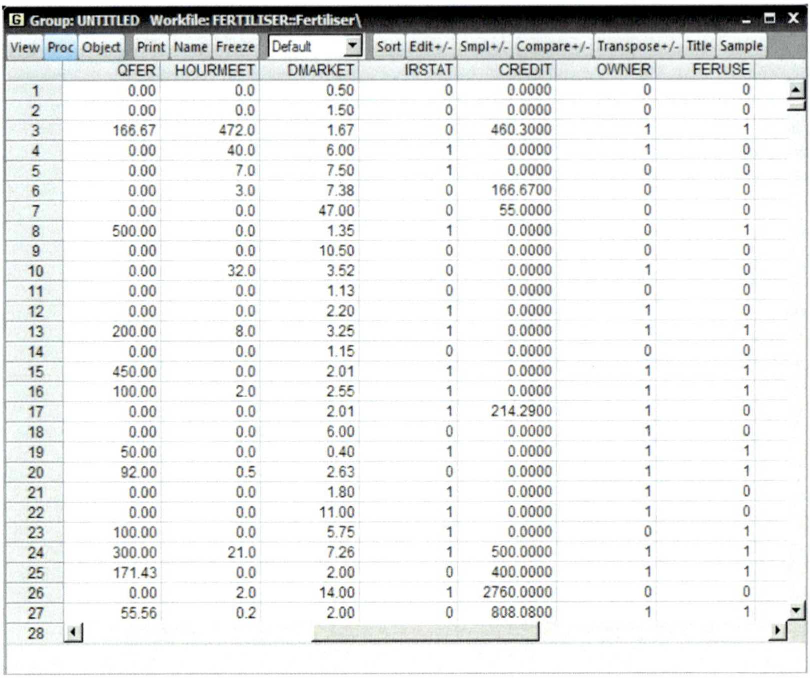

Group: UNTITLED Workfile: FERTILISER::Fertiliser\

View | Proc | Object | Print | Name | Freeze | Default | Sort | Edit+/- | Smpl+/- | Compare+/- | Transpose+/- | Title | Sample

	QFER	HOURMEET	DMARKET	IRSTAT	CREDIT	OWNER	FERUSE
1	0.00	0.0	0.50	0	0.0000	0	0
2	0.00	0.0	1.50	0	0.0000	0	0
3	166.67	472.0	1.67	0	460.3000	1	1
4	0.00	40.0	6.00	1	0.0000	1	0
5	0.00	7.0	7.50	1	0.0000	0	0
6	0.00	3.0	7.38	0	166.6700	0	0
7	0.00	0.0	47.00	0	55.0000	0	0
8	500.00	0.0	1.35	1	0.0000	0	1
9	0.00	0.0	10.50	0	0.0000	0	0
10	0.00	32.0	3.52	0	0.0000	1	0
11	0.00	0.0	1.13	0	0.0000	0	0
12	0.00	0.0	2.20	1	0.0000	1	0
13	200.00	8.0	3.25	1	0.0000	1	1
14	0.00	0.0	1.15	0	0.0000	0	0
15	450.00	0.0	2.01	1	0.0000	1	1
16	100.00	2.0	2.55	1	0.0000	1	1
17	0.00	0.0	2.01	1	214.2900	1	0
18	0.00	0.0	6.00	0	0.0000	1	0
19	50.00	0.0	0.40	1	0.0000	1	1
20	92.00	0.5	2.63	0	0.0000	1	1
21	0.00	0.0	1.80	1	0.0000	1	0
22	0.00	0.0	11.00	1	0.0000	1	0
23	100.00	0.0	5.75	1	0.0000	0	1
24	300.00	21.0	7.26	1	500.0000	1	1
25	171.43	0.0	2.00	0	400.0000	1	1
26	0.00	2.0	14.00	1	2760.0000	0	0
27	55.56	0.2	2.00	0	808.0800	1	1
28							

Fig. 9.4 The raw data file in EViews

File...
  Open...
    Foreign Data as Workfile

to generate Fig. 9.4. To run the logit model involving the five independent variables on the previous page, click:

Quick
  Estimate Equation

which produces the Equation Estimation dialogue box of Fig. 9.5.

Note in Fig. 9.5 that in the Method box, the option BINARY—Binary choice (logit, probit, extreme value) has been chosen and that the readings from the first to the 491st will be used in the analysis. It is now necessary to enter the names of the study variables, starting with the dependent variable **FERUSE**. As shown in Fig. 9.6, we have also included a possible intercept term which is always denoted in EViews by the variable name **C**.

Make sure in Fig. 9.6 that the logit process is selected under the heading 'Binary estimation method', since a process called the probit method is the EViews default. Clicking the OK button generates the results shown in Fig. 9.7.

Adopting a one-tailed test (i.e. a significance level of 5% or 0.05) all five of the independent variables and the intercept term are statistically significant from zero

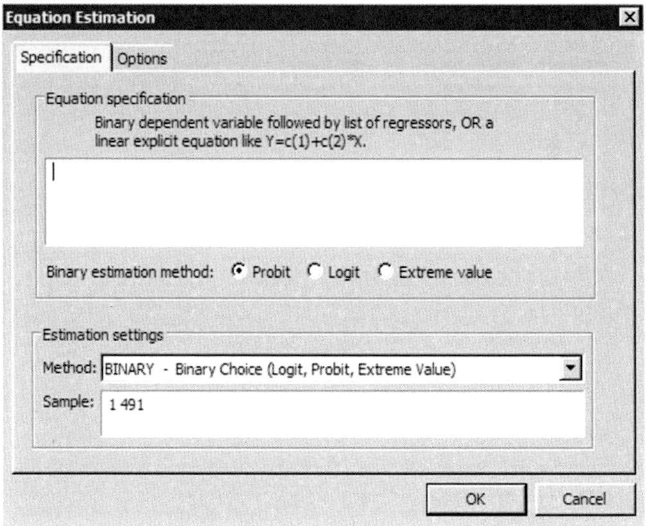

Fig. 9.5 The Equation Estimation dialogue box in EViews

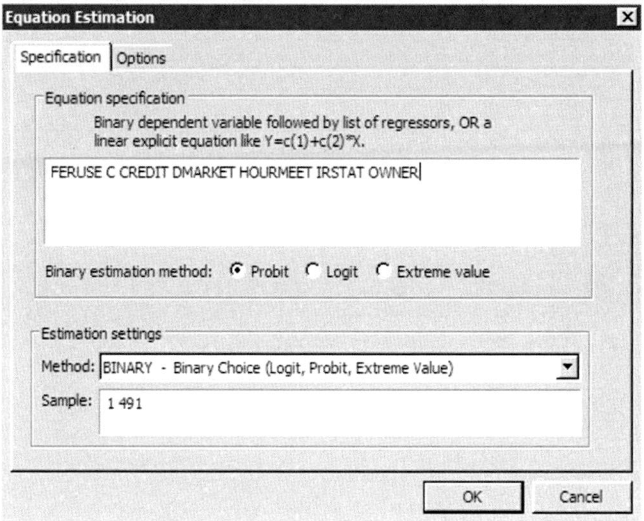

Fig. 9.6 The completed Equation Estimation dialogue box

(i.e. we reject H_0: a particular gradient is zero). The distance to the nearest market (**DMARKET**) has a negative impact on the likelihood that fertiliser is used as we would expect, since increased distances to the market create increased transport costs to the farmer which acts as a deterrent. The gradient of **IRSTAT** is positive. Farmers already using modern methods like irrigation are more likely to use fertiliser.

Equation: UNTITLED Workfile: FERTILISER::Fertiliser\

View | Proc | Object | Print | Name | Freeze | Estimate | Forecast | Stats | Resids

Dependent Variable: FERUSE
Method: ML - Binary Probit (Newton-Raphson / Marquardt steps)
Date: 03/11/18 Time: 20:36
Sample: 1 491
Included observations: 491
Convergence achieved after 4 iterations
Coefficient covariance computed using observed Hessian

Variable	Coefficient	Std. Error	z-Statistic	Prob.
C	-0.708130	0.120114	-5.895473	0.0000
CREDIT	0.000190	6.80E-05	2.797810	0.0051
DMARKET	-0.026717	0.012134	-2.201891	0.0277
HOURMEET	0.016196	0.006744	2.401687	0.0163
IRSTAT	0.909921	0.123017	7.396687	0.0000
OWNER	0.333171	0.125732	2.649855	0.0081

McFadden R-squared	0.152332	Mean dependent var	0.458248
S.D. dependent var	0.498762	S.E. of regression	0.448540
Akaike info criterion	1.193639	Sum squared resid	97.57636
Schwarz criterion	1.244920	Log likelihood	-287.0385
Hannan-Quinn criter.	1.213777	Deviance	574.0770
Restr. deviance	677.2429	Restr. log likelihood	-338.6215
LR statistic	103.1660	Avg. log likelihood	-0.584600
Prob(LR statistic)	0.000000		

Obs with Dep=0	266	Total obs	491
Obs with Dep=1	225		

Fig. 9.7 Fitting the logit model in EViews

Underneath the gradients in Fig. 9.7, the **LR (likelihood ratio) statistic** is reported. This tests the null hypothesis that all the gradients are zero. Naturally, we wish to reject this null as indeed we do here. LR = 103.16, with p = 0.000 which is very highly significant. McFadden's pseudo r^2 does lie between 0 and 1, but is not comparable with the coefficient of determination in multivariate regression. (It is in fact used to see how the percentage explanation increases if another variable(s) is entered into the logit model).

From Fig. 9.7, the equation of the logit model is:

$$\ln\left(\frac{P}{1-p}\right) = -0.708130 + 0.000190(\text{CREDIT}) - 0.026717(\text{DMARKET}) + \ldots$$
$$+ 0.333171(\text{OWNER})$$

For example, if for one particular farmer, HOURMEET = 30, DMARKET = 6, CREDIT = 200, IRSTAT = 1 and OWNER = 1, then from the above equation:

Equation: UNTITLED Workfile: FERTILISER::Fertiliser\

View | Proc | Object | Print | Name | Freeze | Estimate | Forecast | Stats | Resids

Expectation-Prediction Evaluation for Binary Specification
Equation: UNTITLED
Date: 03/11/18 Time: 21:09
Success cutoff: C = 0.5

	Estimated Equation			Constant Probability		
	Dep=0	Dep=1	Total	Dep=0	Dep=1	Total
P(Dep=1)<=C	181	65	246	266	225	491
P(Dep=1)>C	85	160	245	0	0	0
Total	266	225	491	266	225	491
Correct	181	160	341	266	0	266
% Correct	68.05	71.11	69.45	100.00	0.00	54.18
% Incorrect	31.95	28.89	30.55	0.00	100.00	45.82
Total Gain*	-31.95	71.11	15.27			
Percent Gain**	NA	71.11	33.33			

	Estimated Equation			Constant Probability		
	Dep=0	Dep=1	Total	Dep=0	Dep=1	Total
E(# of Dep=0)	168.04	98.15	266.19	144.11	121.89	266.00
E(# of Dep=1)	97.96	126.85	224.81	121.89	103.11	225.00
Total	266.00	225.00	491.00	266.00	225.00	491.00
Correct	168.04	126.85	294.89	144.11	103.11	247.21
% Correct	63.17	56.38	60.06	54.18	45.82	50.35
% Incorrect	36.83	43.62	39.94	45.82	54.18	49.65
Total Gain*	9.00	10.55	9.71			
Percent Gain**	19.63	19.48	19.56			

*Change in "% Correct" from default (constant probability) specification
**Percent of incorrect (default) prediction corrected by equation

Fig. 9.8 The Expectation-Prediction Table

$$\ln\left(\frac{P}{1-p}\right) = 0.89854,$$

$$\frac{P}{1-P} = e^{0.89854} = 2.456$$

$$\mathrm{P} = 2.456 - 2.456\mathrm{P}$$

$$3.456\mathrm{P} = 2.456$$

Whereby P = 0.71

There is an over 71% chance that this particular farmer is a fertiliser user. Further assessment may be made of the logit model's adequacy. In Fig. 9.7, click:

View...
  Expectation-Prediction Evaluation

which produces the results in Fig. 9.8. EViews will ask you to specify a prediction cut-off value lying between 0 and 1. We shall select a probability of 0.5 for this cut-off value. Each observation will be classified as having a predicted probability that lies above or below this cut-off i.e. if the probability >0.5, we will assume that it is more likely that a farmer is in the FERUSE = 1 group. In Fig. 9.8, attention should be paid to the top left-hand corner table under the heading 'Estimated Equation'. Two hundred and sixty-six farmers are in the dependent variable = 0 category i.e. the FERUSE = 0 group—they do not use fertiliser. One hundred and eighty-three of these farmers were predicted by the logit model to have a probability of fertiliser use below the cut-off probability of 0.5. Hence, 181 (68.05%) of the farmers in the FERUSE = 0 group were predicted correctly. Therefore, as also shown in Fig. 9.8, 31.95% of the FERUSE = 0 group were incorrectly classified by the logit model. Similarly, there were 225 farmers observed to be in the dependent variable = 1 category i.e. FERUSE = 1. 160 (71.11%) farmers had probabilities above the cut-off point of 0.5 and were consequently correctly classified. Overall, 341 farmers (181 + 160) have been correctly classified into their FERUSE = 0 or FERUSE = 1 groups. This is an overall success rate of 342 out of 491 farmers or 69.45%.

You can also obtain information as to whether each individual farmer was correctly grouped or not. Click:

View...
  Actual, Fitted, Residual...
    Actual, Fitted, Residual, Table

to produce Fig. 9.9. Farmer 1 is in the FERUSE = 0 group. The probability of his using fertiliser is forecasted by the logit model to be 0.22717, which is lower than the cut-off of 0.5, so this farmer is correctly classified. However, farmer 4 is also in the FERUSE = 0 group, but the logit model estimates that the probability of his using fertiliser is 0.84673, so the model predicts that this individual should be in the FERUSE = 1 group—an incorrect forecasted classification.

As a final assessment of the adequacy of the logit model, there is the **Hosmer-Lemeshow (HL) goodness of fit test**, which is accessed via:

View...
  Goodness-of-Fit Test (Hosmer-Lemeshow)...

To cut a very long story short, the HL test has as its null H_0: the model adequately predicts group membership and the null is rejected if the associated level of significance is less than 5% or 0.05. In the above example, it would be found that HL = 8.3028 with significance 0.4045 (See Fig. 9.10), so the null would not be rejected and the logit model deemed an adequate representation for the data.

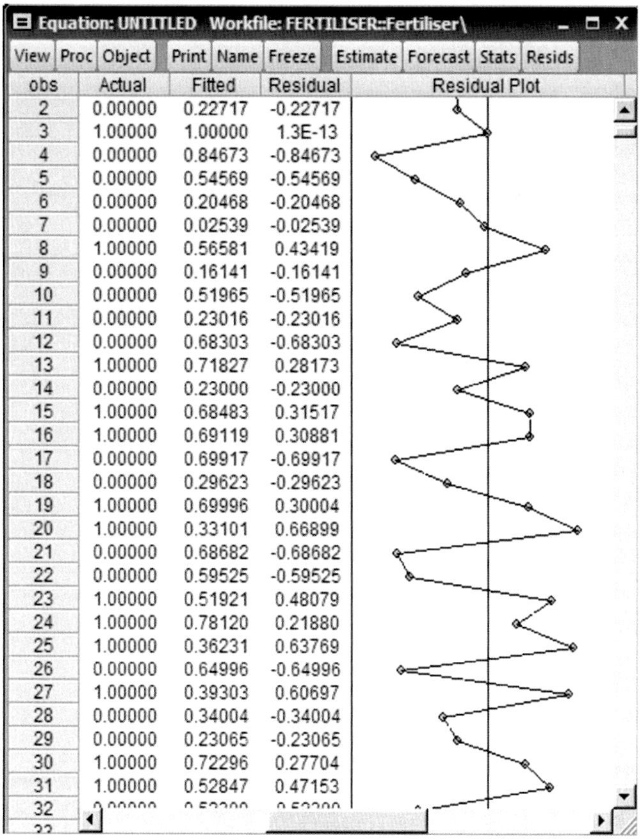

obs	Actual	Fitted	Residual	Residual Plot
2	0.00000	0.22717	-0.22717	
3	1.00000	1.00000	1.3E-13	
4	0.00000	0.84673	-0.84673	
5	0.00000	0.54569	-0.54569	
6	0.00000	0.20468	-0.20468	
7	0.00000	0.02539	-0.02539	
8	1.00000	0.56581	0.43419	
9	0.00000	0.16141	-0.16141	
10	0.00000	0.51965	-0.51965	
11	0.00000	0.23016	-0.23016	
12	0.00000	0.68303	-0.68303	
13	1.00000	0.71827	0.28173	
14	0.00000	0.23000	-0.23000	
15	1.00000	0.68483	0.31517	
16	1.00000	0.69119	0.30881	
17	0.00000	0.69917	-0.69917	
18	0.00000	0.29623	-0.29623	
19	1.00000	0.69996	0.30004	
20	1.00000	0.33101	0.66899	
21	0.00000	0.68682	-0.68682	
22	0.00000	0.59525	-0.59525	
23	1.00000	0.51921	0.48079	
24	1.00000	0.78120	0.21880	
25	1.00000	0.36231	0.63769	
26	0.00000	0.64996	-0.64996	
27	1.00000	0.39303	0.60697	
28	0.00000	0.34004	-0.34004	
29	0.00000	0.23065	-0.23065	
30	1.00000	0.72296	0.27704	
31	1.00000	0.52847	0.47153	
32	[illegible]	[illegible]	[illegible]	

Fig. 9.9 Forecasted group membership for each farmer

Equation: UNTITLED Workfile: FERTILISER::Fertiliser\

View | Proc | Object | Print | Name | Freeze | Estimate | Forecast | Stats | Resids

Goodness-of-Fit Evaluation for Binary Specification
Andrews and Hosmer-Lemeshow Tests
Equation: UNTITLED
Date: 03/11/18 Time: 21:18
Grouping based upon predicted risk (randomize ties)

	Quantile of Risk		Dep=0		Dep=1		Total	H-L
	Low	High	Actual	Expect	Actual	Expect	Obs	Value
1	0.0147	0.2000	39	41.1392	10	7.86080	49	0.69339
2	0.2033	0.2288	42	38.3664	7	10.6336	49	1.58579
3	0.2292	0.2693	35	37.1751	14	11.8249	49	0.52734
4	0.2710	0.3348	36	34.0718	13	14.9282	49	0.35816
5	0.3358	0.4956	28	30.3880	21	18.6120	49	0.49406
6	0.4980	0.5605	25	22.6367	24	26.3633	49	0.45860
7	0.5608	0.6064	25	20.5790	24	28.4210	49	1.63745
8	0.6069	0.6765	17	17.1387	32	31.8613	49	0.00173
9	0.6779	0.7062	10	15.1153	39	33.8847	49	2.50336
10	0.7069	1.0000	9	9.57636	41	40.4236	50	0.04291
		Total	266	266.187	225	224.813	491	8.30280

H-L Statistic	8.3028	Prob. Chi-Sq(8)	0.4045
Andrews Statistic	13.6664	Prob. Chi-Sq(10)	0.1888

Fig. 9.10 Goodness-of-Fit Test (Hosmer-Lemeshow test)

Chapter 10
Vector Autoregression (VAR) Model

In the regression analysis endogenous and exogenous variables are used alongside each other. An endogenous variable is a variable (generated by a statistical model), which is explained by the relationships between functions within the model. For example, the equilibrium price of a good in a supply and demand model is endogenous because it is set by a producer in response to consumer demand. If the general movement of one variable can be expected to produce a particular result in the other, though not necessarily in the same direction, as long as the change is correlating, it will be considered endogenous. In contrast to endogenous variables, exogenous variables are considered independent. This means one variable within the formula does not directly correlate, to a change in the other, such as personal income and colour preference, or rainfall and gas prices.

Causal modelling is used by economists to explain outcomes or dependent variables and find out to which extent a result can be attributed to an endogenous or exogenous cause. This is important for econometrics and economic modelling since it can show a variable causes a particular effect. Classical simultaneous equation models comprised of m endogenous (i.e. dependent) variables end up with m equations, one for each endogenous variable. Each equation may have one or more endogenous variables and some exogenous variables.

The problem of identification needs to be dealt with before estimating these types of equations. In achieving this, often arbitrary restrictions are imposed by excluding some variables from an equation, which may be present in the other equations in the system. Sims (1980) criticized severely this approach. He argued that if there are m endogenous variables, they should all be treated on an equal basis. Particularly, there should not be any distinction between endogenous and exogenous variables. So each equation should have the same number of regressors.

It is for this reason that Sims (1980) developed the VAR model.

A. Aljandali, M. Tatahi, *Economic and Financial Modelling with EViews*, Statistics and Econometrics for Finance, https://doi.org/10.1007/978-3-319-92985-9_10

10.1 The VAR Methodology

Let us consider a system of two in which there is an assumption of an existing relation between three and six-months T-bills illustrated using the following equations:

$$TB3_t = A_1 + \sum_{j=1}^{j=p} B_j TB3_{t-j} + \sum_{j=1}^{j=p} C_j TB6_{t-j} + u_{1t}$$

$$TB6_t = A_2 + \sum_{j=1}^{j=p} D_j TB3_{t-j} + \sum_{j=1}^{j=p} E_j TB6_{t-j} + u_{2t}$$

TB3 and *TB6* are 3 and 6-months T-bill rates.

The U_{1t} and U_{2t} are white noise error terms. We call them **impulses (urges)** or **innovations** or **shock**s in the language of VAR.

The bivariate VAR given above illustrates several features:

1. The above system shows **a simultaneous type equation** system. Each equation contains only its own lagged values and the lagged values of the other variables in the system. The current values of the two variables are included on the right-hand side of these equations.
2. We use the same number of lagged terms in each equation.
3. The bivariate VAR system given above is known as a VAR (p) model. It has p lagged values of each variable on the right-hand side. When we have only one lagged value of each variable on the right-hand side, it would be a VAR (1) model; if two-lagged terms, it would be a VAR (2) model; and so on.
4. The VAR system can be extended to several variables. Each equation in the system then will contain p lagged values of each variable on the right-hand side of each equation.
5. In the two-variable system there can be at most one *cointegrating*, or *equilibrium*, relationship between them. If we have a three-variable VAR system, there can be at most two cointegrating relationships between the three variables. In general, an n-variable VAR system can have at most (n − 1) cointegrating relationships. Johansen's methodology, a utility available in EViews and other computer packages, can help to find out how many cointegrating relationships exist among n variables.
6. If after running the Johansen cointegration test we find out that the variables are **not cointegrated**, we can develop and **unrestricted VAR** model or other models such as those in time series data analysis. If the variables are cointegrated, we can develop an ECM model which we can then call **a restricted VAR model or a VECM model**.

Note: The number of lagged terms can be found by using the Akaike or Schwarz information criteria.

There are three possibilities:

1. Both TB3 and TB6 time series are individually I(0), or stationary. In that case we can estimate each equation by OLS.
2. Both TB3 and TB6 are I(1) then we can take the first differences of the two variables, which, as we know, are stationary. Here again we can use OLS to estimate each equation individually.
3. If the two series are I(1), but are **cointegrated, then we have to use the error correction mechanism (ECM).** The ECM combines the long-run equilibrium with short-run dynamics to reach that equilibrium. Since we are dealing with more than one variable in a VAR system, the multivariate counterpart of ECM is known as **the vector error correction model (VECM).**

A critical requirement of VAR is that he time series under consideration must be stationary.

10.2 The Estimation Process

Step One: Check the stationarity of the time series:
Time Series One: *TB3* 3-months T-bill rates (Informal methods)

(a) **Plot the series**:

Go to *Quick* > *Graph* and write TB3 and Choose ***Line Graph***—OK.

According to Fig. 10.1, the TB3 series appears to be a random walk with drift: it has moved up and down for some time, and it trended downward latter. It seems to be a non-stationary process.

(b) **The Correlogram Test for the TB3 Series**:

Double click the TB3 from the Workfile to get the data of the series TB3.

Go to *View* > *Correlogram*—Select a correlogram of the Level and then click OK.

The Correlogram Test:

(a) H_0: TB3 has no unit root
(b) H_1: TB3 has a unit root
(c) $\alpha = 5\%$ or 0.05

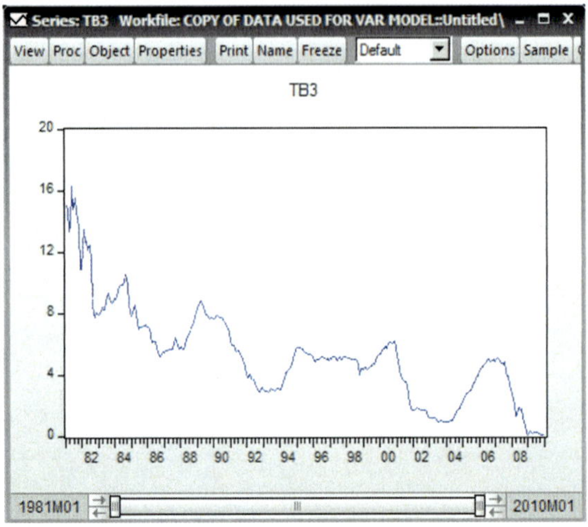

Fig. 10.1 Plot of TB3 series overtime

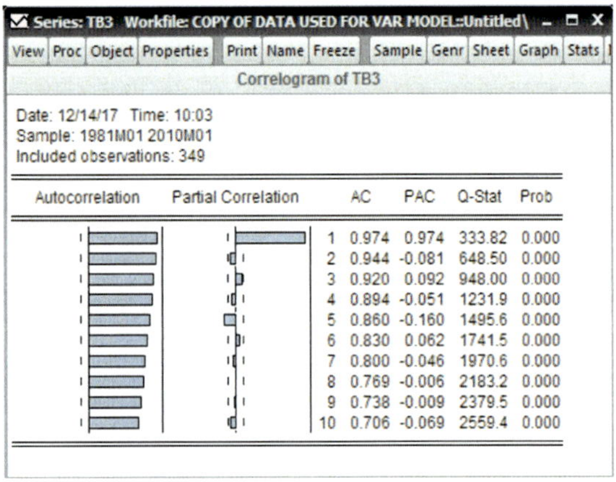
Series: TB3 Workfile: COPY OF DATA USED FOR VAR MODEL::Untitled\

View Proc Object Properties Print Name Freeze Sample Genr Sheet Graph Stats

Correlogram of TB3

Date: 12/14/17 Time: 10:03
Sample: 1981M01 2010M01
Included observations: 349

Autocorrelation	Partial Correlation		AC	PAC	Q-Stat	Prob
		1	0.974	0.974	333.82	0.000
		2	0.944	-0.081	648.50	0.000
		3	0.920	0.092	948.00	0.000
		4	0.894	-0.051	1231.9	0.000
		5	0.860	-0.160	1495.6	0.000
		6	0.830	0.062	1741.5	0.000
		7	0.800	-0.046	1970.6	0.000
		8	0.769	-0.006	2183.2	0.000
		9	0.738	-0.009	2379.5	0.000
		10	0.706	-0.069	2559.4	0.000

Fig. 10.2 Correlogram of TB3

There are no test statistics to be calculated. We look at the Prob column in Fig. 10.2.

(d) The p-values are = 0.000.
(e) Prob = 0.000 < 0.05 therefore reject the H_0
(f) We conclude that TB3 is not stationary

Formal method:

Augmented Dickey-Fuller (ADF) Test for TB3:

Before running the test in EViews, we need to decide whether to include a trend or not.

<u>If the plot of the series is trended, include a trend with the intercept.</u>

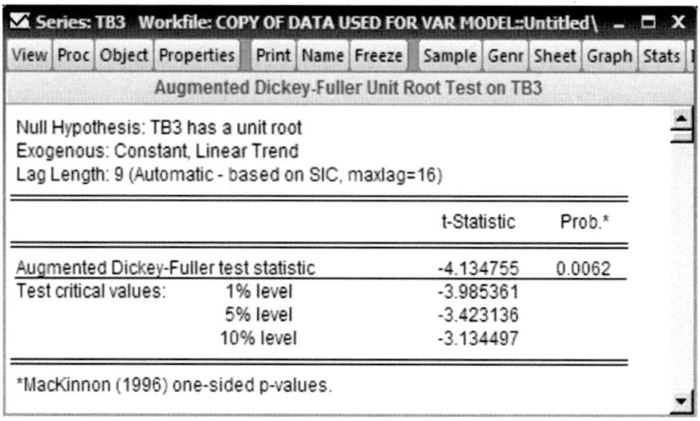

Series: TB3 Workfile: COPY OF DATA USED FOR VAR MODEL::Untitled\

View | Proc | Object | Properties | Print | Name | Freeze | Sample | Genr | Sheet | Graph | Stats

Augmented Dickey-Fuller Unit Root Test on TB3

Null Hypothesis: TB3 has a unit root
Exogenous: Constant, Linear Trend
Lag Length: 9 (Automatic - based on SIC, maxlag=16)

		t-Statistic	Prob.*
Augmented Dickey-Fuller test statistic		-4.134755	0.0062
Test critical values:	1% level	-3.985361	
	5% level	-3.423136	
	10% level	-3.134497	

*MacKinnon (1996) one-sided p-values.

Fig. 10.3 TB3 unit root test

If the plot of the series is not trended, do not include a trend. Include only the intercept.

Follow these steps in EViews: Double click the variable, TB3, from the Workfile. From the Window containing the data of TB3, choose View-Unit Root test. From Test type, choose Augmented Dickey-Fuller. From Test for unit root in, choose level. From Include in test equation, choose Trend and Intercept if the plot of the series indicates it is trended. Otherwise choose only the Intercept. And then click OK.

1. H_0: TB3 has a unit root and is not stationary
2. H_1: TB3 has no unit root and is stationary.
3. $\alpha = 5\%$ or 0.05

According to Fig. 10.3, ADF test statistic is −4.134755.

4. The p-values = 0.0062.
5. Prob = 0.0062 < 0.05 therefore we reject the H_0
6. TB3 does not have a unit root and is stationary.

Time Series Two: *TB6* 6-month T-bill rates
Informal methods:

(a) **Plot the series**:

Go to **Quick > Graph**-write, TB6-Choose **Line Graph-OK.**

According to Fig. 10.4, TB6 appears to be trended down and non-stationary.

Fig. 10.4 Plot of TB6

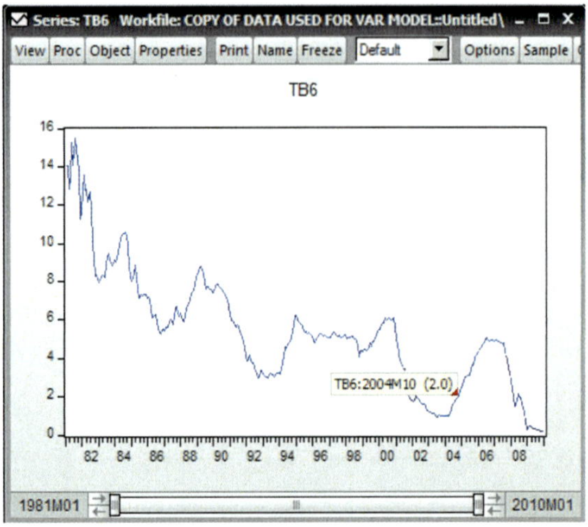

(b) **The Correlogram Test for TB6 Series**:

Double click on TB6 form the Workfile

> Go to **View > Correlogram**—Select a correlogram of the Level and then press **OK**.

The Correlogram Test:
Note:

> The null hypothesis and the alternative hypothesis for the correlogram and for the formal test of the Augmented Dickey-Fuller test are completely opposite.

(a) H_0: TB6 has no unit root
(b) H_1: TB6 has a unit root
(c) $\alpha = 5\%$ or 0.05

There are no test statistics to calculate. We look at the Prob column in Fig. 10.5.

(d) The p-values = 0.000.
(e) Prob = 0.000 < 0.05 therefore we reject the H_0.
(f) TB6 is not stationary

Formal method:
ADF Test for TB6:

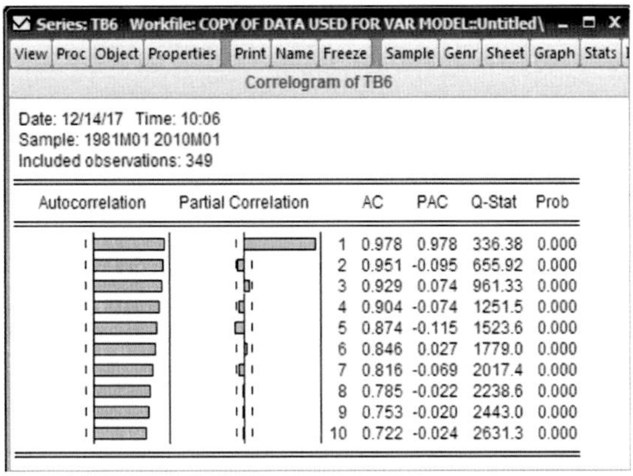

Series: TB6 Workfile: COPY OF DATA USED FOR VAR MODEL::Untitled\

View | Proc | Object | Properties | Print | Name | Freeze | Sample | Genr | Sheet | Graph | Stats

Correlogram of TB6

Date: 12/14/17 Time: 10:06
Sample: 1981M01 2010M01
Included observations: 349

Autocorrelation	Partial Correlation		AC	PAC	Q-Stat	Prob
		1	0.978	0.978	336.38	0.000
		2	0.951	-0.095	655.92	0.000
		3	0.929	0.074	961.33	0.000
		4	0.904	-0.074	1251.5	0.000
		5	0.874	-0.115	1523.6	0.000
		6	0.846	0.027	1779.0	0.000
		7	0.816	-0.069	2017.4	0.000
		8	0.785	-0.022	2238.6	0.000
		9	0.753	-0.020	2443.0	0.000
		10	0.722	-0.024	2631.3	0.000

Fig. 10.5 Correlogram of TB6

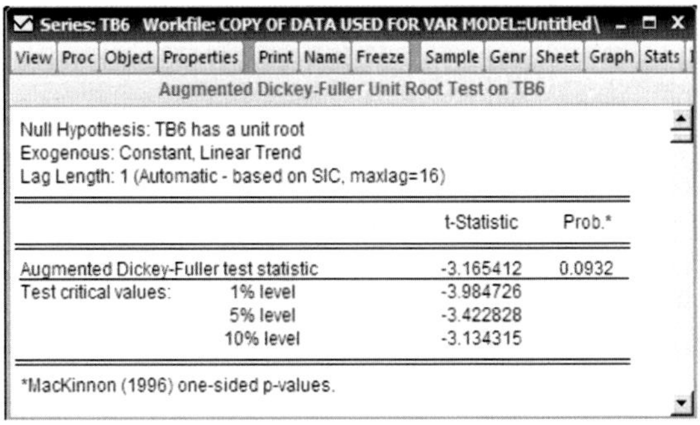

Series: TB6 Workfile: COPY OF DATA USED FOR VAR MODEL::Untitled\

View | Proc | Object | Properties | Print | Name | Freeze | Sample | Genr | Sheet | Graph | Stats

Augmented Dickey-Fuller Unit Root Test on TB6

Null Hypothesis: TB6 has a unit root
Exogenous: Constant, Linear Trend
Lag Length: 1 (Automatic - based on SIC, maxlag=16)

		t-Statistic	Prob.*
Augmented Dickey-Fuller test statistic		-3.165412	0.0932
Test critical values:	1% level	-3.984726	
	5% level	-3.422828	
	10% level	-3.134315	

*MacKinnon (1996) one-sided p-values.

Fig. 10.6 TB6 unit root test

Follow these steps in EViews: Double click the variable, TB6, from the Workfile. From the Window containing the data of TB6, choose View-Unit Root test. From Test type, choose Augmented Dickey-Fuller. From Test for unit root in, choose level. From Include in test equation, choose Trend and Intercept if the plot of the series indicates it is trended. Otherwise choose only the Intercept. And then click OK.

1. H_0: TB6 has a unit root and is not stationary
2. H_1: TB6 has no unit root and is stationary.
3. $\alpha = 5\%$ or 0.05

According to Fig. 10.6, ADF test statistic is -3.165412.

4. The p-values = 0.0932.
5. Prob = 0.0932 > 0.05 therefore we reject the H_0.
6. TB6 has a unit root and is not stationary.

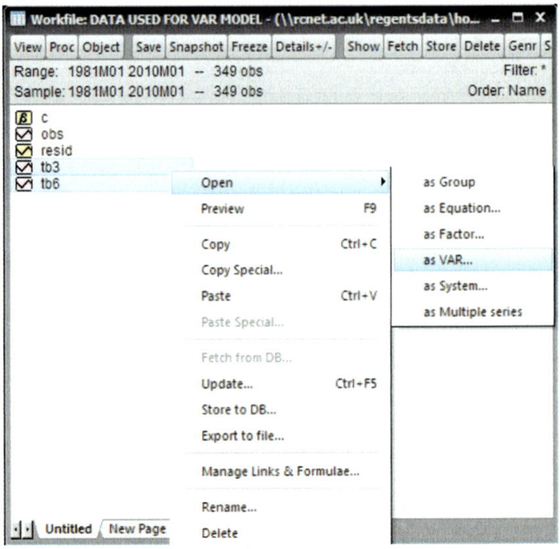

Fig. 10.7 The VAR method in EViews

What is found above indicates that TB3, is stationary but TB6, is not.

Now we have to run the cointegration test for the VAR model known as **Johansen Cointegration Test**. We are presented with two options:

1. If the series are conitegrated then we can proceed and use the VECM model in other term a **Restricted VAR Model.**
2. If the series are not cointegrated (meaning there is no long-run relationship between them) we should follow a model which is known as: **Unrestricted VAR Model**.

Step Two: The Lag Length:

Follow this approach step by step in order to find how many lags should be included in this VAR model processes.

Highlight both variables. Right click the highlight > open > as VAR as shown in Fig. 10.7.

You will get the window shown in Fig. 10.8:

Go to View > Lag Structure > Lag Length Criteria as shown in Fig. 10.9

Choose lag 8 and click OK to get the output of Fig. 10.10

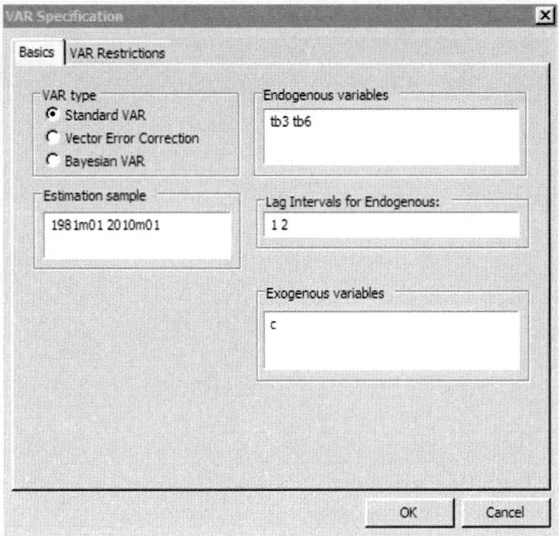

Fig. 10.8 The VAR specification window

The stars of columns 3–7 indicate the number of length suggested by each criterion: LR, FPE, and AIC suggest eight lags and SC and HQ suggest three lags. We follow the one suggested by SC (Schwarz information criterion) and we choose three lags (studies show this criterion is used widely).

Step Three: Johansen Cointegration Test

Go to Quick > Group Statistics > Johansen Cointegration Test as shown in Fig. 10.11.
You will then be prompted to write the name of the variables of your model as shown in Fig. 10.12. It should be noted that the order is not important here. Enter the name of your variables and click OK.

In Fig. 10.13, we are prompted to specify the characteristics of the Johansen cointegration Test—whether the result should include a trend as well as an intercept or not. We decide based on our judgement regarding the graph of the series. The graphs of the series above show they don't have deterministic trends. So, we choose option 3 as indicated above. If we want to add extra exogenous variables we should write the name of them in the dialog box: Exog Variables. For the Lag Intervals part we choose an interval equal to the lag length we found above, three lags. So, we choose an interval of 1 and 2, which means 1 + 2 = 3. Click OK to get output of Fig. 10.14

The Cointegration Test:

1. H_0: The series are not cointegrated,
2. H_1: The series are cointegrated,
3. $\alpha = 5\%$ or 0.05,

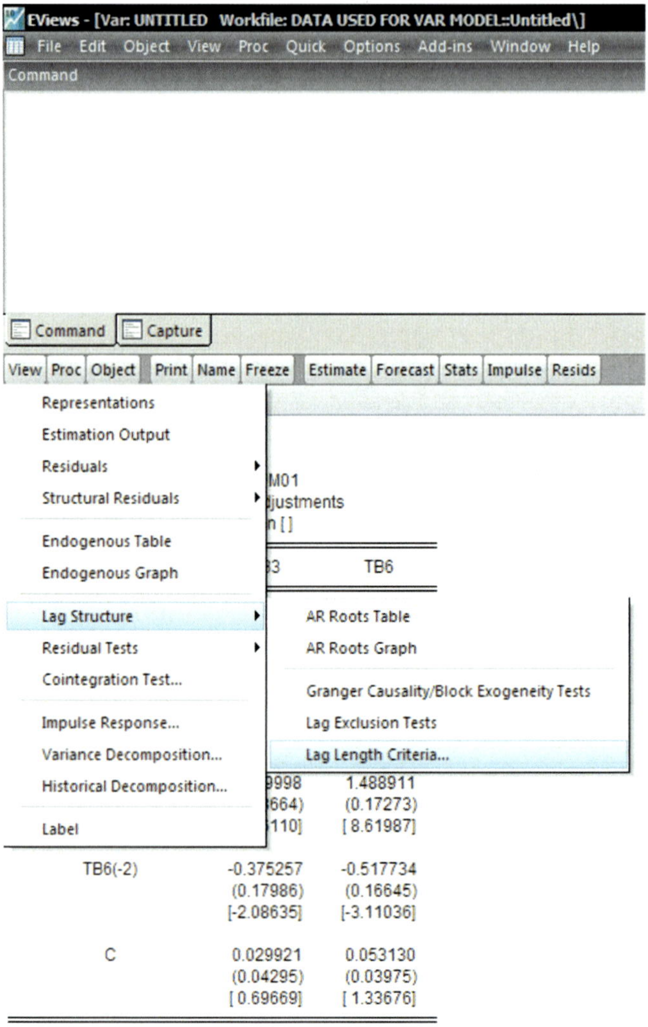

Fig. 10.9 Lag structure

Two Unrestricted Cointegration Rank Tests: Trace and Maximum Eigenvalue tests are reported in this output.

We pay attention to two statements in this output:

1. Trace test indicates 1 cointegrating eqn(s) at the 0.05 level.
2. Max-eigenvalue test indicates 1 cointegrating equ(s) at the 0.05 level.

These two statements help to reject the H_0 and conclude that the series are cointegrated.

Var: UNTITLED Workfile: DATA USED FOR VAR MODEL::Untitled\

View | Proc | Object | Print | Name | Freeze | Estimate | Forecast | Stats | Impulse | Resids

VAR Lag Order Selection Criteria
Endogenous variables: TB3 TB6
Exogenous variables: C
Date: 12/18/17 Time: 08:20
Sample: 1981M01 2010M01
Included observations: 341

Lag	LogL	LR	FPE	AIC	SC	HQ
0	-720.4864	NA	0.237332	4.237457	4.259931	4.246411
1	190.2517	1805.452	0.001163	-1.080655	-1.013232	-1.053793
2	236.4522	91.04605	0.000908	-1.328165	-1.215793	-1.283395
3	248.8227	24.23320	0.000865	-1.377259	-1.219939*	-1.314580*
4	251.2035	4.635874	0.000873	-1.367762	-1.165493	-1.287175
5	254.3717	6.132116	0.000877	-1.362884	-1.115666	-1.264389
6	262.2958	15.24384	0.000857	-1.385899	-1.093732	-1.269495
7	270.0300	14.78811	0.000839	-1.407801	-1.070685	-1.273489
8	277.8062	14.77703*	0.000821*	-1.429948*	-1.047884	-1.277728

* indicates lag order selected by the criterion
LR: sequential modified LR test statistic (each test at 5% level)
FPE: Final prediction error
AIC: Akaike information criterion
SC: Schwarz information criterion
HQ: Hannan-Quinn information criterion

Fig. 10.10 VAR Model output

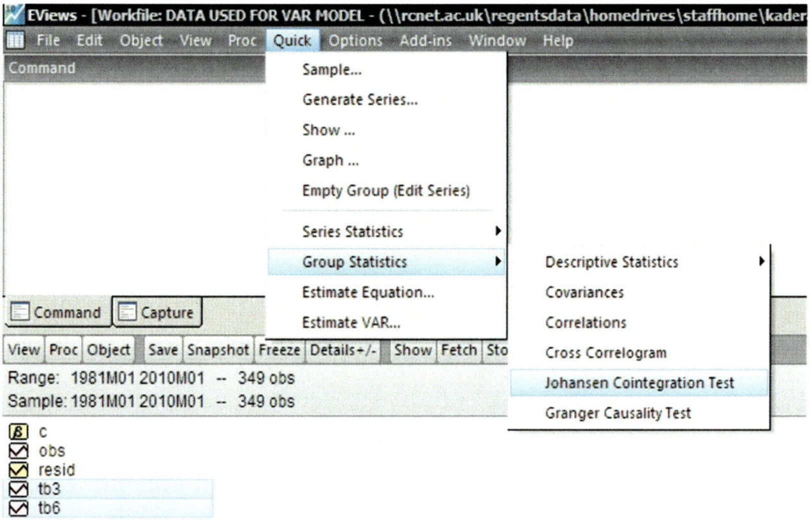

Fig. 10.11 Johansen Cointegration test in EViews

Since the series are cointegrated we can now run a VECM model.

<u>Step Four: The VECM Model</u>

Since the variables are cointegrated we will be estimating a restricted VAR model as mentioned earlier.

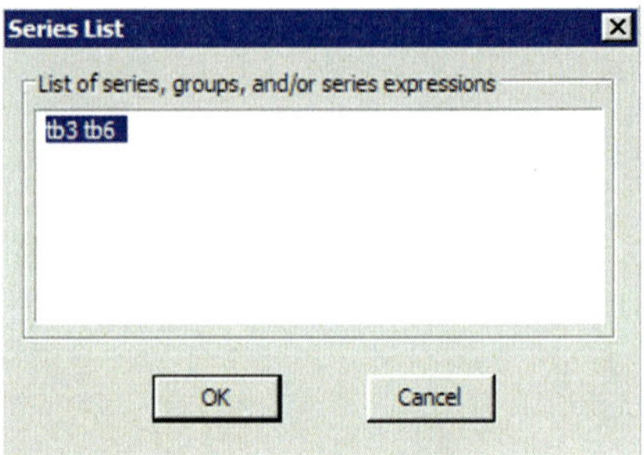

Fig. 10.12 Series list

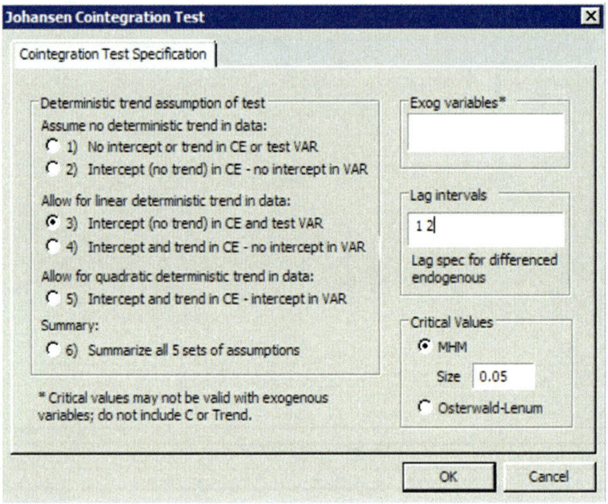

Fig. 10.13 Cointegration test specification

Go to Quick > Estimate VAR as shown in Fig. 10.15.
From the VAR type list in Fig. 10.16 choose Vector Error Correction because series were cointegrated. In the Endogenous Variables part write the name of the variables of the model, in this case: TB3 and TB6 (Do not use the first differences of the variables because the first differences of the variables will be used by the EViews automatically). You can change the number of lags from the Lag Intervals. Then click OK which generates output of Fig. 10.17.

The output in Fig. 10.17 has put the dependent variables on a row and independent variables on a column. Each independent variable has got three numbers. The first number is the coefficient of the independent variable. The second number, inside the smaller bracket, is the standard error of the coefficient. The third number, inside the bigger bracket, is the t-statistics.

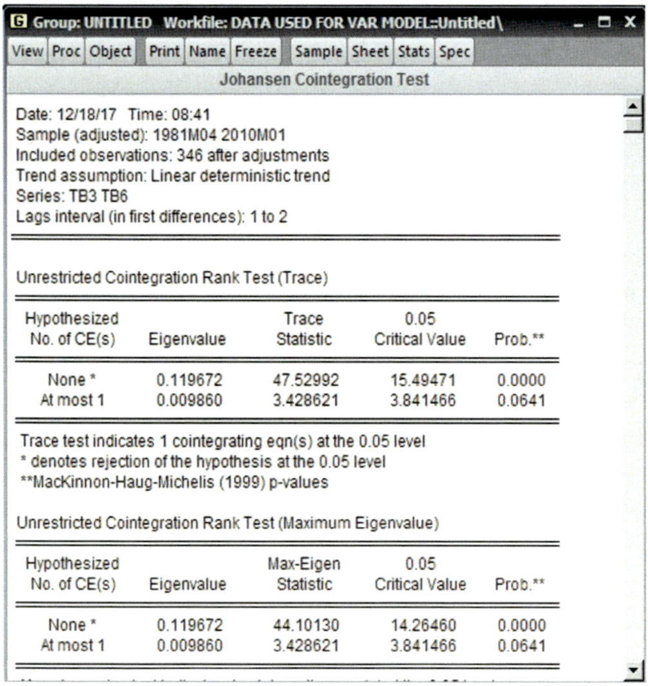

Group: UNTITLED Workfile: DATA USED FOR VAR MODEL::Untitled\

View | Proc | Object | Print | Name | Freeze | Sample | Sheet | Stats | Spec

Johansen Cointegration Test

Date: 12/18/17 Time: 08:41
Sample (adjusted): 1981M04 2010M01
Included observations: 346 after adjustments
Trend assumption: Linear deterministic trend
Series: TB3 TB6
Lags interval (in first differences): 1 to 2

Unrestricted Cointegration Rank Test (Trace)

Hypothesized No. of CE(s)	Eigenvalue	Trace Statistic	0.05 Critical Value	Prob.**
None *	0.119672	47.52992	15.49471	0.0000
At most 1	0.009860	3.428621	3.841466	0.0641

Trace test indicates 1 cointegrating eqn(s) at the 0.05 level
* denotes rejection of the hypothesis at the 0.05 level
**MacKinnon-Haug-Michelis (1999) p-values

Unrestricted Cointegration Rank Test (Maximum Eigenvalue)

Hypothesized No. of CE(s)	Eigenvalue	Max-Eigen Statistic	0.05 Critical Value	Prob.**
None *	0.119672	44.10130	14.26460	0.0000
At most 1	0.009860	3.428621	3.841466	0.0641

Fig. 10.14 Johansen cointegration output

Fig. 10.15 VAR estimation

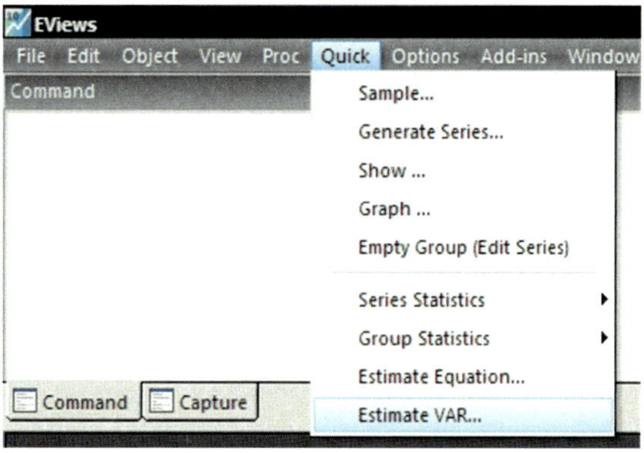

To get the p-value for each coefficient do as follows:

Go to the Proc > Make System > Order by Variables as shown in Fig. 10.18

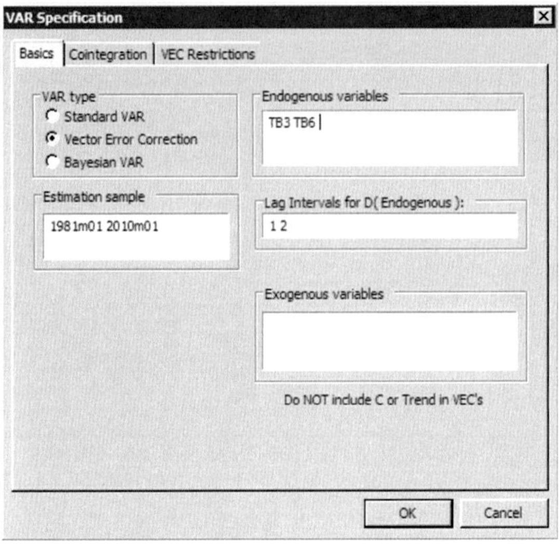

Fig. 10.16 VAR specification: Vector Error Correction

You will get the output of Fig. 10.19:

This output, which is known as a **system equation**, gives 12 system coefficients, 6 coefficients for each equation (the intercepts are included). The names: C(1), C(2), C(3) and so on are used for the coefficients.

In order to find the p-values for all coefficients do as follows:

Go to Estimate and from the list of the Estimation Methods choose Ordinary Least Squares as shown in Fig. 10.20. Click OK

You will get output of Fig. 10.21 which gives you the p-values for all coefficients.

From two provided equations in the second half of this output it can be identified that which coefficient belongs to which variable.

Start reading these two equations from the end and move to the beginning. Coefficients C(6) and C(12) are the intercepts of these two equations, respectively.

Step Five: The Coefficient Test

We need to run a hypothesis testing for each coefficient. This will give us the idea which variable has effected the dependent variable of each model and which one has not.

To make sure there is a long-run relationship between TB3 and TB6 in model one and a similar relationship between TB6 and TB3 in the model two we run two hypothesises testing for the coefficients C(1) and C(7). These two coefficients are the ECM_{t-1} coefficient in model one and two, respectively.

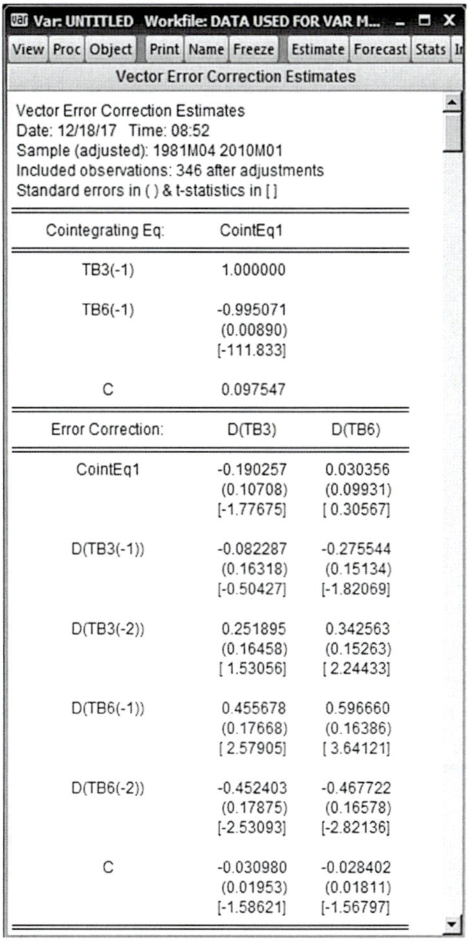

Vector Error Correction Estimates
Date: 12/18/17 Time: 08:52
Sample (adjusted): 1981M04 2010M01
Included observations: 346 after adjustments
Standard errors in ( ) & t-statistics in [ ]

Cointegrating Eq:	CointEq1	
TB3(-1)	1.000000	
TB6(-1)	-0.995071 (0.00890) [-111.833]	
C	0.097547	
Error Correction:	**D(TB3)**	**D(TB6)**
CointEq1	-0.190257 (0.10708) [-1.77675]	0.030356 (0.09931) [0.30567]
D(TB3(-1))	-0.082287 (0.16318) [-0.50427]	-0.275544 (0.15134) [-1.82069]
D(TB3(-2))	0.251895 (0.16458) [1.53056]	0.342563 (0.15263) [2.24433]
D(TB6(-1))	0.455678 (0.17668) [2.57905]	0.596660 (0.16386) [3.64121]
D(TB6(-2))	-0.452403 (0.17875) [-2.53093]	-0.467722 (0.16578) [-2.82136]
C	-0.030980 (0.01953) [-1.58621]	-0.028402 (0.01811) [-1.56797]

Fig. 10.17 Vector Error Correction Estimates

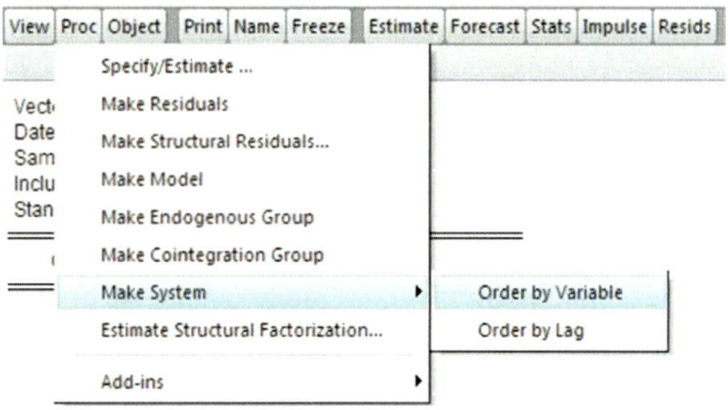

Fig. 10.18 Order by variable option

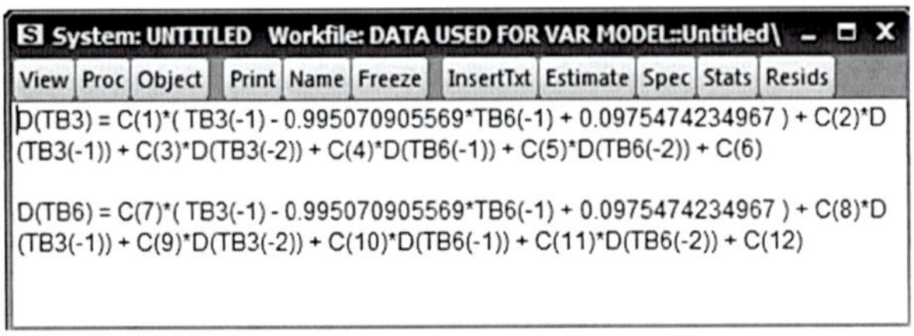

Fig. 10.19 Order by variable output

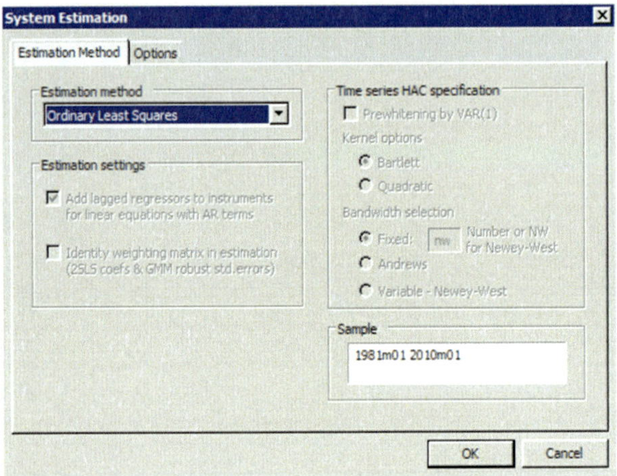

Fig. 10.20 System estimation

Teat for C(1) the coefficient of ECM_{t-1} of Model One:

1. $H_0 : C(1) = 0$
2. $H_1 : C(1) \neq 0$
3. $\alpha = 5\ \% \quad or \ 0.05$
4. $t_{\beta_2} = \dfrac{CoefficientC(1) - C(1)inTHEH_0}{S.E.C(1)} = \dfrac{-190257 - 0}{0.107082} = -1.776747$
5. $p - value = 0.0761$
6. p-value of 0.0761 > 0.05. We cannot reject the H_0. The test is not significant.
7. This coefficient is not statistically significant. It means that there is no long-term relationship between the TB3 and TB6.

Teat for C(7) the coefficient of ECM_{t-1} of Model Two:

1. $H_0 : C(7) = 0$
2. $H_1 : C(7) \neq 0$
3. $\alpha = 5\ \% \ or 0.05$
4. $t_{\beta_2} = \dfrac{CoefficientC(7) - C(7)inTHEH_0}{S.E.C(7)} = \dfrac{0.030356 - 0}{0.099311} = 0.305669$

Fig. 10.21 Least squares output

System: UNTITLED Workfile: DATA USED FOR VAR MODEL::Untitled\

View | Proc | Object | Print | Name | Freeze | InsertTxt | Estimate | Spec | Stats | Resids

System: UNTITLED
Estimation Method: Least Squares
Date: 12/18/17 Time: 09:00
Sample: 1981M04 2010M01
Included observations: 346
Total system (balanced) observations 692

	Coefficient	Std. Error	t-Statistic	Prob.
C(1)	-0.190257	0.107082	-1.776747	0.0761
C(2)	-0.082287	0.163182	-0.504266	0.6142
C(3)	0.251895	0.164577	1.530559	0.1263
C(4)	0.455678	0.176684	2.579049	0.0101
C(5)	-0.452403	0.178750	-2.530926	0.0116
C(6)	-0.030980	0.019531	-1.586207	0.1132
C(7)	0.030356	0.099311	0.305669	0.7600
C(8)	-0.275544	0.151340	-1.820687	0.0691
C(9)	0.342563	0.152635	2.244331	0.0251
C(10)	0.596660	0.163863	3.641208	0.0003
C(11)	-0.467722	0.165779	-2.821356	0.0049
C(12)	-0.028402	0.018114	-1.567971	0.1174

Determinant residual covariance 0.001287

Equation: D(TB3) = C(1)*(TB3(-1) - 0.995070905569*TB6(-1) + 0.0975474234967) + C(2)*D(TB3(-1)) + C(3)*D(TB3(-2)) + C(4) *D(TB6(-1)) + C(5)*D(TB6(-2)) + C(6)
Observations: 346

R-squared	0.128282	Mean dependent var	-0.038439
Adjusted R-squared	0.115463	S.D. dependent var	0.382314
S.E. of regression	0.359566	Sum squared resid	43.95771
Durbin-Watson stat	1.909156		

5. $p - value = 0.7600$
6. p-value of 0.7600 > 0.05. We cannot reject the H_0. The test is not significant.
7. This coefficient is not statistically significant. It means that there is no long-term relationship between the TB6 and TB3.

Coefficients C(1) and C(7) are the error correction mechanism (ECM_{t-1}) of equation one and two, respectively. These two coefficients are not statistically significant; since their p-values (0.0761 and 0.7600) both are bigger than 0.05. This result indicates that a long-run relationship does not exist in the relation between TB3 and TB6.

If we run a hypothesis testing for the other coefficients one by one similar to the two hypothesises testing above it can be confirmed that c(4), c(5), c(9), c(10) and c (11) are statistically significant since there p-values: 0.0101, 0.0116, 0.0251, 0.0003, and 0.0049, respectively are smaller than 0.05.

For Model One Since:

1. C(4) is statistically significant: DTB6(−1) effects DTB3.
2. C(5) is statistically significant: DTB6(−2) effects DTB3.

For Model Two Since:

1. C(9) is statistically significant: DTB3(−2) effects DTB6.
2. C(10) is statistically significant: DTB6(−1) effects DTB6.
3. C(11) is statistically significant: DTB6(−2) effects DTB6.

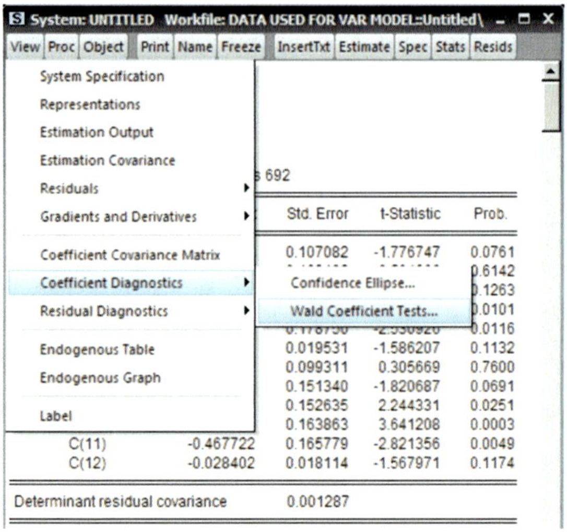

Fig. 10.22 Coefficient diagnostics

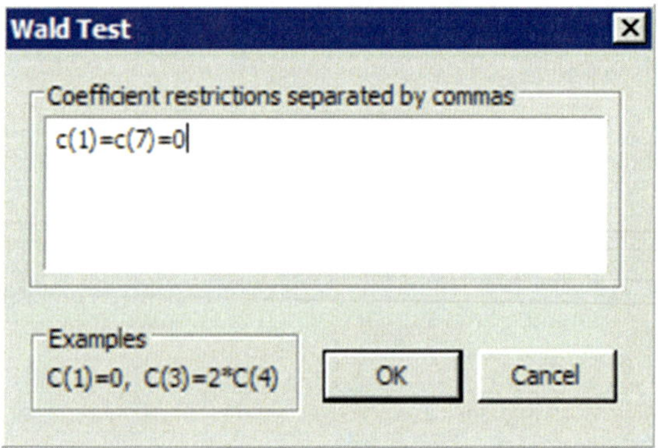

Fig. 10.23 Wald test

Step Six: The Wald Test

Assume we run a joint significant test for C(1) and C(7). We follow the steps below:

From the above output in EViews go to the View > Coefficient Diagnostics > Wald Coefficient Tests as shown in Fig. 10.22. Write the statement of the H_0 below. Click OK to generate output of Fig. 10.23.

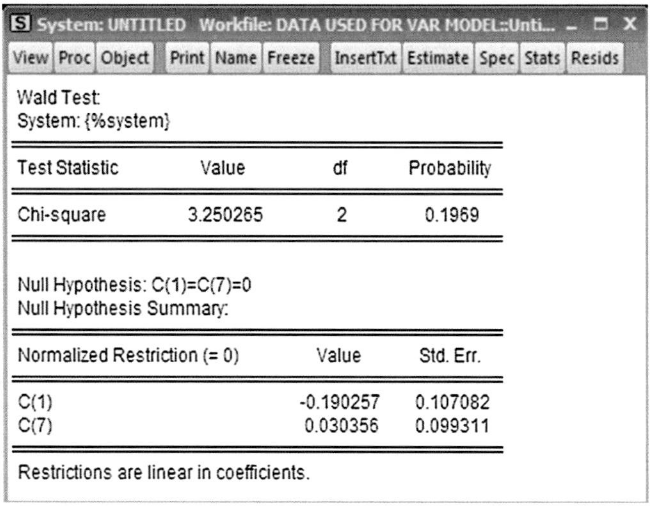

System: UNTITLED Workfile: DATA USED FOR VAR MODEL::Unti...

View | Proc | Object | Print | Name | Freeze | InsertTxt | Estimate | Spec | Stats | Resids

Wald Test:
System: {%system}

Test Statistic	Value	df	Probability
Chi-square	3.250265	2	0.1969

Null Hypothesis: C(1)=C(7)=0
Null Hypothesis Summary:

Normalized Restriction (= 0)	Value	Std. Err.
C(1)	-0.190257	0.107082
C(7)	0.030356	0.099311

Restrictions are linear in coefficients.

Fig. 10.24 Wald test output

1. H_0: C(1) = C(7) = 0,
2. H_1: C(1) and C(7) are not zero,
3. $\alpha = 5\%$ or 0.05,

According to Fig. 10.24, the Chi-square test statistic is: 3.250265.

4. The p-value is: 0.1969,
5. 0.1969 > 0.05. We cannot reject the H_0. The test is not significant.
6. The join coefficient test is not statistically significant.

Step Seven: Residuals Tests - The Portmanteau Autocorrelation Test

Go to the View of the VECM output of the EViews > Residual Diagnostics > Portmanteau Autocorrelation Test as shown in Fig. 10.25. For the number of lags you can choose 3. Then click OK to generate output of Fig. 10.26.

1. H_0: There is no problem of serial correlation
2. H_1: There is a problem of serial correlation
3. $\alpha = 5\%$ or 0.05
4. According to Fig. 10.27, for one lag the p-value is: 0.9746, for two legs the p-value is: 09514 and for three lags the p-value is: 0.9821.

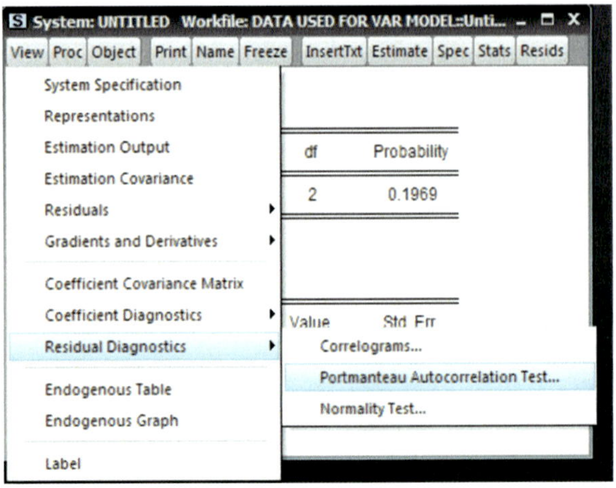

Fig. 10.25 Residual diagnostics

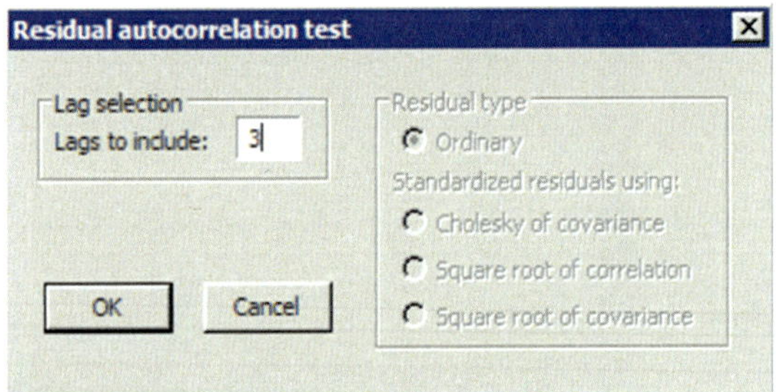

Fig. 10.26 Residual autocorrelation test

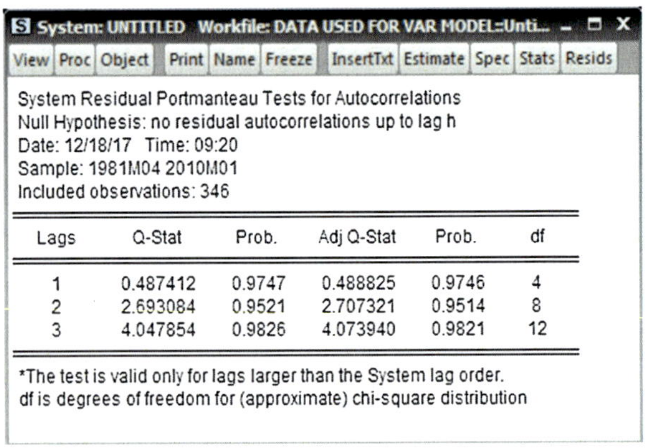
System: UNTITLED Workfile: DATA USED FOR VAR MODEL::Unti...

View Proc Object Print Name Freeze InsertTxt Estimate Spec Stats Resids

System Residual Portmanteau Tests for Autocorrelations
Null Hypothesis: no residual autocorrelations up to lag h
Date: 12/18/17 Time: 09:20
Sample: 1981M04 2010M01
Included observations: 346

Lags	Q-Stat	Prob.	Adj Q-Stat	Prob.	df
1	0.487412	0.9747	0.488825	0.9746	4
2	2.693084	0.9521	2.707321	0.9514	8
3	4.047854	0.9826	4.073940	0.9821	12

*The test is valid only for lags larger than the System lag order.
df is degrees of freedom for (approximate) chi-square distribution

Fig. 10.27 Residual Portmanteau Tests for Autocorrelations

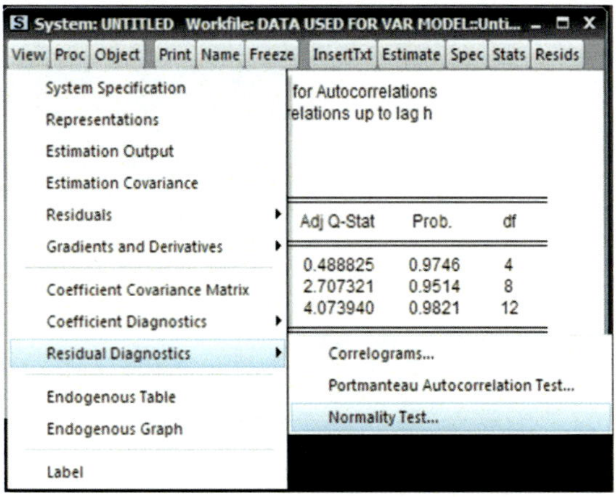

Fig. 10.28 Residual Diagnostic: Normality Test

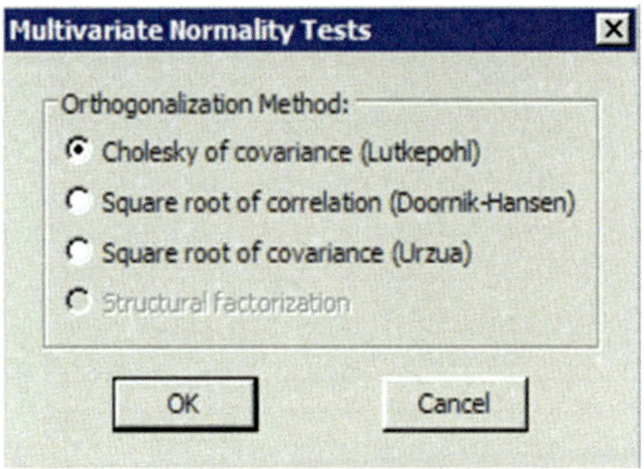

Fig. 10.29 Multivariate Normality Tests

5. ALL p-values are bigger than 0.05. We cannot reject the H_0. The test is not significant.
6. These results show that there is no residual autocorrelations.

The Normality Test:

Go to the View of the VECM output of the EViews > Residual Diagnostics > Normality Test as shown in Fig. 10.28
Then choose Cholesky of Covariance (lutkepohl) as shown in Fig. 10.29. Then click OK which generates output in Fig. 10.30.

1. H_0: Residuals are multivariate normal,
2. H_A: Residuals are not multivariate normal.
3. $\alpha = 5\%$ or 0.05
4. The p-values for both components and for the joint of Jarque-Bera Test are 0.0000.

System: UNTITLED Workfile: DATA USED FOR VAR MODEL::Unti...

View | Proc | Object | Print | Name | Freeze | InsertTxt | Estimate | Spec | Stats | Resids

System Residual Normality Tests
Orthogonalization: Cholesky (Lutkepohl)
Null Hypothesis: residuals are multivariate normal
Date: 12/18/17 Time: 09:24
Sample: 1981M04 2010M01
Included observations: 346

Component	Skewness	Chi-sq	df	Prob.
1	-1.360991	106.8158	1	0.0000
2	-0.087683	0.443360	1	0.5055
Joint		107.2592	2	0.0000

Component	Kurtosis	Chi-sq	df	Prob.
1	18.67092	3540.412	1	0.0000
2	7.453558	285.9427	1	0.0000
Joint		3826.355	2	0.0000

Component	Jarque-Bera	df	Prob.
1	3647.228	2	0.0000
2	286.3861	2	0.0000
Joint	3933.614	4	0.0000

Fig. 10.30 System residuals Normality tests

5. Since p-values are smaller than 0.05, we reject the H_0. The test is significant.
6. These results show that residuals are not multivariate normal.

Step Eight: The Granger Causality Test

Go to the View from the first VECM output (The one is shown below) > Lag Structure > Granger Causality/Block Erogeneity Tests as shown in Fig. 10.31.

Which generates the output in Fig. 10.32
Two tables, one for each model, are given.

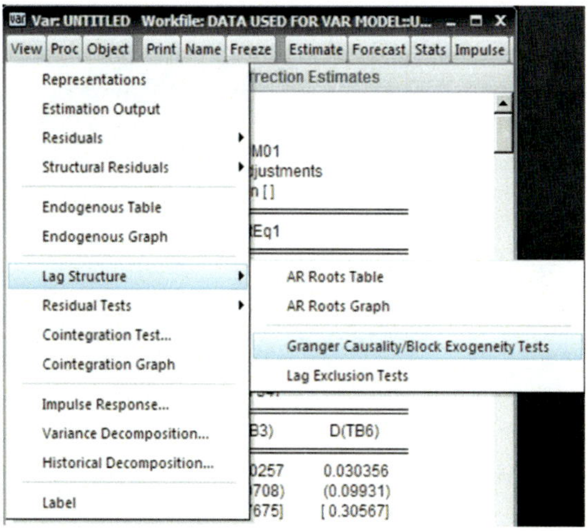

Fig. 10.31 Granger Causality/Block Exogeneity Tests

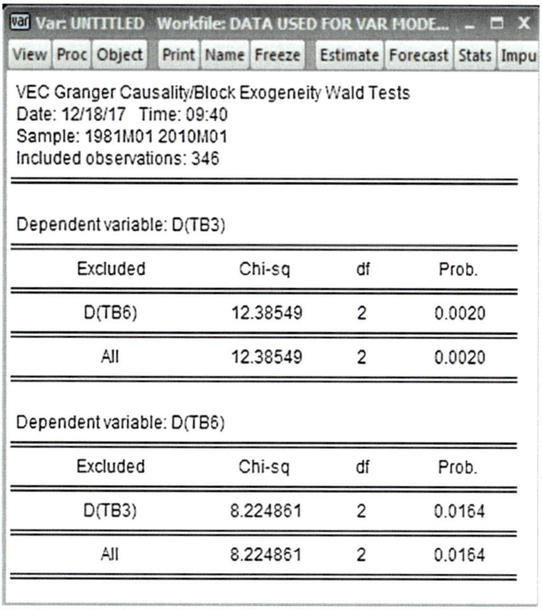

Var: UNTITLED Workfile: DATA USED FOR VAR MODE...

View Proc Object Print Name Freeze Estimate Forecast Stats Impu

VEC Granger Causality/Block Exogeneity Wald Tests
Date: 12/18/17 Time: 09:40
Sample: 1981M01 2010M01
Included observations: 346

Dependent variable: D(TB3)

Excluded	Chi-sq	df	Prob.
D(TB6)	12.38549	2	0.0020
All	12.38549	2	0.0020

Dependent variable: D(TB6)

Excluded	Chi-sq	df	Prob.
D(TB3)	8.224861	2	0.0164
All	8.224861	2	0.0164

Fig. 10.32 VEC Granger Causality/Block Exogeneity Wald Tests

Results from Table One:

Probabilities of DTB6 and ALL are smaller than 0.05. We can claim that DTB6 and all other variables of the Model One granger cause the DTB3, the dependent variable.

Results from Table Two:

Probabilities of DTB3 and ALL are smaller than 0.05. We can claim that DTB3 and all other variables of the Model Two granger cause the DTB6, the dependent variable.

Appendix 10.1: The Wald Test

This test is available in EViews and is used for testing restrictions on parameters, especially those derived from regression models. For example, suppose you are conducting a regression analysis where the dependent variable is the price of tea. Further, suppose you believe that the price of tea at time t depends on the prices of coffee at times t − 1 and t − 2, then the model to be examined would be:

$$\text{TEA PRICE}_t = \alpha + \beta_1 \text{COFFEE PRICE}_{t-1} + \beta_2 \text{COFFEE PRICE}_{t-2}.$$

If you believe that the coffee prices at times t − 1 and t − 2 had an equal impact on the price of tea at time t, then you would want to test the parameter restriction $H_0 : \beta_1 = \beta_2$ or equivalently that $H_0 : \beta_1 - \beta_2 = 0$. Parameter restrictions can take a variety of forms and there may be more than one of them. For example you could test the restrictions that $\alpha = 0$ **and** that $\beta_1 - \beta_2 = 0$ in the above regression. This would be your null hypothesis. In this case, it is called a ***composite hypothesis***, since it consists of more than one part. Testing the above is performed via the Wald test and its associated F statistic.

I mention this test because it has an application that might be of interest to you in the combination forecasting part of your thesis. Of course, most authors use some measure(s) of forecasting adequacy when using regression, ARIMA etc.—measures such as MAPE and RMSE. However, not so many authors test their forecasts for ***unbiassedness*** which is a property that in my view is just as important as a low MAPE.

If a statistic is biased then it is not estimating a population parameter efficiently. There will be consistent error involved in the estimation process. It may be shown that the sample mean is an unbiased estimator of the population mean; the sample regression gradient is an unbiased estimator of the population regression gradient. However, such "logical" rules do not always apply. For example, the sample variance s^2 is a biased estimator of the population variance σ^2 because it consistently underestimates the latter's value. In fact an unbiased estimator of the population variance is $\widehat{\sigma}^2 = \dfrac{ns^2}{n-1}$.

Now consider your forecasts $\widehat{Y}_t$ and your observed values Y_t recorded over the hold back period. Suppose you regress Y_t against $\widehat{Y}_t$:

$$Y_t = \alpha + \beta \widehat{Y}_t.$$

The composite hypothesis H_0: $\alpha = 0$ and $\beta = 1$ is a sufficient for $\widehat{Y}_t$ to be an unbiased estimator of Y_t. You perform this regression in EViews via:

Quick

Estimate Equation

and the form of the regression would be Y C YHAT. Having run this, EViews will refer to the coefficient α as C(1) and β as C(2). Once the regression results have been generated, click the 'View' button, then:

Coefficient tests

Wald – Coefficient Restrictions

and type in C(1) = 0, C(2) = 1 in the box provided. (**Note:** a comma must separate each restriction). You will be given the value of the F statistic associated with the Wald test and its significance level. You reject H_0 if the significance is less than 0.05, since this is a one tailed test. Acceptance of the null indicates that the forecasts in question are unbiased estimators of Y_t. (**Note:** EViews also generates a chi-square statistic associated with the Wald test. Should this latter statistic contradict the F statistic, the user **must** opt for F since it is more sensitive to the sample size in that one of its two degrees of freedom depends on the size of the sample).

Of course, one could apply the Wald test to the forecasts derived from the individual models and/or the combined models (average method and variance-covariance method).

Chapter 11
Panel Data Analysis

The regression models discussed so far primarily used either cross-sectional or time series data. Each of these types of data has its exclusive features. This chapter discusses **panel data regression models** using the same group of entities like individuals, firms, states, countries, **over time**. Panel data has many advantages over pure cross-sectional data or pure time series data. The advantages are as follows:

1. Since panel data deals with individuals, firms, states, countries and so on over time, there is bound to be ***heterogeneity*** in these units, which may be often unobservable. The panel data estimation techniques can take such heterogeneity into account by allowing for ***subject-specific variables***. The term ***subject*** includes micro units such individuals, firms or states.
2. By combining time series of cross-sectional observations, panel data gives "***more informative data, more variability, less collinearity among variables, more degrees of freedom and more efficiency***".
3. By studying the repeated cross-sections of observations, panel data are better suited to study ***the dynamics of change***. For example, unemployment, job turnover, duration of unemployment, and labor mobility are better studied with panel data.
4. Panel data can better ***detect and measure effects that cannot be observed in pure cross-sectional or time series data***. For example, the effects of minimum wage laws on employment and earnings can be better studied if we follow successive waves of increases in federal and/or state minimum wages.
5. Phenomena such as ***economies of scale and technological change*** can be better studied by panel data than by pure cross-sectional or pure time series data.

We call a panel data ***a balanced panel*** if the number of time observations is the same for each individual. If that was not the case we call it ***an unbalanced panel***.

We call a panel data ***a short panel when the number of cross-sectional or individual units N is greater than the number of time periods, T***.

A. Aljandali, M. Tatahi, *Economic and Financial Modelling with EViews*, Statistics and Econometrics for Finance, https://doi.org/10.1007/978-3-319-92985-9_11

In ***a long panel, on the other hand, T is greater than N***.

To estimate a model with panel data there are five options:

1. ***Individual time series functions*** can be used to estimate models by OLS.
2. ***Cross-sectional functions*** one for each year can be used to estimate models by OLS.
3. ***Pooled OLS function***: We can pool all observations, time series and cross-sectional, and estimate a "grand" function, neglecting the dual nature of time series and cross-sectional data. But, such a pooling assumes that the coefficients of the function remain constant across time and cross-section. The pooled OLS estimation is also known as the *constant coefficient model*, for we are assuming that coefficients across time and cross-section remain the same.
4. ***Fixed effects least-squares dummy variable (LSDV) model***: As in option 3, we pool all observations, but allow each individual to have his or her individual intercept dummy. A variant of this is the **within estimator**, which we will explain shortly.
5. ***The random effects model***: Instead of allowing each individual to have their own (fixed) intercept value as in LSDV, we assume that the intercept values of all individuals are random drawings from a much larger population of individuals.

We will discuss options 3, 4 and 5 sequentially.
Before we run these three options we need to examine two points:

1. Are panel data variables stationary?
2. Are they cointegrated?

11.1 Panel Stationary Approach

We have two panel data in this example. We convert them to the log or ln (natural log) through the use of one of the approaches below:

Note: We treat the log or ln data of the panel data as our original data.

Go to Quick-Generate Series and write LGDP = Log(GDP). Click OK.
Go to Quick-Generate Series and write LGEX = Log(GEX). Click OK.

If you want to have natural logs of the data do as follows:

Go to Quick-Generate Series and write LGDP = @Log(GDP). Click OK.
Go to Quick-Generate Series and write LGEX = @Log(GEX). Click OK.

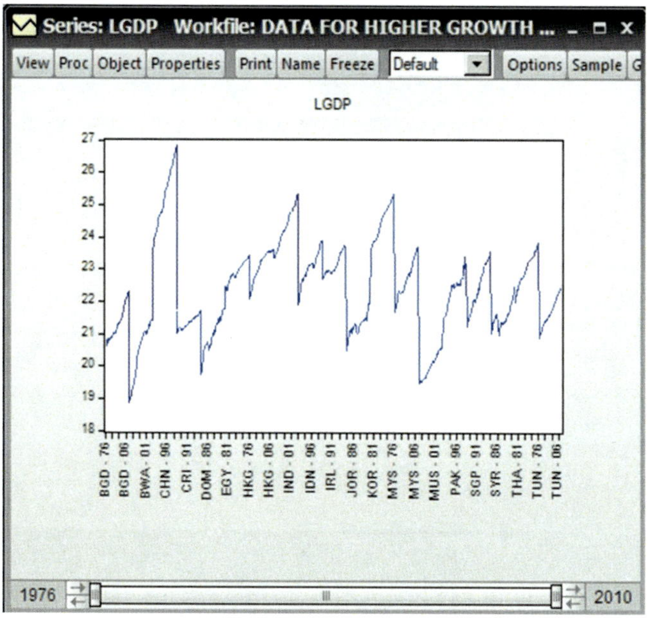

Fig. 11.1 Plot of LGDP series

11.1.1 The LGDP Panel Data

Figure 11.1 presents the graph of the series. In order to make sure the LGDP panel data is stationary we need to run a unit root test as follow:

Open the LGDP data. From the View choose: **Unit Root Test** and then click **OK**.

The Test:

1. H_0: The panel data of LGDP has a unit root,
2. H_1: The panel data LGDP is stationary,
3. $\alpha = 5\%$ or 0.05,

According to Fig. 11.2, the four test statistics are equivalent to −1.67631, 3.64252, 24.5320 and 35.8464.

4. The p-values of these test statistics are 0.0468, 0.9999, 0.9553 and 0.5694.
5. The first p-value is smaller than 0.05 but the rest are bigger. As the majority is bigger than 0.05 cannot reject the H_0. The test is not significant.
6. The LGDP Panel data is not stationary.

Fig. 11.2 Panel unit root test on LGDP

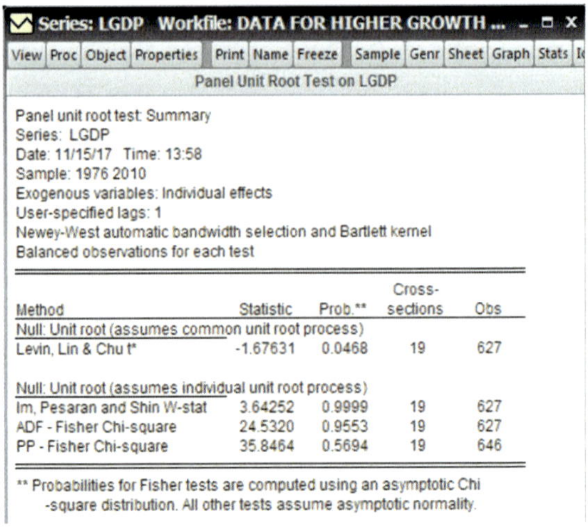
Series: LGDP Workfile: DATA FOR HIGHER GROWTH ...

Panel Unit Root Test on LGDP

Panel unit root test: Summary
Series: LGDP
Date: 11/15/17 Time: 13:58
Sample: 1976 2010
Exogenous variables: Individual effects
User-specified lags: 1
Newey-West automatic bandwidth selection and Bartlett kernel
Balanced observations for each test

Method	Statistic	Prob.**	Cross-sections	Obs
Null: Unit root (assumes common unit root process)				
Levin, Lin & Chu t*	-1.67631	0.0468	19	627
Null: Unit root (assumes individual unit root process)				
Im, Pesaran and Shin W-stat	3.64252	0.9999	19	627
ADF - Fisher Chi-square	24.5320	0.9553	19	627
PP - Fisher Chi-square	35.8464	0.5694	19	646

** Probabilities for Fisher tests are computed using an asymptotic Chi -square distribution. All other tests assume asymptotic normality.

Fig. 11.3 Plot of LGEX

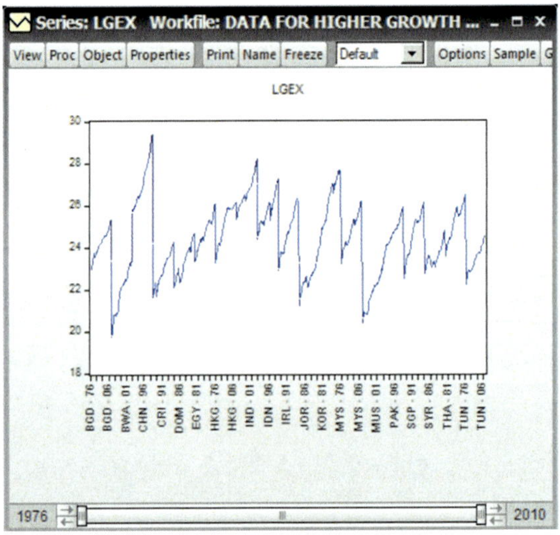

11.1.2 The LGEX Panel Data

Figure 11.3 shows the graph of the LGEX:

The data seems to be stationary as the line of the series has crossed the zero line many times.

However, in order to make sure the LGEX panel data is stationary or not we need to run a unit root test as follows:

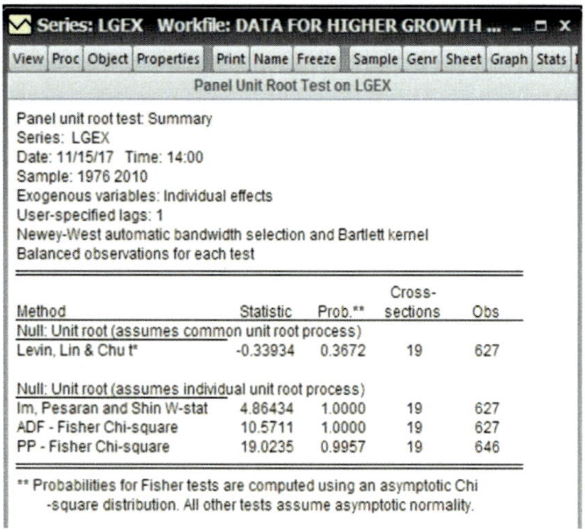
Series: LGEX Workfile: DATA FOR HIGHER GROWTH ...

View Proc Object Properties Print Name Freeze Sample Genr Sheet Graph Stats

Panel Unit Root Test on LGEX

Panel unit root test: Summary
Series: LGEX
Date: 11/15/17 Time: 14:00
Sample: 1976 2010
Exogenous variables: Individual effects
User-specified lags: 1
Newey-West automatic bandwidth selection and Bartlett kernel
Balanced observations for each test

Method	Statistic	Prob.**	Cross-sections	Obs
Null: Unit root (assumes common unit root process)				
Levin, Lin & Chu t*	-0.33934	0.3672	19	627
Null: Unit root (assumes individual unit root process)				
Im, Pesaran and Shin W-stat	4.86434	1.0000	19	627
ADF - Fisher Chi-square	10.5711	1.0000	19	627
PP - Fisher Chi-square	19.0235	0.9957	19	646

** Probabilities for Fisher tests are computed using an asymptotic Chi -square distribution. All other tests assume asymptotic normality.

Fig. 11.4 Panel unit root test on LGEX

Open the LGEX data. From the View choose: Unit Root Test and then click OK which generates output of Fig. 11.4.

The Test:

1. H_0: The panel data of LGEX has a unit root,
2. H_1: The panel data of LGEX is stationary.
3. $\alpha = 5\%$ or 0.05

According to Fig. 11.4, the four test statistics are equivalent to: −0.33934, 4.86434, 10.5711 and 19.235.

4. The p-values of these test statistics are 0.3672, 1.0000, 1.0000 and 0.9957.
5. All p-values are bigger than 0.05. We cannot reject the H_0. The test is not significant.
6. The LGEX Panel data is not stationary.

11.1.3 The First Difference of the Data

In order to find the first differences of the two panel data we follow these steps:

Go to Quick-Generate Series-write DLGDP = LGDP − LGDP(−1) and then OK.
Go to Quick-Generate Series-write DLGEX = LGEX − LGEX(−1) and the OK.

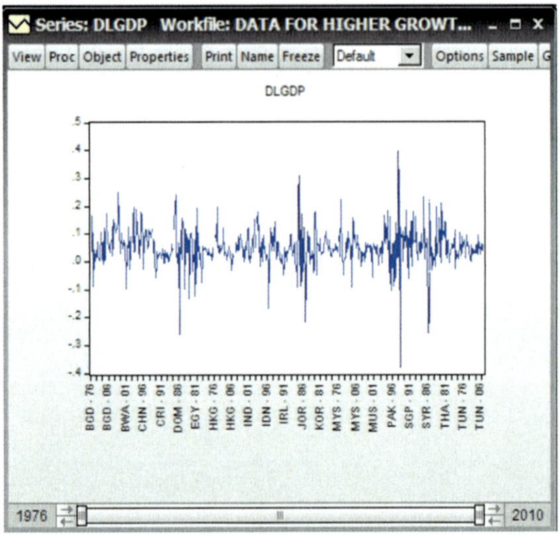

Fig. 11.5 Plot of DGDP

11.1.4 The DLGDP Panel Data

Figure 11.5, presents the graph of the DGDP:

The data seems to be stationary as the line of the series has crossed the zero line many times.

However, in order to make sure the DLGDP panel data is stationary or not we need to run a unit root test as follow:

Open the DLGDP data. From the View choose: Unit Root Test and then click OK which generates Fig. 11.6.

The Test:

1. H_0: The panel data of DLGDP has a unit root.
2. H_1: The panel data of DLGDP is stationary.
3. $\alpha = 5\%$ or 0.05.

The output shows four test statistics: −6.61669, −8.00521, 136.808 and 293.287.

4. The p-values of these test statistics are all 0.0000.
5. Prob = 0.0000 < 0.05 therefore we reject the H_0
6. The DLGDP Panel data is stationary.

Fig. 11.6 Panel unit Root Test on DGDP

Series: DLGDP Workfile: DATA FOR HIGHER GROWT...

View | Proc | Object | Properties | Print | Name | Freeze | Sample | Genr | Sheet | Graph | Stats

Panel Unit Root Test on DLGDP

Panel unit root test: Summary
Series: DLGDP
Date: 11/15/17 Time: 14:03
Sample: 1976 2010
Exogenous variables: Individual effects
User-specified lags: 1
Newey-West automatic bandwidth selection and Bartlett kernel
Balanced observations for each test

Method	Statistic	Prob.**	Cross-sections	Obs
Null: Unit root (assumes common unit root process)				
Levin, Lin & Chu t*	-6.61669	0.0000	19	608
Null: Unit root (assumes individual unit root process)				
Im, Pesaran and Shin W-stat	-8.00521	0.0000	19	608
ADF - Fisher Chi-square	136.808	0.0000	19	608
PP - Fisher Chi-square	293.287	0.0000	19	627

** Probabilities for Fisher tests are computed using an asymptotic Chi-square distribution. All other tests assume asymptotic normality.

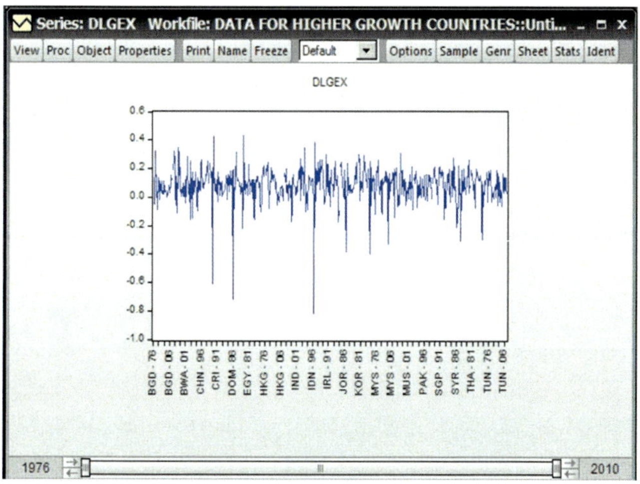

Fig. 11.7 Plot of DLGEX series

11.1.5 The DLGEX Panel Data

Figure 11.7 presents the DLGEX

The data seems to be stationary as the line of the series has crossed the zero line many times. However, in order to make sure the DLGEX panel data is stationary or not we need to run a unit root test as follow:

Open the DLGEX data. From the View choose: Unit Root Test and then click OK which generates output of Fig. 11.8

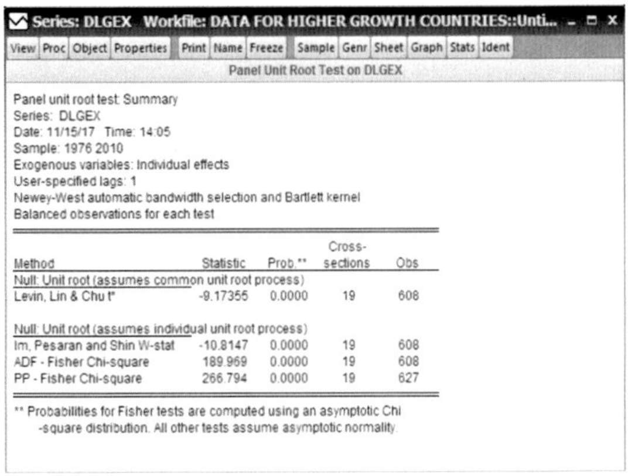

Series: DLGEX Workfile: DATA FOR HIGHER GROWTH COUNTRIES::Unti...

View Proc Object Properties Print Name Freeze Sample Genr Sheet Graph Stats Ident

Panel Unit Root Test on DLGEX

Panel unit root test: Summary
Series: DLGEX
Date: 11/15/17 Time: 14:05
Sample: 1976 2010
Exogenous variables: Individual effects
User-specified lags: 1
Newey-West automatic bandwidth selection and Bartlett kernel
Balanced observations for each test

Method	Statistic	Prob.**	Cross-sections	Obs
Null: Unit root (assumes common unit root process)				
Levin, Lin & Chu t*	-9.17355	0.0000	19	608
Null: Unit root (assumes individual unit root process)				
Im, Pesaran and Shin W-stat	-10.8147	0.0000	19	608
ADF - Fisher Chi-square	189.969	0.0000	19	608
PP - Fisher Chi-square	266.794	0.0000	19	627

** Probabilities for Fisher tests are computed using an asymptotic Chi -square distribution. All other tests assume asymptotic normality.

Fig. 11.8 Panel Unit Root Test on DLGEX

The Test:

1. H_0: The panel data of DLGEX has a unit root.
2. H_1: The panel data of DLGEX is stationary.
3. $\alpha = 5\%$ or 0.05.

The output shows four test statistics: −9.17355, −108147, 189.969 and 266.794.

4. The p-values of these test statistics are all 0.0000.
5. Prob = 0.0000 < 0.05 therefore we reject the H_0
6. The DLGEX Panel data is stationary.

We run a regression by using the original panel data of this model (Fig. 11.9): $lGDP_{it} = \alpha_i + \beta_i lGEX_{it} + \varepsilon_{it}$

After running this model by using the OLS method save the residual by generating a variable call it, ECM. Do as follows:

After having the regression output: Go to **Quick-Generate Series**-write ECM = RESID and then click OK.

Which generates the graph of the Residual-ECM in Fig. 11.10:

The data seems to be stationary as the line of the series has crossed the zero line many times.

However, in order to make sure the ECM is stationary, we need to run a unit root test as follows:

Open the ECM data. From the View choose: Unit Root Test and then click OK which generates output in Fig. 11.11

Fig. 11.9 Cointegration test

Equation: UNTITLED Workfile: DATA FOR HIGHE...

View | Proc | Object | Print | Name | Freeze | Estimate | Forecast | Stats | Resids

Dependent Variable: LGDP
Method: Panel Least Squares
Date: 11/15/17 Time: 14:07
Sample: 1976 2010
Periods included: 35
Cross-sections included: 19
Total panel (balanced) observations: 665

Variable	Coefficient	Std. Error	t-Statistic	Prob.
C	2.263669	0.255219	8.869527	0.0000
LGEX	0.825913	0.010446	79.06733	0.0000

R-squared	0.904117	Mean dependent var	22.39565
Adjusted R-squared	0.903972	S.D. dependent var	1.456159
S.E. of regression	0.451240	Akaike info criterion	1.249370
Sum squared resid	134.9987	Schwarz criterion	1.262903
Log likelihood	-413.4156	Hannan-Quinn criter.	1.254614
F-statistic	6251.643	Durbin-Watson stat	0.053854
Prob(F-statistic)	0.000000		

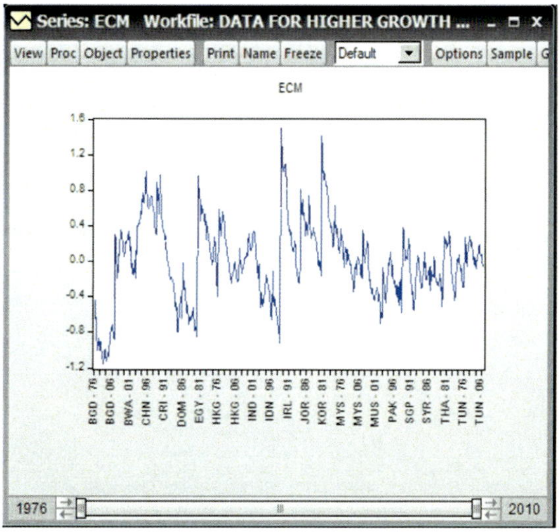

Fig. 11.10 Plot of the ECM residuals

The Test:

1. H_0: The residuals are not stationary
2. H_1: The residuals are stationary.
3. $\alpha = 5\%$ or 0.05.
4. Four out of six probabilities of these six tests are smaller than 0.05.
5. As most of the p-values are smaller than 0.05, we reject the H_0
6. The ECM data is stationary (Figs. 11.12, 11.13, 11.14, 11.15 and 11.16).

Fig. 11.11 The Levin, Lin & Chu Unit Root Test on ECM

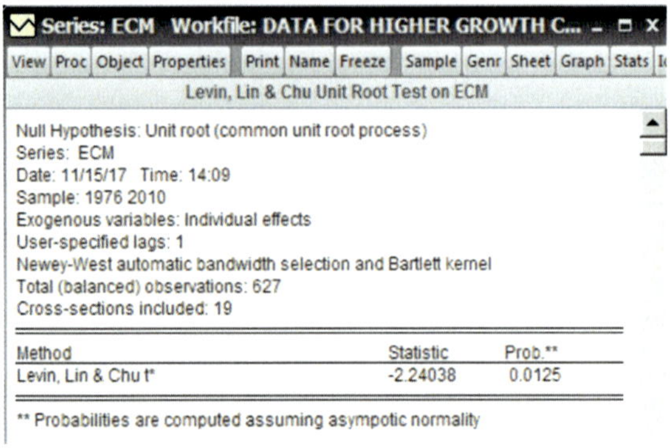

Series: ECM Workfile: DATA FOR HIGHER GROWTH C...

View | Proc | Object | Properties | Print | Name | Freeze | Sample | Genr | Sheet | Graph | Stats

Levin, Lin & Chu Unit Root Test on ECM

Null Hypothesis: Unit root (common unit root process)
Series: ECM
Date: 11/15/17 Time: 14:09
Sample: 1976 2010
Exogenous variables: Individual effects
User-specified lags: 1
Newey-West automatic bandwidth selection and Bartlett kernel
Total (balanced) observations: 627
Cross-sections included: 19

Method	Statistic	Prob.**
Levin, Lin & Chu t*	-2.24038	0.0125

** Probabilities are computed assuming asympotic normality

Fig. 11.12 The Breitung Unit Root Test on ECM

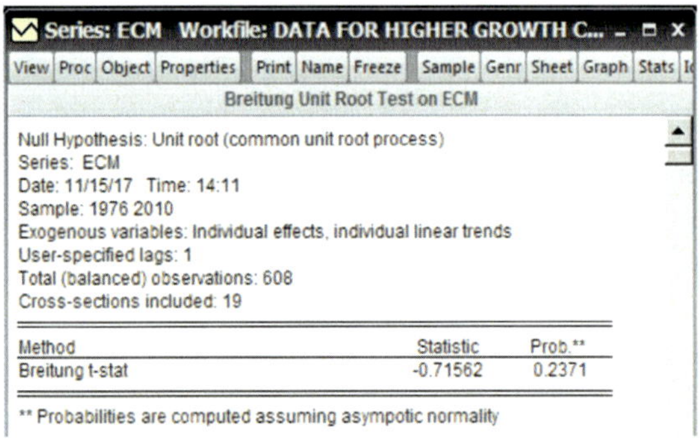

Series: ECM Workfile: DATA FOR HIGHER GROWTH C...

View | Proc | Object | Properties | Print | Name | Freeze | Sample | Genr | Sheet | Graph | Stats

Breitung Unit Root Test on ECM

Null Hypothesis: Unit root (common unit root process)
Series: ECM
Date: 11/15/17 Time: 14:11
Sample: 1976 2010
Exogenous variables: Individual effects, individual linear trends
User-specified lags: 1
Total (balanced) observations: 608
Cross-sections included: 19

Method	Statistic	Prob.**
Breitung t-stat	-0.71562	0.2371

** Probabilities are computed assuming asympotic normality

Fig. 11.13 The IM, Pesaran and Shin Unit Root Test on ECM

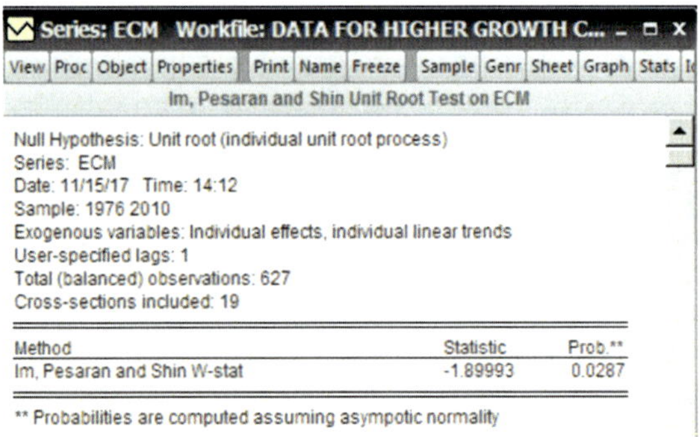

Series: ECM Workfile: DATA FOR HIGHER GROWTH C...

View | Proc | Object | Properties | Print | Name | Freeze | Sample | Genr | Sheet | Graph | Stats

Im, Pesaran and Shin Unit Root Test on ECM

Null Hypothesis: Unit root (individual unit root process)
Series: ECM
Date: 11/15/17 Time: 14:12
Sample: 1976 2010
Exogenous variables: Individual effects, individual linear trends
User-specified lags: 1
Total (balanced) observations: 627
Cross-sections included: 19

Method	Statistic	Prob.**
Im, Pesaran and Shin W-stat	-1.89993	0.0287

** Probabilities are computed assuming asympotic normality

11.2 The Panel ECM Model

The cointegration test showed the residual is stationary. This indicates that there is a long-run relationship between the variables of this model. This result allows us to run a panel ECM model.

For this ECM model we can now run three panel data regression models as discussed above: model 3, 4 and 5.

Fig. 11.14 The ADF Fisher Unit Root Test on ECM

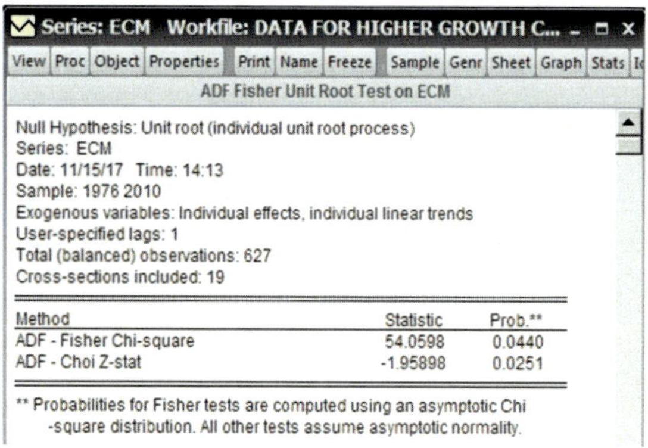
Series: ECM Workfile: DATA FOR HIGHER GROWTH C...

View Proc Object Properties Print Name Freeze Sample Genr Sheet Graph Stats

ADF Fisher Unit Root Test on ECM

Null Hypothesis: Unit root (individual unit root process)
Series: ECM
Date: 11/15/17 Time: 14:13
Sample: 1976 2010
Exogenous variables: Individual effects, individual linear trends
User-specified lags: 1
Total (balanced) observations: 627
Cross-sections included: 19

Method	Statistic	Prob.**
ADF - Fisher Chi-square	54.0598	0.0440
ADF - Choi Z-stat	-1.95898	0.0251

** Probabilities for Fisher tests are computed using an asymptotic Chi -square distribution. All other tests assume asymptotic normality.

Fig. 11.15 The Phillips-Perron Fisher Unit Root Test on ECM

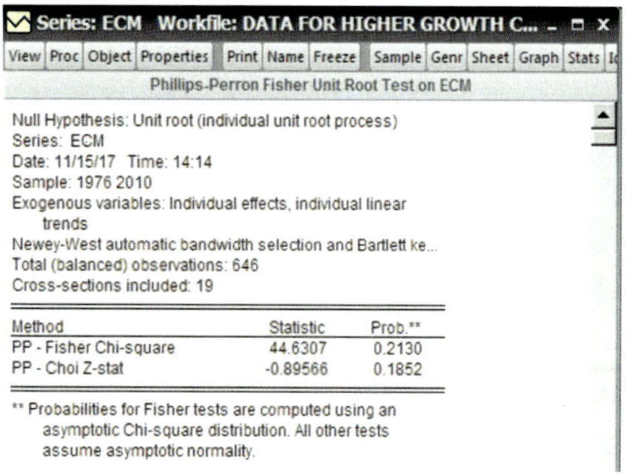
Series: ECM Workfile: DATA FOR HIGHER GROWTH C...

View Proc Object Properties Print Name Freeze Sample Genr Sheet Graph Stats

Phillips-Perron Fisher Unit Root Test on ECM

Null Hypothesis: Unit root (individual unit root process)
Series: ECM
Date: 11/15/17 Time: 14:14
Sample: 1976 2010
Exogenous variables: Individual effects, individual linear trends
Newey-West automatic bandwidth selection and Bartlett ke...
Total (balanced) observations: 646
Cross-sections included: 19

Method	Statistic	Prob.**
PP - Fisher Chi-square	44.6307	0.2130
PP - Choi Z-stat	-0.89566	0.1852

** Probabilities for Fisher tests are computed using an asymptotic Chi-square distribution. All other tests assume asymptotic normality.

Fig. 11.16 The Hadri Unit Root Test on ECM

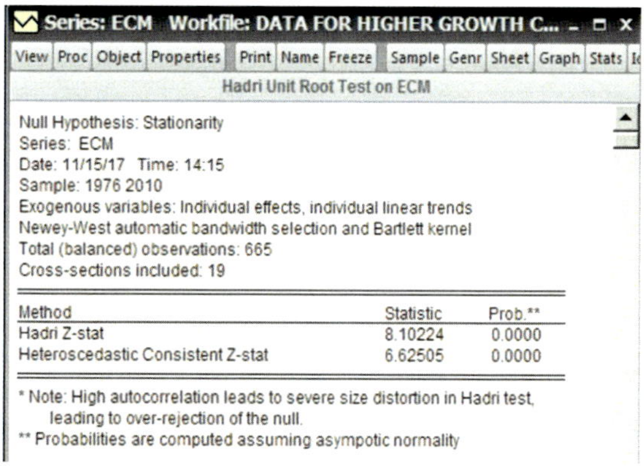
Series: ECM Workfile: DATA FOR HIGHER GROWTH C...

View Proc Object Properties Print Name Freeze Sample Genr Sheet Graph Stats

Hadri Unit Root Test on ECM

Null Hypothesis: Stationarity
Series: ECM
Date: 11/15/17 Time: 14:15
Sample: 1976 2010
Exogenous variables: Individual effects, individual linear trends
Newey-West automatic bandwidth selection and Bartlett kernel
Total (balanced) observations: 665
Cross-sections included: 19

Method	Statistic	Prob.**
Hadri Z-stat	8.10224	0.0000
Heteroscedastic Consistent Z-stat	6.62505	0.0000

* Note: High autocorrelation leads to severe size distortion in Hadri test, leading to over-rejection of the null.
** Probabilities are computed assuming asympotic normality

11.2.1 Pooled OLS regression

Consider the following relationship.

$$LGDP_{it} = \beta_1 + \beta_2 LGEX_{it} + \varepsilon_{it}$$

It is assumed the government expenditure is main drive of economic growth. Other variables are assumed to be less important. We have put two subscripts on the variables: i, representing the cross-section unit, and t, the time.

It is also assumed that the error term satisfies the usual classical assumptions.

The ECM Model we will estimate is:

$$DLGDP_{it} = \beta_1 + \beta_2 DGP_{it-1} + \beta_3 DGEX_{it} + \beta_4 DGEX_{it-1} + \beta_5 ECM_{it-1} + \epsilon_{it}$$

We assume the pooling of the data is valid.

Using *EViews*, we obtained the results of the below output:

Go to Quick and choose Estimate Equation. DLGDP C DLGDP(−1) DLGEX DLGEX(−1) ECM(−1). Then click, OK which generates output of Fig. 11.17.

The Result:

1. If we run hypothesis testings, one by one, for the coefficients of variables: DLGDP(−1), DLGEX, DLGEX(−1) and ECM(−1) it can be seen that all the coefficients, including the intercept, except for DLGDP(−1), are statistically significant since their probabilities are smaller than 0.05.

Fig. 11.17 Panel Least Squares

Equation: UNTITLED Workfile: DATA FOR HIGHER GROWTH COUNTRI...

View | Proc | Object | Print | Name | Freeze | Estimate | Forecast | Stats | Resids

Dependent Variable: DLGDP
Method: Panel Least Squares
Date: 01/23/17 Time: 12:42
Sample (adjusted): 1978 2010
Periods included: 33
Cross-sections included: 19
Total panel (balanced) observations: 627

Variable	Coefficient	Std. Error	t-Statistic	Prob.
C	0.030219	0.003752	8.054479	0.0000
DLGDP(-1)	0.046662	0.039238	1.189215	0.2348
DLGEX	0.126615	0.022578	5.607943	0.0000
DLGEX(-1)	0.089529	0.022823	3.922756	0.0001
ECM(-1)	-0.013294	0.005829	-2.280720	0.0229

R-squared	0.095810	Mean dependent var	0.050546
Adjusted R-squared	0.089996	S.D. dependent var	0.066775
S.E. of regression	0.063699	Akaike info criterion	-2.661354
Sum squared resid	2.523803	Schwarz criterion	-2.625939
Log likelihood	839.3343	Hannan-Quinn criter.	-2.647595
F-statistic	16.47719	Durbin-Watson stat	2.028725
Prob(F-statistic)	0.000000		

2. Since the coefficient of the ECM(−1) is negative (−0.013294) and it is statistically significant it can be concluded that there is a long run relationship between these variables.
3. The R^2 of the model is very low (only 0.095810). The F statistic's probability shows the joint coefficient of the model is statistically significant.
4. The Durbin-Watson Stat shows the model does not have serial correlation problem (2.028725).

11.2.2 The Fixed Effects Least Squares Dummy Variable (LSDV) Model

The data we use looks like this screen shot:

1	COUNTRY	YEAR	GDP	GEX
2	BGD	2010	4.88E+09	1E+11
3	BGD	2009	4.48E+09	8.94E+10
4	BGD	2008	4.23E+09	7.96E+10
5	BGD	2007	4.08E+09	6.84E+10
6	BGD	2006	3.84E+09	6.19E+10
7	BGD	2005	3.62E+09	6.03E+10
8	BGD	2004	3.36E+09	5.66E+10
9	BGD	2003	3.04E+09	5.19E+10
10	BGD	2002	2.68E+09	4.76E+10
11	BGD	2001	2.25E+09	4.7E+10
12	BGD	2000	2.15E+09	4.71E+10
13	BGD	1999	2.13E+09	4.57E+10
14	BGD	1998	2.12E+09	4.41E+10
15	BGD	1997	1.87E+09	4.23E+10
16	BGD	1996	1.82E+09	4.07E+10
17	BGD	1995	1.83E+09	3.79E+10
18	BGD	1994	1.79E+09	3.38E+10
19	BGD	1993	1.73E+09	3.32E+10
20	BGD	1992	1.55E+09	3.17E+10
21	BGD	1991	1.4E+09	3.1E+10
22	BGD	1990	1.37E+09	3.01E+10
23	BGD	1989	1.36E+09	2.68E+10
24	BGD	1988	1.35E+09	2.56E+10
25	BGD	1987	1.34E+09	2.38E+10
26	BGD	1986	1.32E+09	2.12E+10
27	BGD	1985	1.22E+09	2.16E+10
28	BGD	1984	1.2E+09	1.97E+10
29	BGD	1983	1.16E+09	1.72E+10
30	BGD	1982	1.13E+09	1.81E+10
31	BGD	1981	1.13E+09	1.98E+10
32	BGD	1980	1.08E+09	1.81E+10
33	BGD	1979	1.11E+09	1.56E+10
34	BGD	1978	9.98E+08	1.33E+10
35	BGD	1977	1.09E+09	9.63E+09
36	BGD	1976	9.21E+08	1.01E+10
37	BWA	2010	2E+09	1.49E+10
38	BWA	2009	1.99E+09	1.15E+10
39	BWA	2008	1.97E+09	1.35E+10
40	BWA	2007	1.8E+09	1.24E+10

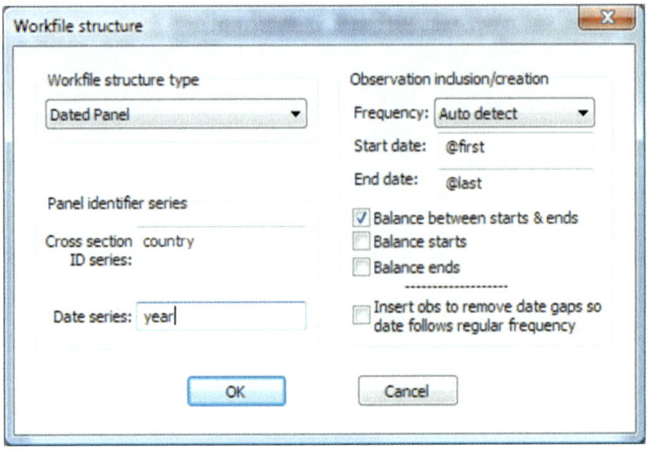

Fig. 11.18 Dated Panel

The name of countries, the years from 1976 to 2010 and the data of GDP and GEX are given in column 1–4, respectively.

Go to Proc > Structure/Resize Current Page. Change the Workfile structure to "Dated Panel". Enter Country as Cross section ID. Enter Year as Date series as shown in Fig. 11.18.

After doing this the workfile headline changes to this:

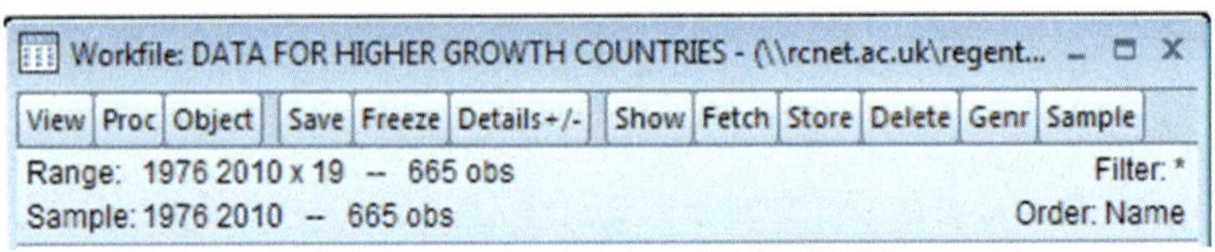

In order to take into consideration **the heterogeneity** that may exist among all countries **we allow each country to have its own intercept**, as in the following equation.

$$DLGDP_{it} = \beta_{1i} + \beta_2 DGP_{it-1} + \beta_3 DGEX_{it} + \beta_4 DGEX_{it-1} + \beta_5 ECM_{it} + \epsilon_{it}$$

This equation is called the **fixed effects regression model** (**FEM**).

The added subscript i to the intercept indicates that the intercept of the all countries can be different. The difference may be due special features of each country, such as economic endowments or other economic features.

The term "fixed effects" implies that each individual intercept, although different from the intercepts of the others, **does not vary over time** and it is **time-invariant**.

If the expression of the intercept included t and it was written as, $\boldsymbol{\beta 1_{it}}$, in the above model, the intercept of each individual would be **time-variant**. But note that in the above model we assumed that the slope coefficients are time-invariant.

We follow this approach by introducing **differential intercept dummies.** We modify the above equation as follows:

$$DLGDP_{it} = \beta_1 + \beta_2 D_{2i} + \beta_3 D_{3i} + \ldots + \beta_{18} D_{18i} + \beta_{19} DGP_{it-1} + \beta_{20} DGEX_{it} + \beta_{21} DGEX_{it-1} + \beta_{22} ECM_{it} + \epsilon_{it}$$

where $D_{2\mathrm{i}} = 1$ for country 2, 0 otherwise; $D_{3\mathrm{i}} = 1$ for country 3, 0 otherwise; and so on.

It is important to note that we have used only 18 dummies to represent 19 countries to avoid the **dummy variable trap** (perfect collinearity). In this case the 18 dummies will represent the **differential intercept dummy coefficients**—that is, they will show by how much the intercept coefficient of the country that is assigned a dummy variable will differ from the benchmark category. We are treating the first country as the **benchmark** or **reference category**, although any country can be chosen for that purpose.

Go to Quick choose Estimate Equation. Now you must have the Panel Data option on the top of the estimation window. Write the model equation in the window box. Choose Panel Options from the screen in Fig. 11.19.
Click panel data and chose "Fixed" as shown in Fig 11.20 then hit OK to generate output of Fig 11.21

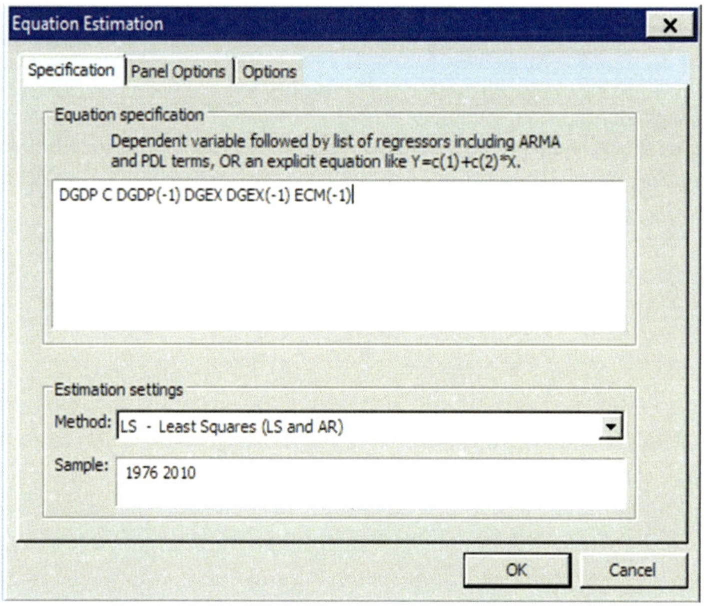

Fig. 11.19 Least Squares

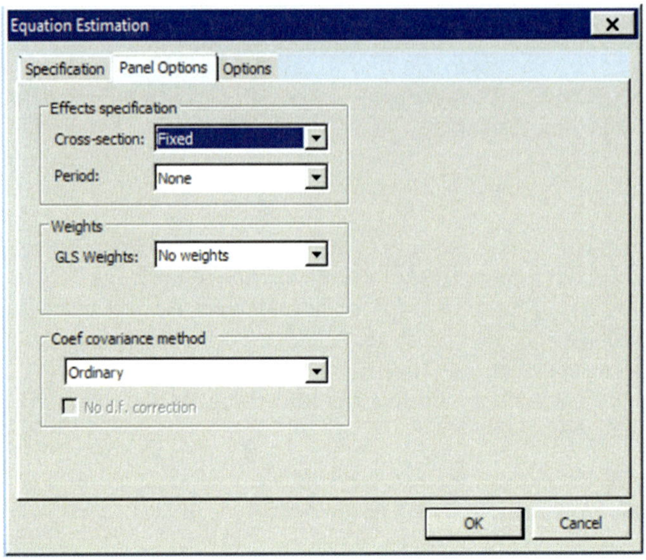

Fig. 11.20 Panel Options

Notes:

The first thing to notice about the results in the above output is that the output does not produce the values of the country differential intercept coefficients, although they are taken into account in the estimation processes. However, the differential intercept coefficients can be easily obtained if we want to do it manually. Secondly, if you compare the OLS pooled regression results with the FEM results, you will see substantial differences between the two, not only in the values of the coefficients, but also in their signs.

These results, therefore, cast doubt on the pooled OLS estimates. If you examine the country differential intercept dummies, you will find that several of them are statistically highly significant, suggesting that the pooled estimates hide the heterogeneity among the 19 countries.

The Result:

1. If we run hypothesis tastings, one by one, for the coefficients of variables: DLGDP(−1), DLGEX, DLGEX(−1) and ECM(−1) it can be seen that all the

Equation: UNTITLED Workfile: DATA FOR HIGHER...

View | Proc | Object | Print | Name | Freeze | Estimate | Forecast | Stats | Resids

Dependent Variable: DLGDP
Method: Panel Least Squares
Date: 11/15/17 Time: 14:25
Sample (adjusted): 1978 2010
Periods included: 33
Cross-sections included: 19
Total panel (balanced) observations: 627

Variable	Coefficient	Std. Error	t-Statistic	Prob.
C	0.033046	0.003711	8.904219	0.0000
DLGDP(-1)	-0.003634	0.039588	-0.091799	0.9269
DLGEX	0.133556	0.022340	5.978476	0.0000
DLGEX(-1)	0.079359	0.022660	3.502208	0.0005
ECM(-1)	-0.043340	0.010571	-4.099899	0.0000

Effects Specification

Cross-section fixed (dummy variables)

R-squared	0.161122	Mean dependent var	0.050546
Adjusted R-squared	0.130567	S.D. dependent var	0.066775
S.E. of regression	0.062263	Akaike info criterion	-2.678912
Sum squared resid	2.341502	Schwarz criterion	-2.516006
Log likelihood	862.8388	Hannan-Quinn criter.	-2.615621
F-statistic	5.273172	Durbin-Watson stat	2.001710
Prob(F-statistic)	0.000000		

Fig. 11.21 Panel Least Squares

coefficients, including the intercept, except for DLGDP(−1), are statistically significant since their probabilities are smaller than 0.05.

2. Since the coefficient of the ECM(−1) is negative (−0.095049) and it is statistically significant it can be concluded that there is a long run relationship between these variables.
3. The R^2 of this model is still low (0.197174). The F statistic's probability shows the joint coefficient of the model is statistically significant.
4. The Durbin-Watson Stat shows the model does not have serial correlation problem (1.989628).

11.2.3 Model Testing

We can provide a test to find out if the fixed effects model is better than the OLS pooled model. Since the pooled model neglects the heterogeneity effects that are explicitly taken into account in the fixed effects model, **the pooled model is a restricted** version **of the fixed effects model**.

As a result of this we call the pooled model a restricted model and the fixed effects model an unrestricted model.

Which one passes the below test should be used.

The test statistics used for the comparison between these two models is based on the below F test:

$$F = \frac{\left(R_{ur}^2 - R_r^2\right)/m}{\left(1 - R_{ur}^2\right)/(n - k)}$$

where R_{ur}^2 and R_r^2 are unrestricted and restricted coefficients of determination, m is the number of parameters omitted from the restricted model (18 here), n is the number of observations in the sample, and k is the number of parameters estimated in the unrestricted regression (here a total of 22). The restricted and unrestricted R^2 values are obtained from above EViews outputs, respectively.

The Test:

1. H_0: The restricted model-the pooled model is a better model,
2. H_a: The unrestricted model-the fixed effects model is a better model
3. $\alpha = 5\%$ or 0.05.
4. $F = \frac{\left(R_{ur}^2 - R_r^2\right)/m}{\left(1 - R_{ur}^2\right)/(n - k)}$

   $F = \frac{(0.197174 - 0.095810)/18}{(1 - 0.197174)/(627 - 22)} = \frac{0.005631}{0.001326} = 4.2466$
5. $V_1 = K - 1 = 22 - 1 = 21$ and $V_2 = n - k = 627 - 22 = 605$. The **Critical F-Value** for $V_1 = 21$ and $V_2 = 605$ is: 1.59
6. **Decision:** 4.2466 > 1.59 therefore reject the H_0 in favour of the alternative.
7. **The fixed effects model is a better model.**

11.2.4 Limitations of the Fixed Effects LSDV Model

Although easy to implement, the LSDV model has the following limitations:

1. Every additional dummy variable will cost an additional degree of freedom. Therefore, if the sample is not very large, introducing too many dummies will leave few observations to do meaningful statistical analysis.
2. Too many additive and multiplicative dummies may lead to the possibility of multicollinearity, which make precise estimation of one or more parameters difficult.
3. To obtain estimates with desirable statistical properties, we need to pay careful attention to the error term of the model. The statistical results presented above are based on the assumption that the error term follows the classical assumptions,

namely $u_{it} \sim N(0, \sigma^2)$. Since the index i refers to cross-sectional observation and t to time series observations, the classical assumption regarding U_{it} may have to be modified. There are several possibilities:

(a) We can assume that the error variance is the same for all cross-sectional units or we can assume that the error variance is heteroscedastic.
(b) For each subject, we can assume that there is no autocorrelation over time or we can assume autocorrelation of the AR (1) type.
(c) At any given time, we can allow the error term of individual #1 to be non-correlated with the error term for say, country #2, or we can assume that there is such correlation.

11.3 The Random Effects Model (REM) or Error Components Model (ECM)

In the fixed effects model it is assumed that the county specific coefficient β_{1i} is fixed for each subject and it is time-invariant. In the random effects model it is assumed that β_{1i} is a random variable with a mean value of β_1 (no i subscript here) and the intercept of any cross-section unit is expressed as:

$$\beta_{1i} = \beta_1 + \varepsilon_i$$

where ε_i is a random error term with mean 0 and variance σ^2.

In terms of our model, this means that the 19 counties included in our sample and they have a common mean value for the intercept ($=\beta_1$). Differences in the country values of the intercept for each country are reflected in the error term ε_i

Therefore, we can write the above model as:

$$DLGDP_{it} = \beta_1 + \beta_2 DGP_{it-1} + \beta_3 DGEX_{it} + \beta_4 DGEX_{it-1} + \beta_5 ECM_{it-1} + w_{it}$$
$$w_{it} = \varepsilon_i + u_{it}$$

The composite error term w_{it} has two components: ε_i, which is the cross-section or country-specific error component and u_{it}, which is the combined time series and cross-section error component.

Now you can see why the REM model is also called an **error components model (ECM)**: the composite error term consists of two (or more) error components. The usual assumptions of ECM are that:

$$\varepsilon_i \sim N\left(0, \sigma_\varepsilon^2\right)$$
$$u_{it} \sim N\left(0, \sigma_u^2\right)$$
$$E(\varepsilon_i u_{it}) = 0; \quad E\left(\varepsilon_i \varepsilon_j\right) = 0 \ (i \neq j)$$
$$E(u_{it} u_{is}) = E\left(u_{it} u_{ij}\right) = E\left(u_{it} u_{js}\right) = 0 \ (i \neq j; t \neq s)$$

That is, country error components are not correlated with each other and are not autocorrelated across both cross-section and time series units.

It is also critical to note that w_{it} is not correlated with any of the explanatory variables included in the model.

Since ε_i is a part of w_{it}, it is possible that the latter is correlated with one or more regressors. If that turns out to be the case, the REM will result in inconsistent estimation of the regression coefficients.

The **Hausman test**, which will be explained shortly, will show in a given application if w_{it} is correlated with the regressors—that is, whether REM is the appropriate model.

Go to Quick choose Estimate Equation. Now you must have the Panel Data option on the top of the estimation window. Write the model equation in the window box. Choose Panel Options from screen in Fig. 11.22
Click panel data and chose Fixed effect from Effects Specification as shown in Fig. 11.23. Then hit OK in Fig. 11.23 and to generate ouput of Fig. 11.24.

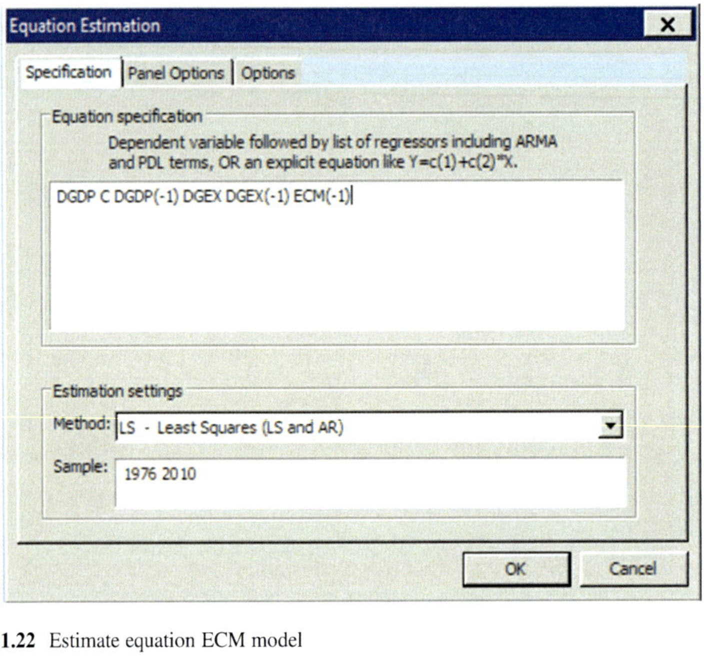

Fig. 11.22 Estimate equation ECM model

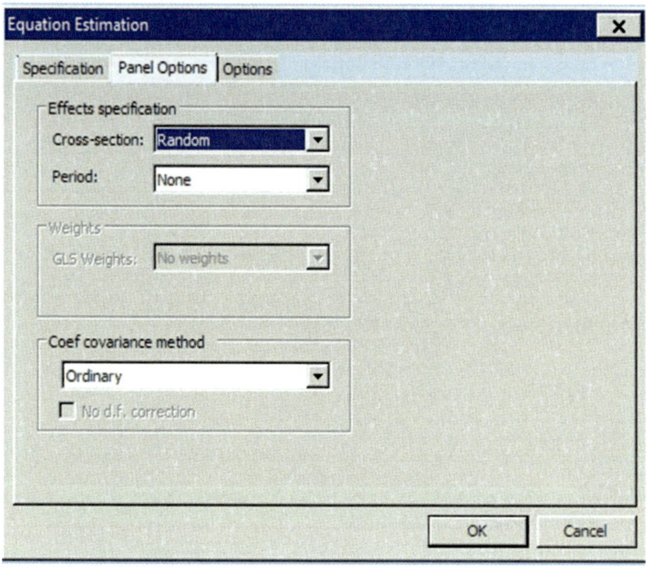

Fig. 11.23 Panel Options

11.4 Fixed Effects Model vs. Random Effects Model

Comparing the fixed effect estimators and the random effects estimators given above, you will see substantial differences between the two. So which model is better in the present example: fixed effects or random effects?

The answer to this question depends on the assumption we make about the likely correlation between the cross-section specific error component ε_i and the X regressors.

If it assumed that $\varepsilon_i$and the regressors are uncorrelated, REM may be appropriate, but if they are correlated, FEM may be appropriate.

Which Model Is an Appropriate Model?

We use the **Hausman Test** to answer this question.

Go to View > Fixed/Random Effects Testing > Correlated Random Effects-Hausman Test as shown in Fig. 11.25

Equation: UNTITLED Workfile: DATA FOR HIGHER...

View | Proc | Object | Print | Name | Freeze | Estimate | Forecast | Stats | Resids

Dependent Variable: DLGDP
Method: Panel EGLS (Cross-section random effects)
Date: 11/15/17 Time: 14:51
Sample (adjusted): 1978 2010
Periods included: 33
Cross-sections included: 19
Total panel (balanced) observations: 627
Swamy and Arora estimator of component variances

Variable	Coefficient	Std. Error	t-Statistic	Prob.
C	0.030219	0.003667	8.240265	0.0000
DLGDP(-1)	0.046662	0.038353	1.216646	0.2242
DLGEX	0.126615	0.022069	5.737297	0.0000
DLGEX(-1)	0.089529	0.022308	4.013239	0.0001
ECM(-1)	-0.013294	0.005697	-2.333328	0.0199

Effects Specification

	S.D.	Rho
Cross-section random	0.000000	0.0000
Idiosyncratic random	0.062263	1.0000

Weighted Statistics

R-squared	0.095810	Mean dependent var	0.050546
Adjusted R-squared	0.089996	S.D. dependent var	0.066775
S.E. of regression	0.063699	Sum squared resid	2.523803
F-statistic	16.47719	Durbin-Watson stat	2.028725
Prob(F-statistic)	0.000000		

Unweighted Statistics

R-squared	0.095810	Mean dependent var	0.050546
Sum squared resid	2.523803	Durbin-Watson stat	2.028725

Fig. 11.24 Panel EGLS output

The Test:

1. H_0: Random-effects model is appropriate
2. H_1: Fixed-effects model is appropriate
3. $\alpha = 5\%$ or 0.05,

   According to Fig. 11.26, the Chi-sq Statistic = 80.555399.

4. p-value = 0.0000.
5. Prob = 0.0000 < 0.05. We reject the H_0. The test is significant.
6. Fixed-effects model is appropriate.

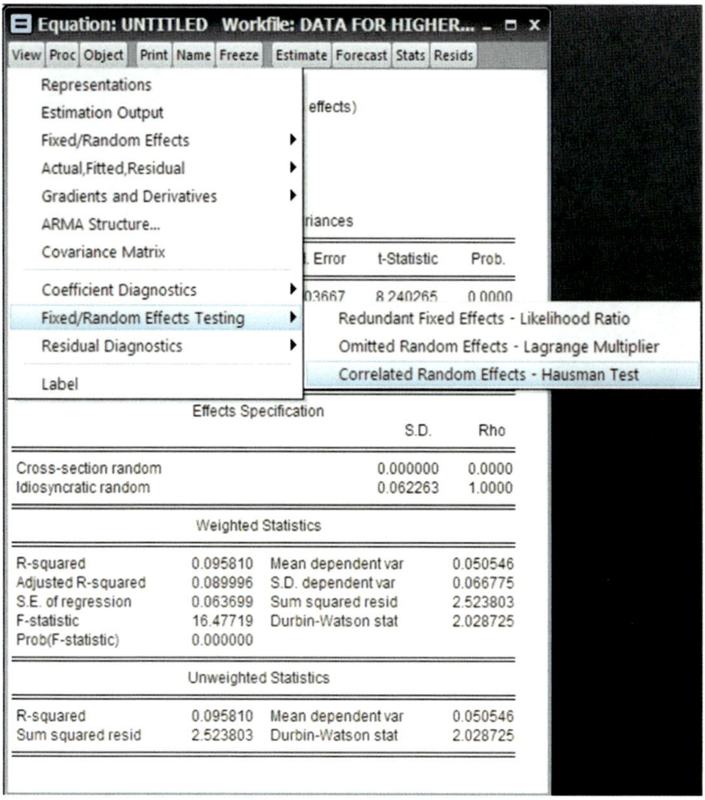

Fig. 11.25 Correlated Random Effects—Hausman Test

11.5 The Final Result

Earlier we found that the fixed effects model was better than the pooled model (restricted model). This result was found through running a test. Now based one the last test (the Hausman Test) it became clear that the fixed effect model is a better model than the random effects model. So, it can be concluded that the fixed effects model is an appropriate model for the above relationship. We follow the result found for the fixed effects model.

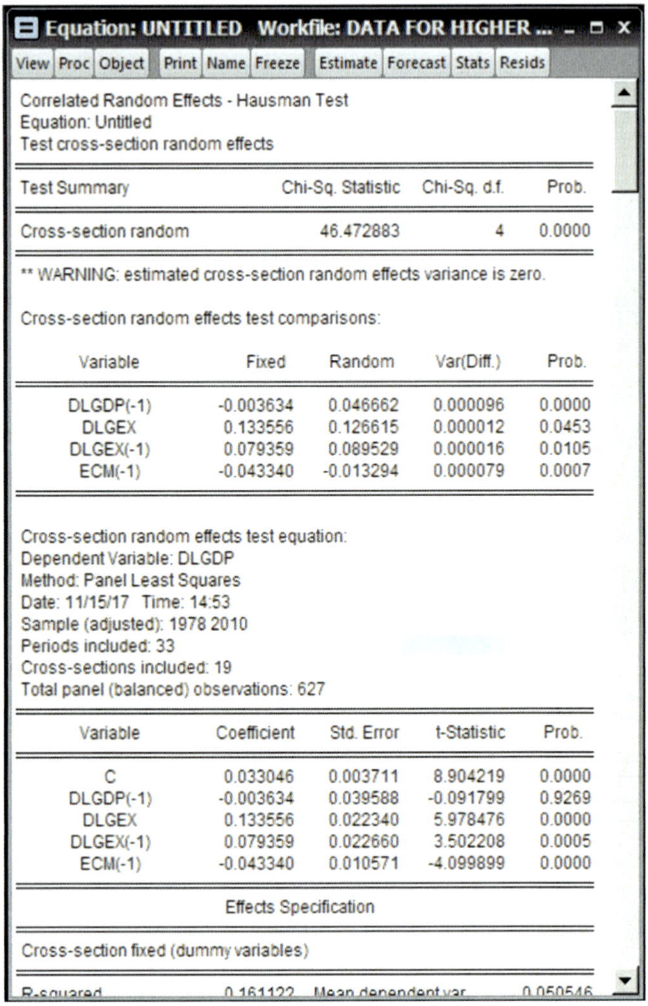

Equation: UNTITLED Workfile: DATA FOR HIGHER ...

View | Proc | Object | Print | Name | Freeze | Estimate | Forecast | Stats | Resids

Correlated Random Effects - Hausman Test
Equation: Untitled
Test cross-section random effects

Test Summary	Chi-Sq. Statistic	Chi-Sq. d.f.	Prob.
Cross-section random	46.472883	4	0.0000

** WARNING: estimated cross-section random effects variance is zero.

Cross-section random effects test comparisons:

Variable	Fixed	Random	Var(Diff.)	Prob.
DLGDP(-1)	-0.003634	0.046662	0.000096	0.0000
DLGEX	0.133556	0.126615	0.000012	0.0453
DLGEX(-1)	0.079359	0.089529	0.000016	0.0105
ECM(-1)	-0.043340	-0.013294	0.000079	0.0007

Cross-section random effects test equation:
Dependent Variable: DLGDP
Method: Panel Least Squares
Date: 11/15/17 Time: 14:53
Sample (adjusted): 1978 2010
Periods included: 33
Cross-sections included: 19
Total panel (balanced) observations: 627

Variable	Coefficient	Std. Error	t-Statistic	Prob.
C	0.033046	0.003711	8.904219	0.0000
DLGDP(-1)	-0.003634	0.039588	-0.091799	0.9269
DLGEX	0.133556	0.022340	5.978476	0.0000
DLGEX(-1)	0.079359	0.022660	3.502208	0.0005
ECM(-1)	-0.043340	0.010571	-4.099899	0.0000

Effects Specification

Cross-section fixed (dummy variables)

R-squared	0.161122	Mean dependent var	0.050546

Fig. 11.26 Correlated Random Effects—Hausman Test

Chapter 12
Capital Asset Pricing Model (CAPM)

The **capital asset pricing model (CAPM)** is the basic theory that links risk and return for all assets—it quantifies the relationship between risk and return. The aim of running the CAPM model is to identify systematic risk. Capital asset pricing model (CAPM) is an attempt to explain and quantify the non-diversifiable type of risk. In other words, it measures how much additional return an investor should expect from taking a little extra risk. Investors demand a premium for bearing risk and therefore the higher the risk of the security, the higher the expected return to encourage investors to buy that security. As investors hold well-diversified portfolios, they are concerned with the non-diversifiable part of the risk of an individual stock. The relevant risk of an individual stock is its contribution to the risk of a well-diversified portfolio. **Nondiversifiable risk** is the relevant portion of an asset's risk attributable to market factors that affect all firms and which cannot be eliminated through diversification. Also called systematic risk. Because any investor can create a portfolio of assets that will eliminate virtually all diversifiable risk, the only relevant risk is the one that is non-diversifiable. The beta coefficient is used as measure to quantify the systematic risk.

The beta coefficient (β) is a relative measure of non-diversifiable risk. It is an index of the degree of movement of an asset's return in response to a change in the market return. Beta is the stock's relative volatility (how the price of a stock moves up or down in to market movements). An asset's historical returns are used to find the asset's beta coefficient given that the beta coefficient for the entire market equals 1.0. All other betas are viewed in relation to this value. The market return is the return on the market portfolio of all traded securities.

12.1 The CAPM Equation

Using the beta coefficient to measure non-diversifiable risk, the capital asset pricing model (CAPM) is given in the following equation:

A. Aljandali, M. Tatahi, *Economic and Financial Modelling with EViews*, Statistics and Econometrics for Finance, https://doi.org/10.1007/978-3-319-92985-9_12

$$E(r_i) = R_F + \beta_i[E(R_m) - R_F]$$

where

r_i = required return on asset i
R_F = risk-free rate of return, commonly measured by the return on a U.S. Treasury bill
β_i = beta coefficient or index of non-diversifiable risk for asset j
R_m = market return; return on the market portfolio of assets

The CAPM can be divided into two parts:

1. The **risk-free rate of return, (R_F)** which is the required return on a risk-free asset, typically a 3-month U.S. Treasury bill.
2. The risk premium.

The $(r_m - R_F)$ portion of the risk premium is called the *market risk premium,* because it represents the premium the investor must receive for taking the average amount of risk associated with holding the market portfolio of assets. The **security market line (SML)** is the depiction of the capital asset pricing model (CAPM) as a graph that reflects the required return in the marketplace for each level of nondiversifiable risk (beta). In a graph, risk as measured by beta, *b,* is plotted on the *x* axis, and required returns, *r,* are plotted on the *y* axis. In practice, we use linear regression to estimate the relation. The output is the best-fitting line that represents the historical relation between the stock and the market. The slope of this line is our estimate of beta and indicates how much the stock's excess return changed for a 1% change in the market's excess return.

For example, consider the daily returns of an individual security *i* which can be calculated as follows:

$$\mathrm{R}_{\mathrm{i,t}} = \ln\left(\mathrm{P}_{\mathrm{i,t+1}}\right) - \ln\left(\mathrm{P}_{\mathrm{i,t}}\right)$$

where P_{it} represents the closing price for security *i* on day *t*. The same method is applied in computing the returns of market indices. We use the market model to find the returns for the security HOGANAS at time *t:*

$$RHOGANAS_t = \beta_0 + \beta_1 ROMX_t + u_t$$

where *RHOGANAS* and *ROMX* are returns on security *HOGANAS* and on the OMX market (Stockholm Market Index) at time *t* respectively. We run the above model to find the two parameters, **α and β by using OLS regression**. The main aim is to find **β,** which indicates the systematic risk attached to the security, HOGANAS. Data values of both variables are uploaded to EViews and the returns are found by converting the data (prices) to natural logs as follows:

LHOGANAS = LOG (HOGANAS)
LOMX = LOG (OMX)

The returns of the above variables are found as follows:

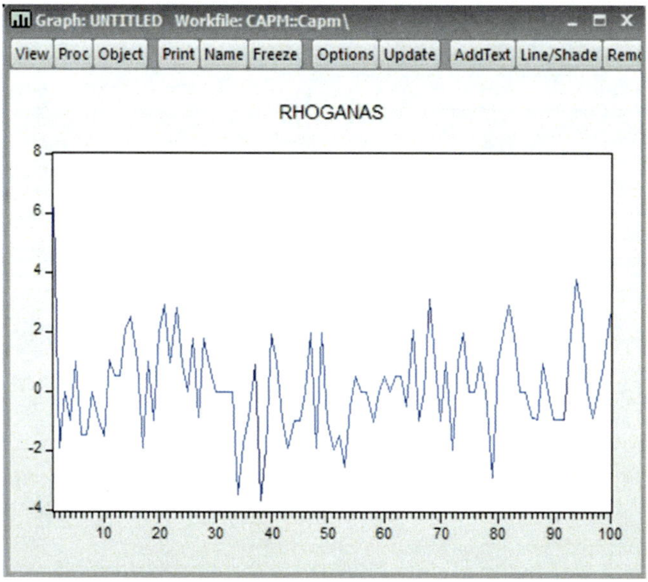

Fig. 12.1 Plot of RHOGANAS series

RHOGANAS = LHOGANAS − LHOGANAS (−1)
ROMX = LOMX − LOMX (−1)

Figure 12.1 shows the series is stationary and I(0).

(a) H_0 : The series has no unit root
(b) H_a : The series has a unit root
(c) $\alpha = 5\%$ or 0.05

There are no test statistics to calculate. We look at the Prob column.

(d) The p-values are bigger than 0.05
(e) As p-values are >0.05, we cannot reject the H_0; we accept the H_0.
(f) The series is stationary.

1. H_0: series has a unit root and it is not stationary
2. H_1: series has no unit root and it is stationary.
3. $\alpha = 5\%$ or 0.05 (Fig. 12.2)

According to Fig. 12.3, ADF test statistic is −9.010727

1. The p-values = 0.0000
2. Prob = 0.0000 < 0.05 therefore we reject the H_0
3. RHOGANAS has no unit root and is stationary.

Figure 12.4 shows that the ROMX series is stationary and I(0).

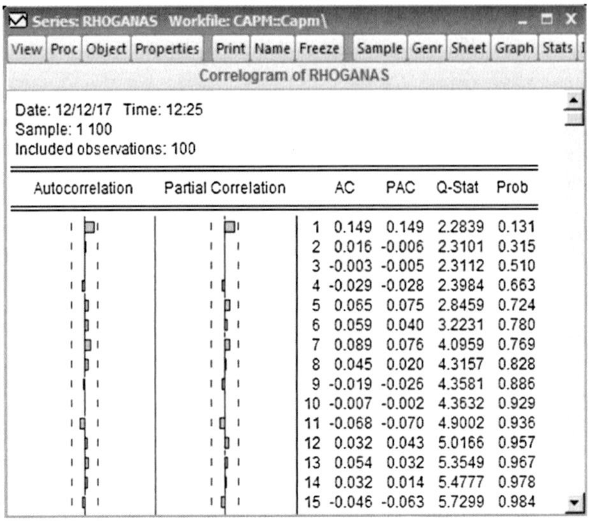

Series: RHOGANAS Workfile: CAPM::Capm\

View | Proc | Object | Properties | Print | Name | Freeze | Sample | Genr | Sheet | Graph | Stats

Correlogram of RHOGANAS

Date: 12/12/17 Time: 12:25
Sample: 1 100
Included observations: 100

Lag	AC	PAC	Q-Stat	Prob
1	0.149	0.149	2.2839	0.131
2	0.016	-0.006	2.3101	0.315
3	-0.003	-0.005	2.3112	0.510
4	-0.029	-0.028	2.3984	0.663
5	0.065	0.075	2.8459	0.724
6	0.059	0.040	3.2231	0.780
7	0.089	0.076	4.0959	0.769
8	0.045	0.020	4.3157	0.828
9	-0.019	-0.026	4.3581	0.886
10	-0.007	-0.002	4.3632	0.929
11	-0.068	-0.070	4.9002	0.936
12	0.032	0.043	5.0166	0.957
13	0.054	0.032	5.3549	0.967
14	0.032	0.014	5.4777	0.978
15	-0.046	-0.063	5.7299	0.984

Fig. 12.2 Correlogram of RHOGANAS

Correlogram of ROMX (Fig. 12.5)

1. H_0: ROMX has no unit root
2. H_1: ROMX has a unit root
3. $\alpha = 5\%$ or 0.05

There are no test statistics to calculate. We look at the Prob column.

1. The p-values are bigger than 0.05
2. As p-values are >0.05, we cannot reject the H_0
3. ROMX is stationary.

Formal Approach

1. H_0: ROMX has a unit root and is not stationary
2. H_1: ROMX has no unit root and is stationary.
3. $\alpha = 5\%$ or 0.05

According to Fig. 12.6, ADF test statistic $= -8.799709$

1. The p-values $= 0.0000$
2. Prob $= 0.0000 < 0.05$ therefore reject the H_0
3. ROMX has no unit root and is stationary

Correlogram of the residual (Fig. 12.8)

1. H_0: The series has no unit root
2. H_a: The series has a unit root
3. $\alpha = 5\%$ or 0.05

Series: RHOGANAS Workfile: CAPM::Capm\

View | Proc | Object | Properties | Print | Name | Freeze | Sample | Genr | Sheet | Graph

Augmented Dickey-Fuller Unit Root Test on RHOGANAS

Null Hypothesis: RHOGANAS has a unit root
Exogenous: Constant
Lag Length: 0 (Automatic - based on SIC, maxlag=12)

		t-Statistic	Prob.*
Augmented Dickey-Fuller test statistic		-9.010727	0.0000
Test critical values:	1% level	-3.497727	
	5% level	-2.890926	
	10% level	-2.582514	

*MacKinnon (1996) one-sided p-values.

Augmented Dickey-Fuller Test Equation
Dependent Variable: D(RHOGANAS)
Method: Least Squares
Date: 12/12/17 Time: 14:47
Sample (adjusted): 2 100
Included observations: 99 after adjustments

Variable	Coefficient	Std. Error	t-Statistic	Prob.
RHOGANAS(-1)	-0.848125	0.094124	-9.010727	0.0000
C	0.106146	0.152371	0.696631	0.4877

R-squared	0.455647	Mean dependent var	-0.035645
Adjusted R-squared	0.450035	S.D. dependent var	2.033406
S.E. of regression	1.507966	Akaike info criterion	3.679396
Sum squared resid	220.5742	Schwarz criterion	3.731822
Log likelihood	-180.1301	Hannan-Quinn criter.	3.700607
F-statistic	81.19320	Durbin-Watson stat	1.765989
Prob(F-statistic)	0.000000		

Fig. 12.3 RHOGANAS unit root test

There are no test statistics to calculate. We look at the Prob column.

4. The p-values are bigger than 0.05
5. As p-values are >0.05, we cannot reject the H_0; we accept the H_0.
6. The series is stationary.

Figure 12.9 shows the residuals are stationary.

The unit root test of the residual (Fig. 12.10)

7. H_0: series has a unit root and it is not stationary
8. H_1: series has no unit root and it is stationary.
9. $\alpha = 5\%$ or 0.05

ADF test statistic = −9.653714

1. The p-values = 0.0000
2. 0.0000 < 0.05. We reject the H_0; we accept the H_a.
   The series has no unit root and it is stationary.

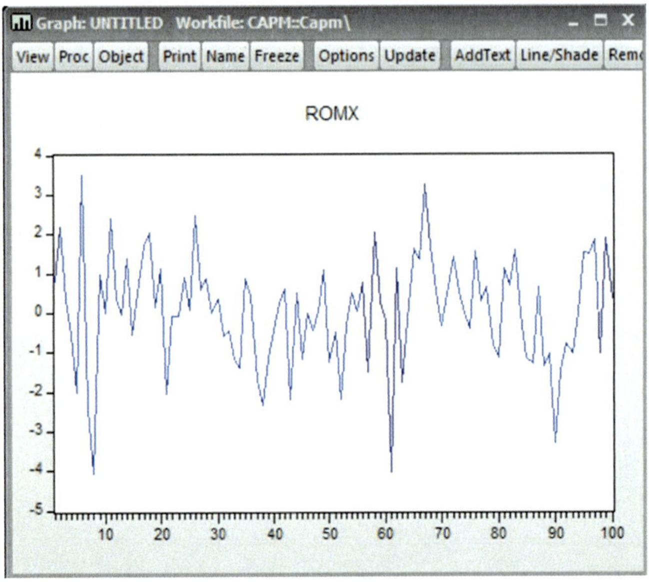

Fig. 12.4 Plot of ROMX series

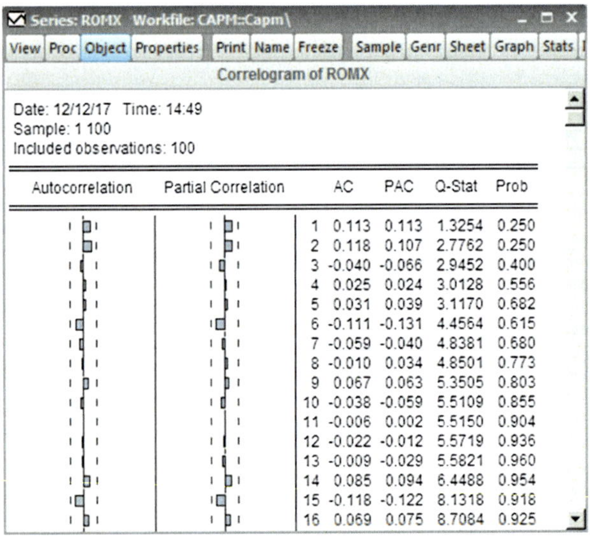

Date: 12/12/17 Time: 14:49
Sample: 1 100
Included observations: 100

	AC	PAC	Q-Stat	Prob
1	0.113	0.113	1.3254	0.250
2	0.118	0.107	2.7762	0.250
3	-0.040	-0.066	2.9452	0.400
4	0.025	0.024	3.0128	0.556
5	0.031	0.039	3.1170	0.682
6	-0.111	-0.131	4.4564	0.615
7	-0.059	-0.040	4.8381	0.680
8	-0.010	0.034	4.8501	0.773
9	0.067	0.063	5.3505	0.803
10	-0.038	-0.059	5.5109	0.855
11	-0.006	0.002	5.5150	0.904
12	-0.022	-0.012	5.5719	0.936
13	-0.009	-0.029	5.5821	0.960
14	0.085	0.094	6.4488	0.954
15	-0.118	-0.122	8.1318	0.918
16	0.069	0.075	8.7084	0.925

Fig. 12.5 Correlogram of ROMX

Series: ROMX Workfile: CAPM::Capm\

View | Proc | Object | Properties | Print | Name | Freeze | Sample | Genr | Sheet | Graph

Augmented Dickey-Fuller Unit Root Test on ROMX

Null Hypothesis: ROMX has a unit root
Exogenous: Constant
Lag Length: 0 (Automatic - based on SIC, maxlag=12)

		t-Statistic	Prob.*
Augmented Dickey-Fuller test statistic		-8.799709	0.0000
Test critical values:	1% level	-3.497727	
	5% level	-2.890926	
	10% level	-2.582514	

*MacKinnon (1996) one-sided p-values.

Augmented Dickey-Fuller Test Equation
Dependent Variable: D(ROMX)
Method: Least Squares
Date: 12/12/17 Time: 14:51
Sample (adjusted): 2 100
Included observations: 99 after adjustments

Variable	Coefficient	Std. Error	t-Statistic	Prob.
ROMX(-1)	-0.886545	0.100747	-8.799709	0.0000
C	0.029665	0.142159	0.208673	0.8351

R-squared	0.443919	Mean dependent var	-0.004874
Adjusted R-squared	0.438186	S.D. dependent var	1.886384
S.E. of regression	1.413925	Akaike info criterion	3.550612
Sum squared resid	193.9209	Schwarz criterion	3.603038
Log likelihood	-173.7553	Hannan-Quinn criter.	3.571824
F-statistic	77.43488	Durbin-Watson stat	2.018622
Prob(F-statistic)	0.000000		

Fig. 12.6 ROMX unit root test

12.2 Residual Analysis

1. **Check for Serial Correlation**

Serial Correlation LM Test

In EViews after running the main regression choose view—Residual Diagnostics—Serial Correlation LM test. The output should look like this output the one reported in Fig. 12.11

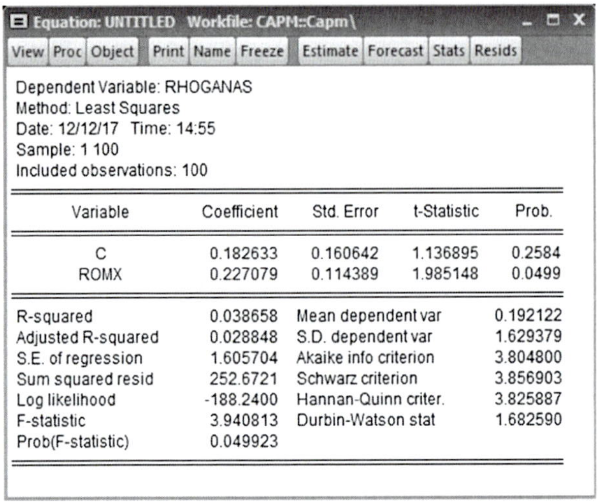
Equation: UNTITLED Workfile: CAPM::Capm\

View | Proc | Object | Print | Name | Freeze | Estimate | Forecast | Stats | Resids

Dependent Variable: RHOGANAS
Method: Least Squares
Date: 12/12/17 Time: 14:55
Sample: 1 100
Included observations: 100

Variable	Coefficient	Std. Error	t-Statistic	Prob.
C	0.182633	0.160642	1.136895	0.2584
ROMX	0.227079	0.114389	1.985148	0.0499

R-squared	0.038658	Mean dependent var	0.192122
Adjusted R-squared	0.028848	S.D. dependent var	1.629379
S.E. of regression	1.605704	Akaike info criterion	3.804800
Sum squared resid	252.6721	Schwarz criterion	3.856903
Log likelihood	-188.2400	Hannan-Quinn criter.	3.825887
F-statistic	3.940813	Durbin-Watson stat	1.682590
Prob(F-statistic)	0.049923		

Fig. 12.7 RHOGANAS Regression output (constant not significant)

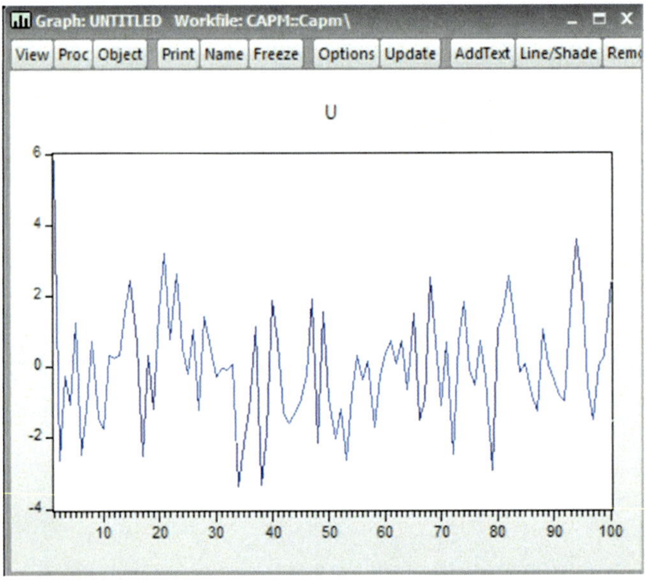

Fig. 12.8 Correlogram of U series

Series: U Workfile: CAPM::Capm\

View | Proc | Object | Properties | Print | Name | Freeze | Sample | Genr | Sheet | Graph | Stats

Correlogram of U

Date: 12/12/17 Time: 14:59
Sample: 1 100
Included observations: 100

Autocorrelation	Partial Correlation		AC	PAC	Q-Stat	Prob
		1	0.080	0.080	0.6613	0.416
		2	-0.032	-0.039	0.7680	0.681
		3	-0.031	-0.025	0.8670	0.833
		4	-0.032	-0.029	0.9763	0.913
		5	0.015	0.018	1.0012	0.962
		6	0.051	0.046	1.2854	0.972
		7	0.126	0.119	3.0332	0.882
		8	0.080	0.066	3.7492	0.879
		9	-0.035	-0.035	3.8885	0.919
		10	0.001	0.020	3.8886	0.952
		11	-0.080	-0.078	4.6238	0.948
		12	0.024	0.033	4.6881	0.968
		13	0.036	0.012	4.8371	0.979
		14	0.042	0.019	5.0448	0.985
		15	-0.053	-0.075	5.3776	0.988

Fig. 12.9 Plot of U residuals

Series: U Workfile: CAPM::Capm\

View | Proc | Object | Properties | Print | Name | Freeze | Sample | Genr | Sheet | Graph | Stats

Augmented Dickey-Fuller Unit Root Test on U

Null Hypothesis: U has a unit root
Exogenous: Constant
Lag Length: 0 (Automatic - based on SIC, maxlag=12)

		t-Statistic	Prob.*
Augmented Dickey-Fuller test statistic		-9.653714	0.0000
Test critical values:	1% level	-3.497727	
	5% level	-2.890926	
	10% level	-2.582514	

*MacKinnon (1996) one-sided p-values.

Augmented Dickey-Fuller Test Equation
Dependent Variable: D(U)
Method: Least Squares
Date: 12/12/17 Time: 16:02
Sample (adjusted): 2 100
Included observations: 99 after adjustments

Variable	Coefficient	Std. Error	t-Statistic	Prob.
U(-1)	-0.918558	0.095151	-9.653714	0.0000
C	-0.056857	0.150260	-0.378388	0.7060

R-squared	0.489995	Mean dependent var	-0.034538
Adjusted R-squared	0.484737	S.D. dependent var	2.082545
S.E. of regression	1.494889	Akaike info criterion	3.661976
Sum squared resid	216.7651	Schwarz criterion	3.714402
Log likelihood	-179.2678	Hannan-Quinn criter.	3.683187
F-statistic	93.19419	Durbin-Watson stat	1.756978
Prob(F-statistic)	0.000000		

Fig. 12.10 U unit root test

Equation: UNTITLED Workfile: CAPM::Capm\

View | Proc | Object | Print | Name | Freeze | Estimate | Forecast | Stats | Resids

Breusch-Godfrey Serial Correlation LM Test:

F-statistic	0.395549	Prob. F(2,96)	0.6744
Obs*R-squared	0.817325	Prob. Chi-Square(2)	0.6645

Test Equation:
Dependent Variable: RESID
Method: Least Squares
Date: 12/12/17 Time: 16:04
Sample: 1 100
Included observations: 100
Presample missing value lagged residuals set to zero.

Variable	Coefficient	Std. Error	t-Statistic	Prob.
C	0.001521	0.161714	0.009407	0.9925
ROMX	-0.010890	0.118413	-0.091968	0.9269
RESID(-1)	0.086804	0.104741	0.828750	0.4093
RESID(-2)	-0.038072	0.104362	-0.364813	0.7161

R-squared	0.008173	Mean dependent var	-4.44E-17
Adjusted R-squared	-0.022821	S.D. dependent var	1.597574
S.E. of regression	1.615701	Akaike info criterion	3.836593
Sum squared resid	250.6070	Schwarz criterion	3.940800
Log likelihood	-187.8296	Hannan-Quinn criter.	3.878767
F-statistic	0.263699	Durbin-Watson stat	1.839912
Prob(F-statistic)	0.851394		

Fig. 12.11 Breusch-Godfrey Serial Correlation LM Test

(a) H_0: No serial correlation
(b) H_1: There is serial correlation
(c) $\alpha = 5\%$ or 0.05
Obs*R-square = 0.817325
(d) The p-value = 0.6645
(e) As 0.6645 > 0.05, we cannot reject the H_0; we accept the H_0.
There is no serial correlation

2. **Check for Heteroscedasticity**
White's Test

White's Test in EViews

The EViews is doing this test directly. Go to the main regression output from EViews. Choose View—Residual Diagnostics—Heteroscedasticity Test—choose White—click OK to generate Fig. 12.12

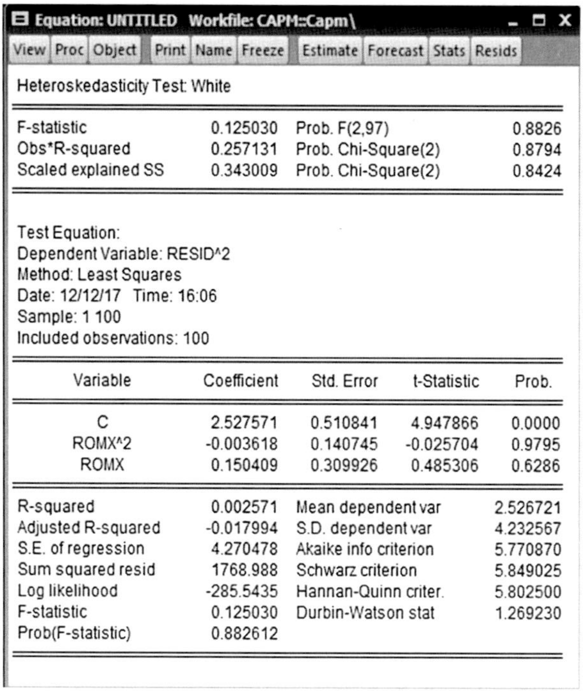

Heteroskedasticity Test: White

F-statistic	0.125030	Prob. F(2,97)	0.8826
Obs*R-squared	0.257131	Prob. Chi-Square(2)	0.8794
Scaled explained SS	0.343009	Prob. Chi-Square(2)	0.8424

Test Equation:
Dependent Variable: RESID^2
Method: Least Squares
Date: 12/12/17 Time: 16:06
Sample: 1 100
Included observations: 100

Variable	Coefficient	Std. Error	t-Statistic	Prob.
C	2.527571	0.510841	4.947866	0.0000
ROMX^2	-0.003618	0.140745	-0.025704	0.9795
ROMX	0.150409	0.309926	0.485306	0.6286

R-squared	0.002571	Mean dependent var	2.526721
Adjusted R-squared	-0.017994	S.D. dependent var	4.232567
S.E. of regression	4.270478	Akaike info criterion	5.770870
Sum squared resid	1768.988	Schwarz criterion	5.849025
Log likelihood	-285.5435	Hannan-Quinn criter.	5.802500
F-statistic	0.125030	Durbin-Watson stat	1.269230
Prob(F-statistic)	0.882612		

Fig. 12.12 The White test statistic in EViews

(a) H_0: The residuals are Homoskedastic
(b) H_1: The residuals are not Homoskedastic
(c) $\alpha = 5\%$ or 0.05
Obs*R-square = 0.257131
(d) The p-value = 0.8794
(e) As 0.8794 > 0.05. We cannot reject the H_0; we accept the H_0.
(f) There is no problem of Heteroscedasticity.

3. **Check for Normality**

Go to View—Residual Diagnostics—Histogram Normality Test—click OK to generate Fig. 12.13

(a) H_0: The residuals are Normally distributed
(b) H_1: The residuals are not Normally distributed
(c) $\alpha = 5\%$ or 0.05

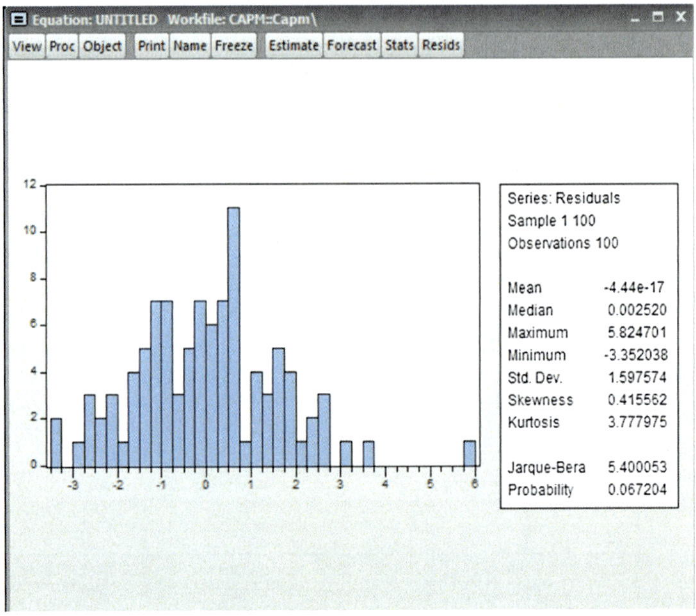

Fig. 12.13 Testing for Normality of the residuals

Jarque-Bera test statistic = 5.400053

(d) p-value = 0.067204
(e) 0.067204 > 0.05 we cannot reject the H_0
(f) The residual is distributed normally.

Our regression model can be written as follows:
RHOGANAS = 0.182633 + 0.227079ROMX

Test for β Coefficient

1. $H_0 : \beta_1 = 0$
2. $H_1 : \beta_1 \neq 0$
3. $\alpha = 5\ \%\ or 0.05$
4. $t_{\beta_1} = \frac{Coefficient \beta_1 - \beta_1 (\beta_1 in THE H_0)}{S.E.(\beta_1)} = \frac{0.227079 - 0}{0114389} = 1.985145$
5. $p - value = 0.0499$
6. We reject the H_0 because 0.0499 < 0.05. The test is significant.
7. This parameter is statistically significant.

Decision Based on the CAPM Model

As $\beta = 0.227079$ is smaller than one, we can conclude that this security has a systematic risk less than the market risk. The market risk is assumed to be equal to 1. As the market return goes up by 1% the expected return of the security, RHOGANAS, goes up by 0.227079%.

References

Aljandali, A. (2016). *Quantitative Analysis & IBM SPSS Statistics: A Guide for Business & Finance Students*. Switzerland: Springer, Cham.

Aljandali, A. (2017). *Multivariate Methods & Forecasting with IBM SPSS Statistics*. Switzerland: Springer, Cham.

Baba, Y., Engle, R. F., Kraft, D. F., & Kroner, K. F. (1989). *Multivariate simultaneous generalised ARCH* (Discussion Paper 89-57). San Diego: Department of Economics, University of California, San Diego.

Baillie, R. T., & Myers, R. J. (1991). Bivariate GARCH estimation of the optimal commodity futures hedge. *Journal of Applied Econometrics, 6*, 109–124.

Bauwens, L., Laurent, S., & Rombouts, J. V. K. (2006). Multivariate GARCH models: A survey. *Journal of Applied Econometrics, 21*(1), 79–109.

Bollerslev, T. (1986). Generalised autoregressive conditional heteroskedasticity. *Journal of Econometrics, 31*, 307–327.

Bollerslev, T., Engle, R. F., & Wooldridge, J. M. (1988). A capital asset pricing model with time-varying covariances. *Journal of Political Economy, 96*, 116–131.

Box, G.E.P., & Pierce, D.A. (1970). Distribution of residual autocorrelations in autoregressive-integrated moving average time series models. *Journal of the American Statistical Association, 65*, 1509–1526.

Brailsford, T. J., & Faff, R. W. (1996). An evaluation of volatility forecasting. *Journal of Banking and Finance, 20*, 419–438.

Brooks, C. (2004). *Introductory econometrics for finance* (Chap. 8). Cambridge, UK: Cambridge University Press.

Brooks, C., Burke, S., & Persand, G. (2003). Multivariate GARCH models: Software choice and estimation issues. In *ISMA Centre Discussion Papers in Finance*, University of Reading, April 2003.

Brooks, C., Henry, O. T., & Persand, G. (2002). Optimal hedging and the value of news. *Journal of Business, 75*(2), 333–352.

Caporin, M., & McAleer, M. (2012). Do we really need both BEKK and DCC? A tale of two multivariate GARCH models. *Journal of Economic Surveys, 26*(4), 736–751.

Chang, C. -L., González-Serrano, L., & Jimenez-Martin, J. -A. (2012). Currency hedging strategies using dynamic multivariate GARCH. *Paper Presented to the International Conference on Risk Modelling and Management*, Madrid, June 2011. Retrieved Oct 23, 2014, from http://www.ucm.es/data/cont/docs/518-2013-11-05-1207.pdf.

A. Aljandali, M. Tatahi, *Economic and Financial Modelling with EViews*, Statistics and Econometrics for Finance, https://doi.org/10.1007/978-3-319-92985-9

De Goeij, P., & Marquering, W. (2004). Modeling the conditional covariance between stock and bond returns: A multivariate GARCH approach. *Journal of Financial Econometrics, 2*(4), 531–564.

Dickey, D., & Fuller, W. A. (1979). Distribution of the estimates for autoregressive time series with a unit root. *Journal of the American Statistical Association, 74*, 427–431.

Engle, R. F. (1982). Autoregressive conditional heteroskedacticity with estimates of the variance of United Kingdom inflation. *Econometrica, 55*, 251–276.

Engle, R. F. (1982). Autoregressive conditional heteroscedasticity with estimates of the variance of United Kingdom inflation. *Econometrica, 50*(4), 987–1007.

Engle, R. (2002). Dynamic conditional correlation: A simple class of multivariate generalized autoregressive conditional heteroscedasticity models. *Journal of Business and Economic Statistics, 20*(3), 339–359.

Engle, R., & Kroner, K. (1995). Multivariate simultaneous GARCH. *Econometric Theory, 11*, 122–150.

Ewing, B. T. (2002). The transmission shocks among S and P 500 indexes. *Applied Financial Economics, 12*, 285–290.

Gujarati, D. N., & Porter, D. C. (2009). *Basic econometrics*. New York: McGraw Hill.

Kroner, K. F., & Claessens, S. (1991). Optimal dynamic hedging portfolios and the currency composition of external debt. *Journal of International Money and Finance, 10*, 131–148.

Ledoit, O., Santa-Clara, P., & Wolf, M. (2003). Flexible multivariate GARCH modelling with an application to international stock markets. *The Review of Economics and Statistics, 85*(3), 735–747.

Lien, D., & Luo, X. (1994). Multiperiod hedging in the presence of conditional heteroscedasticity. *Journal of Futures Markets, 14*, 927–955.

Nelson, D. B. (1991). Conditional heteroskedasticity in asset returns: A new approach. *Econometrica, 59*, 347–370.

Pesaran, M. H., & Pesaran, B. (1997). Working with Microfit 4.0: Interactive Econometric Analysis. Oxford University Press: Oxford.

Pesaran, B., & Pesaran, M. H. (2009). *Time Series Econometrics Using Microfit 5.0*. Oxford, UK: Oxford University Press.

Righia, M. B., & Cerella, P. S. (2012). Multivariate generalized autoregressive conditional heteroscedasticity (GARCH) modelling of sector volatility transmission: A dynamic conditional correlation (DCC) approach. *African Journal of Business Management, 6*(27), 8157–8162.

Silvennoinen, A., & Teräsvirta, T. (2009). Modelling multivariate autoregressive conditional heteroscedasticity with the double smooth transition conditional correlation GARCH model. *Journal of Financial Econometrics, 7*(4), 373–411.

Sims, C. A. (1980, January). Macroeconomics and reality. *Econometrica, 48*(1), 1–48.

Thomas, R. L. (1997). *Modern econometrics*. Upper Saddle River: Prentice Hall.

Verbeek, M. (2004) A Guide to Modern Econometrics. 2nd Edition, Erasmus University Rotterdam, John Wiley & Sons Ltd., Hoboken.

Index

A. Aljandali, M. Tatahi, *Economic and Financial Modelling with EViews*, Statistics and Econometrics for Finance, https://doi.org/10.1007/978-3-319-92985-9

Made in the USA
Las Vegas, NV
16 September 2026

0f185a78-bb10-4838-b4de-05934094ad6cR02